Biomedical Electronics, Noise Shaping ADCs, and Frequency References

Pieter Harpe • Andrea Baschirotto •
Kofi A. A. Makinwa
Editors

Biomedical Electronics, Noise Shaping ADCs, and Frequency References

Advances in Analog Circuit Design 2022

Springer

Editors
Pieter Harpe
Department of Electrical Engineering
Eindhoven University of Technology
Eindhoven, The Netherlands

Andrea Baschirotto
Department of Physics
University of Milan-Bicocca
Milan, Italy

Kofi A. A. Makinwa
Department of Microelectronics
Delft University of Technology
Delft, The Netherlands

ISBN 978-3-031-28911-8 ISBN 978-3-031-28912-5 (eBook)
https://doi.org/10.1007/978-3-031-28912-5

This Springer imprint is published by the registered company Springer Nature Switzerland AG
The registered company address is: Gewerbestrasse 11, 6330 Cham, Switzerland

Preface

This book is part of the Analog Circuit Design series and contains contributions of all 18 speakers of the 30th workshop on Advances in Analog Circuit Design (AACD). The event was organized by John Morrissey, Ivan O'Connell, Nicola Cooney, Paul Hyland, Kapil Bhate and Mary Kent and some students from MCCI – Microelectronic Circuits Centre Ireland, Tyndall National Institute, Cork, Ireland. The sponsors for this workshop were: Qualcomm, Analog Devices, AMD, Qorvo, Macom, Boston Scientific, Bosch, Renesas, Onsemi, u-blox, Infineon, Cadence, IDA Ireland, OTC Ireland, Vishay. The workshop was held in Cork, Ireland, from October 4 to 6, 2022.

ABOUT AACD

The aim of the AACD workshop is to bring together a group of expert designers to discuss new developments and future options. Each workshop is followed by the publication of a book by Springer in their successful series of Analog Circuit Design. This book is the 30th in this series. The book series can be seen as a reference for all people involved in analog and mixed-signal design. The full list of the previous books and topics in the series is included in this book.

ABOUT MCCI

Funded by Enterprise Ireland and the IDA, MCCI's mission is to deliver high impact research for the semiconductor industry and to generate innovative technology. MCCI is a national technology centre that works collaboratively in microelectronics circuit design to improve the performance of mixed-signal circuits required by their industry partners. MCCI's research focus is on mixed-signal, analog and RF circuits. The centre has established itself as a single point of contact in Ireland for access to high-calibre academic research in the field of microelectronics. MCCI is committed to the development of an engineering talent pipeline for the global semiconductor industry. For more information, visit www.mcci.ie

This book comprises three parts, each with six chapters from experts in the field, covering advanced analog and mixed-signal circuit design topics that are considered highly important by the circuit design community:

- Biomedical Electronics
- Noise Shaping ADCs
- Frequency References

We are confident that this book, like its predecessors, proves to be a valuable contribution to our analog and mixed-signal circuit design community.

Eindhoven, The Netherlands — Pieter Harpe
Milan, Italy — Andrea Baschirotto
Delft, The Netherlands — Kofi A. A. Makinwa

The Topics Covered Before in This Series

2021	Online	Analog Circuits for Machine Learning Current, Voltage and Temperature Sensors High-Speed Communication
2019	Milan (Italy)	Next-Generation ADCs High-Performance Power Management Technology Considerations for Advanced Integrated Circuits
2018	Edinburgh (Scotland)	Analog Techniques for Power Constrained Applications Sensors for Mobile Devices Energy Efficient Amplifiers and Drivers
2017	Eindhoven (The Netherlands)	Hybrid ADCs Smart Sensors for the IoT Sub-1V & Advanced Node Analog Circuit Design
2016	Villach (Austria)	Continuous-time $\Sigma\Delta$ Modulators for Transceivers Automotive Electronics Power Management
2015	Neuchâtel (Switzerland)	Efficient Sensor Interfaces Advanced Amplifiers Low Power RF Systems
2014	Lisbon (Portugal)	High-Performance AD and DA Converters IC Design in Scaled Technologies Time-Domain Signal Processing

2013	Grenoble (France)	Frequency References
		Power Management for SoC
		Smart Wireless Interfaces
2012	Valkenburg (The Netherlands)	Nyquist A/D Converters
		Capacitive Sensor Interfaces
		Beyond Analog Circuit Design
2011	Leuven (Belgium)	Low-Voltage Low-Power Data Converters
		Short-Range Wireless Front-Ends
		Power Management and DC-DC
2010	Graz (Austria)	Robust Design
		Sigma Delta Converters
		RFID
2009	Lund (Sweden)	Smart Data Converters
		Filters on Chip
		Multimode Transmitters
2008	Pavia (Italy)	High-Speed Clock and Data Recovery
		High-Performance Amplifiers
		Power Management
2007	Oostende (Belgium)	Sensors, Actuators and Power Drivers for the Automotive and Industrial Environment
		Integrated PAs from Wireline to RF
		Very High Frequency Front Ends
2006	Maastricht (The Netherlands)	High-Speed AD Converters
		Automotive Electronics: EMC issues
		Ultra Low Power Wireless
2005	Limerick (Ireland)	RF Circuits: Wide Band, Front-Ends, DACs
		Design Methodology and Verification of RF and Mixed-Signal Systems
		Low Power and Low Voltage
2004	Montreux (Swiss)	Sensor and Actuator Interface Electronics
		Integrated High-Voltage Electronics and Power Management
		Low-Power and High-Resolution ADCs
2003	Graz (Austria)	Fractional-N Synthesizers
		Design for Robustness
		Line and Bus Drivers

2002	Spa (Belgium)	Structured Mixed-Mode Design Multi-bit Sigma-Delta Converters Short-Range RF Circuits
2001	Noordwijk (The Netherlands)	Scalable Analog Circuits High-Speed D/A Converters RF Power Amplifiers
2000	Munich (Germany)	High-Speed A/D Converters Mixed-Signal Design PLLs and Synthesizers
1999	Nice (France)	XDSL and Other Communication Systems RF-MOST Models and Behavioural Modelling Integrated Filters and Oscillators
1998	Copenhagen (Denmark)	1-Volt Electronics Mixed-Mode Systems LNAs and RF Power Amps for Telecom
1997	Como (Italy)	RF A/D Converters Sensor and Actuator Interfaces Low-Noise Oscillators, PLLs and Synthesizers
1996	Lausanne (Swiss)	RF CMOS Circuit Design Bandpass Sigma Delta and Other Data Converters Translinear Circuits
1995	Villach (Austria)	Low-Noise/Power/Voltage Mixed-Mode with CAD Tools Voltage, Current and Time References
1994	Eindhoven (The Netherlands)	Low-Power Low-Voltage Integrated Filters Smart Power
1993	Leuven (Belgium)	Mixed-Mode A/D Design Sensor Interfaces Communication Circuits
1992	Scheveningen (The Netherlands)	OpAmps ADC Analog CAD

Contents

Part I
Biomedical Electronics

The first part of this book discusses the development of electronics for biomedical applications. Thanks to the advent of technology scaling and integration, wearable devices, implants, large sensor arrays, and new sensing modalities have become possible. This chapter illustrates some of the recent advances in the field in terms of ExG interfaces and neural recording, on-chip (para-)magnetic resonance detection, and electrochemical sensing.

In Chap. 1, Komail Badami and colleagues give an overview of design challenges for high-performance ExG interfaces. Thanks to progress in electronics, these systems have moved from bedside devices in clinical setups to wearable devices in the home environment. Besides discussing challenges, the chapter also introduces and reviews various state-of-the-art solutions.

In Chap. 2, Corentin Pochet and Drew Hall further expand on the development of ExG digitization interfaces. Time-based digitization paradigms leveraging voltage-controlled oscillators (VCOs) are explored to allow for more digital-friendly implementations of the analog-to-digital conversion process. Circuit-level implementations and measurements are shown to highlight the performance that can be attained with the VCO-based approach.

In Chap. 3, Carolina Mora Lopez and Xiaohua Huang present circuits and architectures for neural recording interfaces. First, a general overview of existing techniques is given, and requirements to perform neural recording are discussed. After that, state-of-the-art neural readout chips and neural probes are reviewed, highlighting some of the key innovations and ultimately benchmarking the various architectures.

In Chap. 4, Jens Anders and colleagues discuss recent developments in the field of chip-integrated electronics for nuclear magnetic resonance (NMR) and electron paramagnetic resonance (EPR) for biomedical applications. Besides explaining the physical principles, potential applications and circuit architectures are discussed. Various chip implementation examples are given, and an outlook on future directions is given.

In Chap. 5, Zhongzheng Wang and colleagues provide a detailed analysis of electrochemical sensors, their models, and possible sensing interface architectures.

Various implementation challenges and solutions are discussed, and the chapter concludes with a discussion on future trends.

In Chap. 6, Tim Cummins discusses the opportunities of using CMOS-integrated capacitive sensors for molecular detection. A review of molecular detection methods is given first, using either laboratory setups or CMOS-integrated biosensors. Next, a biosensor approach based on capacitive sensing is described, inherent challenges are discussed, and initial test results of this new concept are given.

Overview of Design Challenges in High-Performance ExG Interfaces

K. Badami, M. Pons Sole, T. Mavrogordatos, A. Fivaz, P. Persechini, O. Chételat, P.-F. Ruedi, and S. Emery

Abstract Progress in CMOS technology and a huge impetus on biomedical research have moved biopotential monitoring systems from bedside devices in clinical setups to wearable devices in the home environment. In spite of the significant progress, design of an ExG readout system poses several circuit- and system-level challenges. The designed circuit should exhibit very large differential offset handling capability yet while amplifying very low-frequency small ExG signals. The system should maintain high immunity to large common-mode signals arising from EMI from the mains supply or from motion artifacts. Further, systems designed for wearable and long-term monitoring should be comfortable for the patient without restricting mobility yet while enabling high noise/power efficiency. An overview of these challenges and state-of-the-art solutions to mitigate them are discussed in this chapter.

1 Progress in Biomedical Systems

Over the last few decades, significant progress has been made in the integration of biomedical readout circuits in CMOS technology for monitoring of biological signals such as electrocardiogram (ECG), electroencephalogram (EEG), and electromyogram (EMG). Continuous and long-term monitoring of such biomedical signals can improve life quality and enable medical cost reduction by timely and on-demand diagnostics. The synergistic combination of (a) power and area efficiency improvement of ExG sensing systems, (b) proliferation of user-friendly electronic systems such as smart watches and mobile phones, and (c) increasing awareness about healthcare and disease prevention has transformed the biomedical sensing systems from bulky bedside devices in clinical setups to comfortable wearable devices in the home environment.

K. Badami (✉) · M. Pons Sole · T. Mavrogordatos · A. Fivaz · P. Persechini · O. Chételat · P.-F. Ruedi · S. Emery
CSEM, Neuchâtel, Switzerland
e-mail: komail.badami@csem.ch

P. Harpe et al. (eds.), *Biomedical Electronics, Noise Shaping ADCs, and Frequency References*, https://doi.org/10.1007/978-3-031-28912-5_1

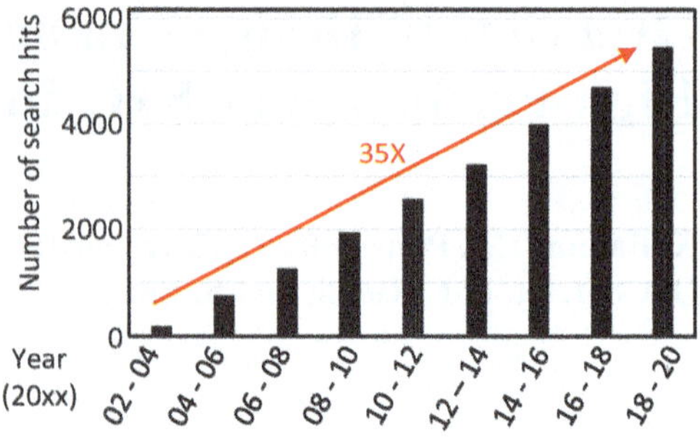

Fig. 1 Number of publications with author keywords containing terms {biomedical, ExG, ECG, EEG, or EMG} on IEEE explore website over consecutive 2-year periods from 2002 to 2020

This progress in biomedical monitoring is evident from the growth of peer-reviewed publications on biomedical signal processing and related topics. An example search was conducted on IEEE Xplore to find the number of publications with author keywords containing the terms {biomedical, ExG, ECG, EEG, or EMG} over consecutive 2-year periods starting from 2002 to 2020. A graphical summary illustrating the growth of research is shown in Fig. 1.

The remainder of this chapter is organized as follows: Sect. 2 briefly discusses different challenges/trade-offs in ExG system design, while Sects. 2.1–2.5 individually detail the five crucial challenges and discuss the state-of-the-art solutions to these challenges. Conclusions are discussed in Sect. 3.

2 Challenges in Design of ExG Readouts

In spite of the significant research in biomedical systems, ExG readout system design still faces a host of conflicting challenges both at system and at circuit level. The space occupied by common ExG signals in the amplitude/frequency domain is shown in Fig. 2, left. Along with the useful ExG signals, the to-be-measured ExG signals are accompanied by a large differential offset and a large in-band common-mode EMI from the mains supply, both of which need to be rejected yet while maintaining sufficient amplification for the useful ExG signals without signal path saturation. Also, the source impedance of the ExG signals, especially from dry electrodes, is quite large and also has a large variable component that varies between different electrode contacts on the same patient. The ExG readout interface should have a sufficiently high input impedance to be able to interface with these large source impedances and also have a high common-mode rejection to limit the impact from EMI from the mains supply. Further, for continuous monitoring in nonclinical setups, the system should be comfortable for the end user without restricting the mobility and yet offer long enough battery lifetime while meeting the stringent noise requirements. An application-aware trade-off has to be made for the above challenges while complying with respective biomedical regulations. An overview

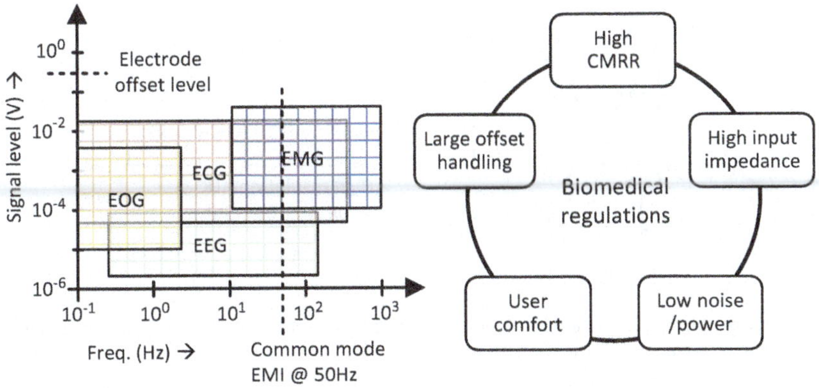

Fig. 2 Overview of design challenges at system and at circuit level for design of ExG readout interfaces

of these challenges is summarized in Fig. 2, right. These challenges and their state-of-the-art solutions are detailed in the subsequent sections.

2.1 Electrode Offsets and CMOS Integration

2.1.1 Problem Description

ExG signals can have large electrode offsets riding on top of small meaningful signals that have to be amplified and measured. The electrode offset levels for ECG signals can be as large as ±300 mV as per IEC 60601-2-25/27/47, while the useful ECG signal is in the range of few 100 μV to few mV, effectively implying up to a 50 dB overhead of electrode offset in signal processing. DC-coupled readouts, as shown in Fig. 3, amplify the electrode offset and the ExG signal together and avoid amplifier saturation with low gain such as ≤5x–20x [1, 2]. The amplified signal along with the electrode offset is then digitized with a high-resolution ADCs as shown in Fig. 3. Very low cutoff HPFs, typically sub-1 Hz, to attenuate the electrode offset are then implemented in the digital domain.

The main limitation of such DC-coupled systems is the large dynamic range of the analog frontend to accommodate the electrode offset in addition to the useful ExG signal. This significantly increases the power consumption of the readout channel and limits operating time in case of battery-based operation.

2.1.2 Existing Solutions

AC-coupled readouts attenuate the electrode offset with an upfront sub-1 Hz HPF in the analog domain before the ExG signal is digitized by the analog frontend. In order

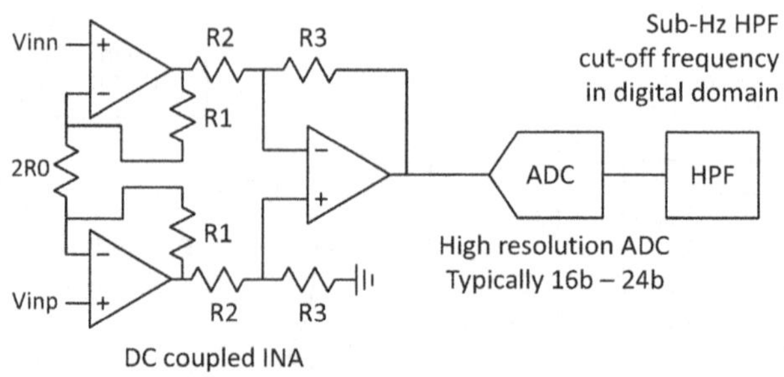

Fig. 3 DC-coupled ExG readouts process electrode offsets and ExG signals through same amplifiers and attenuate electrode offset in a digital HPF

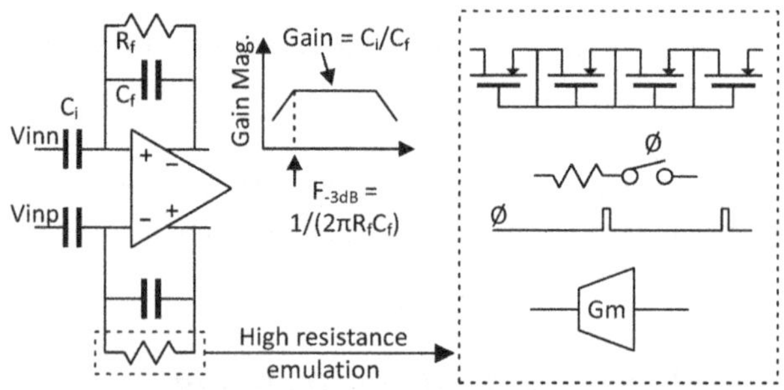

Fig. 4 Some of the options to emulate high-value resistor required in the feedback path of an AC-coupled amplifier

to create a corner frequency of let's say 1 Hz with a generously large capacitor of 100 pF on CMOS, a resistor of about 1.6 GΩ will be required. Biomedical systems with a limited number of channels overcome this limitation by a HPF on the PCB as discussed in [3, 4]. However, for systems with a large number of ExG readouts [5, 6], an external HPF for each channel significantly increases the board area and the BOM, and hence HPF integration solutions on CMOS are sought:

- **High resistor emulation**: Some of the approaches for creating a very low HPF corner on chip by emulating a very high DC resistance are illustrated in Fig. 4 and are discussed further. Pseudo-resistors [7, 8] are essentially FETs biased in cutoff region in order to emulate large resistances required for low HPF corner frequencies. While the pseudo-resistors enable emulation of large resistors of the order of 100s of GΩs, they suffer from ill-controlled process and temperature spread as the resistance depends on leakage current [8]. Multiple solutions exist in literature to improve the pseudo-resistor performance [8]. An example of pseudo-resistor biasing to reduce the PVT spread [5] is shown in Fig. 5. Another alternative is to use a very small gm cell [9] in the feedback path to fix a low cutoff frequency for the high-pass corner. In addition to limited linearity, the offset/noise of the gm cell worsens the signal path. Other options to emulate a

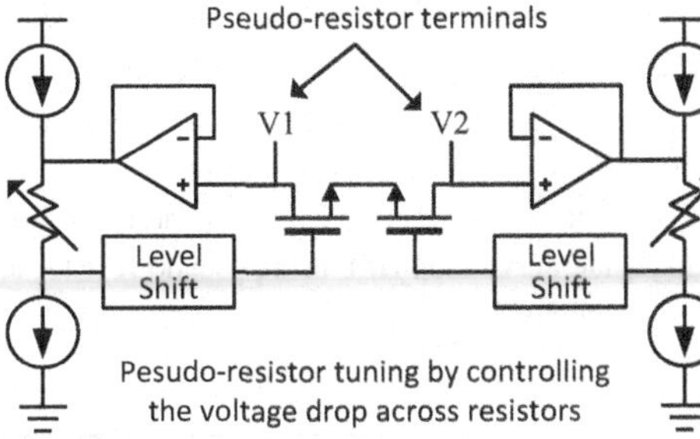

Fig. 5 Example of pseudo-resistor biasing used in a 20-channel EMG recording system [5] to reduce the PVT spread

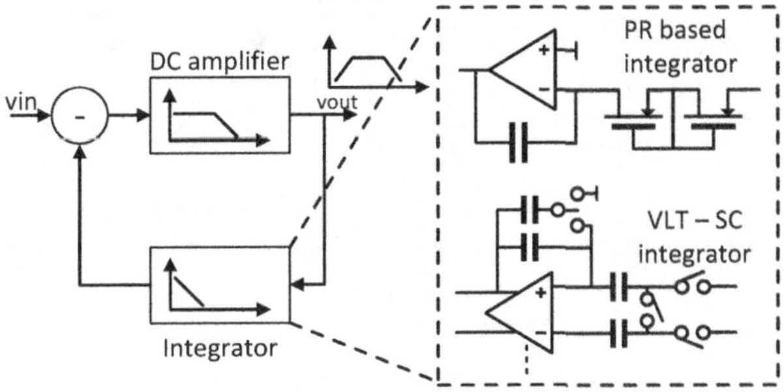

Fig. 6 DC servo loop to create a very low-frequency high-pass corner using pseudo-resistors or very large time constant switched capacitor integrator

large resistor include duty-cycling a smaller resistor [10] as illustrated in Fig. 4 or as a cascade of capacitors sampled on alternating clock phases [11, 12].

- **DC servo loops**: Another option commonly used to create a very low HPF corner is the use of DC servo loops. In such an amplifier, an integrator/LPF is used in the feedback path of a DC-coupled amplifier to effectively create a bandpass response between the input and output of the amplifier [12–15] as shown in the block diagram in Fig. 6. The low-pass cutoff frequency of the feedback path is effectively the high-pass corner frequency of the amplifier. In order to create a very low cutoff frequency in the feedback path, the integrator/LPF uses pseudo-resistors [13] or a switched capacitor network to emulate a large resistor [12]. The use of active elements in the feedback network often worsens the offset and noise of the readout system. An area-efficient, very large time-constant switched capacitor-based integrator [16] is used in a DC servo loop resulting in 0.5 Hz high-pass corner [14].

2.1.3 Summary

A short tabular summary of different approaches to create a very low HPF corner frequency on-chip in the analog domain is provided in Table 1.

Table 1 Summary of solutions for very low high-pass corner frequency

Ref.	Solution adopted for low HPF corner	HPF corner
[7]	Pseudo-resistor for a large resistor emulation	0.014 Hz
[10]	Duty-cycling a smaller resistor	0.2 Hz–4 Hz
[9]	Very low gm cell to emulate a large resistor	1.5 Hz
[17]	DC servo loop HPF using pseudo-resistor using very large time constant integrator	<0.26 Hz
[14]		0.5 Hz

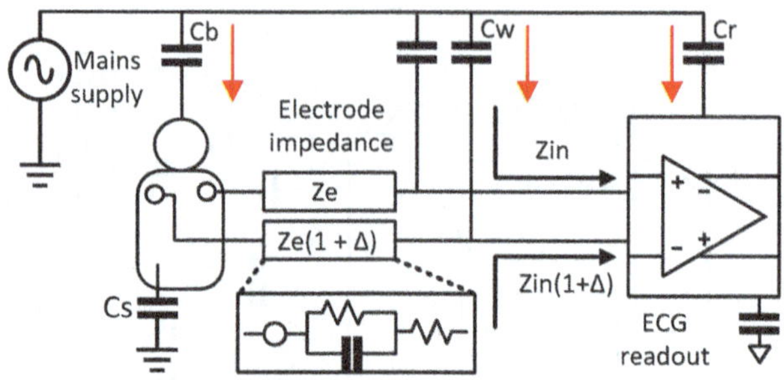

Fig. 7 Coupling of electromagnetic interference at different points to an ECG readout system

2.2 *Power Supply Interference and Common-Mode Rejection*

2.2.1 Problem Description

Common-mode rejection is one of the most important design parameters in an ExG readout system. This is because EMI from mains power supply or cable motion artifacts can result in a relatively large common-mode signal at the input of the ExG readout [18, 19]. This interference can couple to the readout chain through multiple paths such as the patient's body, the recording cables, or the recording instrument as shown in Fig. 7.

Mismatch in the skin-electrode impedance and/or the input impedance of the amplifier results in common-mode to differential conversion of the relatively large common-mode signal and can distort the recorded ExG signal. To limit the impact of common-mode disturbances, for example, ECG standards specify less than 1 mVpp deviation at the output of the recording system while applying 10 Vrms at the input under a very specific test setup specified in IEC 60601-2-25/27. This translates to a CMRR requirement of 90 dB. More details about the test setup for ECG CMRR testing can be found under IEC 60601-2-25/27.

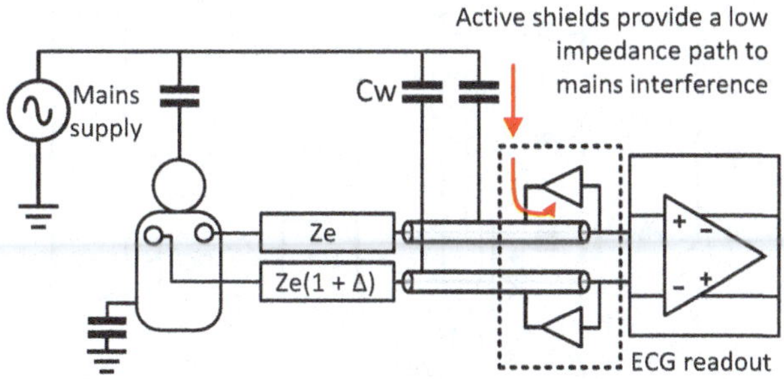

Fig. 8 Active shielding of ECG recording cables to suppress the impact of EM coupling

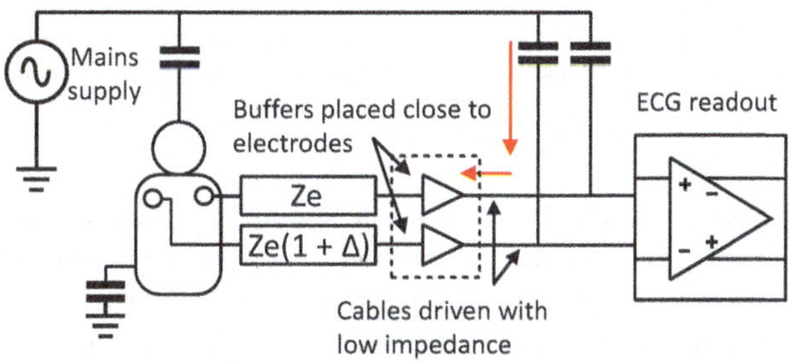

Fig. 9 Active electrodes drive recording cables with very low impedance and hence improve robustness to EMI

2.2.2 Existing Solutions

Techniques at circuit and system level to limit the impact of large common-mode disturbances on the ExG readout system are discussed further:

- **Cable shielding and active electrodes**: The long ExG cables pick up EMI as shown in Fig. 7. To limit the impact of this EMI coupling, not only the recording cables are shielded but the shields also driven by signal buffers as shown in Fig. 8. Further, along with cable shielding, often a buffer or an amplifier is placed in close proximity to the electrode contact or on the electrode. This enables to drive the ExG cables with low impedance, thus further improving the robustness of the cables to EMI pickup as shown in Fig. 9. An eight-channel active electrode-based EEG system achieves 84 dB CMRR and is discussed in [13, 20]. Such active electrodes present a very large input impedance, typically $\geq$100 MΩ, and are well suited for use with dry electrodes as discussed in [17, 21]. Active electrode-based systems increase the cabling complexity for the connected electrodes [13, 17, 21]. Furthermore, while active electrodes improve the immunity of the recording cables to EMI, the mismatch between the buffers/amplifiers on the different electrodes themselves can limit the CMRR of the readout chain [21].
- **Driven body reference**: To limit the impact of common-mode coupling to the patient's body, it is driven with either a fixed reference voltage or by a closed

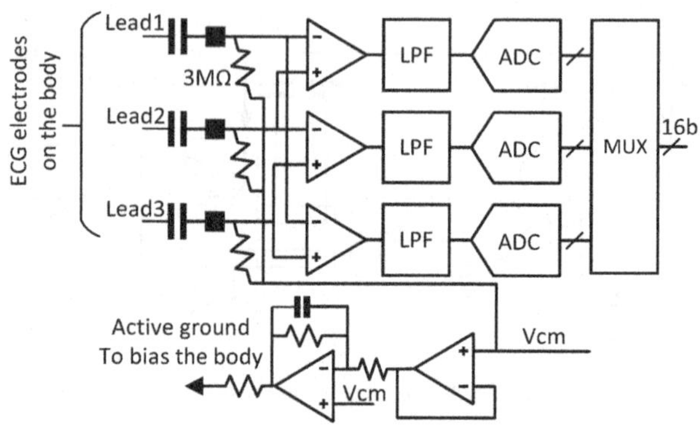

Fig. 10 Driven right leg system [3] for suppressing the EM coupling to the human body in an ECG measurement system. The feedback path provides low impedance for the common-mode interference current

loop system that tracks the common-mode signal from the readout amplifiers and regulates the reference voltage on the body. Such systems in the ECG system's parlance are commonly referred to as driven right leg (DRL) systems [22, 23] and a block diagram is illustrated in Fig. 10. The DRL amplifier provides a low impedance path to the 50/60 Hz common-mode interference signals thereby improving the CMRR [3, 4]. Stability is a major challenge in the design of the DRL systems since multiple low-frequency poles exist in the loop and their location can vary based on skin-electrode impedance [4, 22, 23]. DRL loops need a large few nF capacitor for stability and also need current limiting resistors of few 10's kΩ to limit current in case the DRL loop saturates. An application report [24] discusses a DRL system that achieves a CMRR upto 96 dB. Other options to mitigate the stability issue of the DRL loop include providing the common-mode feedback signal at the input of the readout chain as discussed in [20] or a feed-forward compensation as detailed in [25].

- **Dynamic offset cancellation**: When the common-mode rejection of the ExG readout system is limited by the matching in the readout amplifier, dynamic offset cancellation techniques such as chopping are invariably required to improve the CMRR. Since chopping also improves the flicker noise of the amplifier, a discussion of chopping techniques for ExG readouts is deferred to Sect. 2.5.

2.2.3 Summary

A tabular summary of different state-of-the-art approaches to improve the common-mode rejection of the ExG readout system is provided in Table 2.

Table 2 Summary of solutions to improve CMRR

Ref.	Solutions to improve CMRR	Achieved CMRR
[13, 20]	Signal shielding and active electrodes	84 B
[24]	Driven right leg systems	96 dB
[20]	Common-mode feedback to amplifier input	82 dB
[25]	Feedforward biasing the readout system	66 dB–102 dB
[14]	Dynamic offset cancellation techniques	134 dB

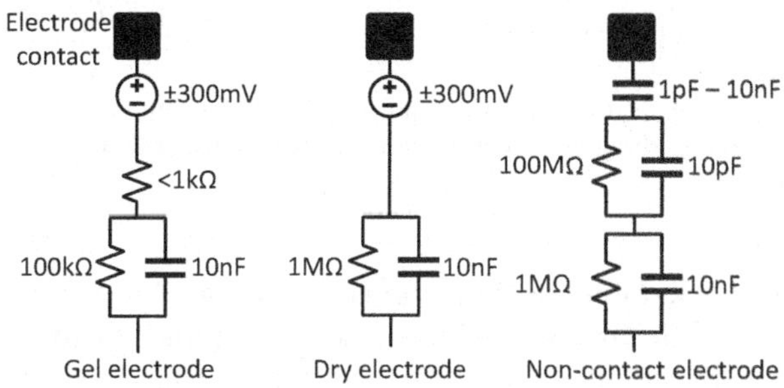

Fig. 11 Typical impedance components of gel and dry contact and dry noncontact electrodes [29]

2.3 High Input Impedance for Use with Dry Electrodes

2.3.1 Problem Description

Gel electrodes provide low impedance contact to the skin, typically ≤10 kΩ [26, 27]. However, their usage is limited to clinical environments as they require skin preparation steps such as shaving, cleanup, and abrasion prior to gel application [26–28]. Furthermore, for long-term monitoring over multiple days, gel would need to be applied every few hours, and prolonged usage of gel can result in skin irritation. This is why for home-based long-term monitoring setups, readout systems that work with dry electrodes are more preferred. The main limitation of dry electrodes is relatively high skin-electrode impedance which can be as high 100 kΩ to few MΩs [26–28]. Typical impedance components of gel, dry, and noncontact electrodes are shown in Fig. 11.

2.3.2 Existing Solutions

ExG readouts based on dry electrodes require very high input impedance analog frontends to accommodate the large source impedance of dry electrodes and to limit the impact of skin-electrode impedance variation (among the different electrodes) on common-mode rejection. Techniques for improving the input impedance of the amplifier are discussed further:

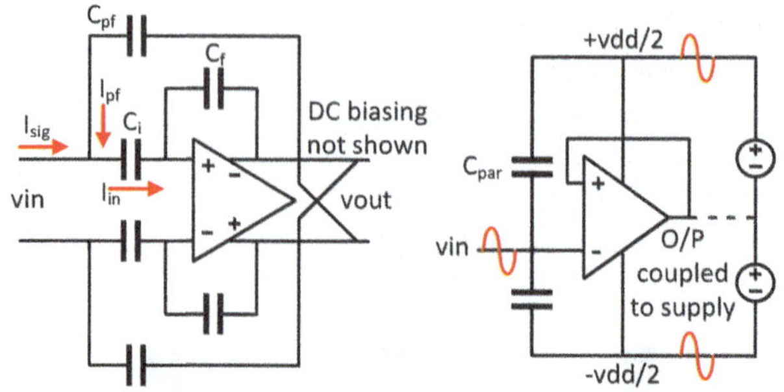

Fig. 12 (left) A positive feedback loop around the amplifier to improve the input impedance and (right) bootstrapping the power supply with the input signal enables to enhance the input impedance of the buffer

- **Positive feedback loop for impedance boosting**: An approach often used to interface dry electrodes with the readout amplifier is to boost the input impedance of the amplifier with an impedance boosting loop [9, 14, 20, 30] as shown in Fig. 12 left. A positive feedback loop around the AC-coupled amplifier compensates for the input current I_{in}, thus reducing the current sourced from the signal source I_{sig} as shown in Fig. 12, left. The feedback from the negative feedback loop still needs to dominate to ensure an overall stable amplifier. A calibration-based approach has been introduced in [30] to maximize the current cancellation and hence maximize the input impedance. This technique has been shown to improve input impedance from a couple of G Ωs to 50 GΩ [30].
- **Power supply bootstrapping**: Power supply bootstrapping can be used to significantly enhance the input impedance of a buffer/amplifier. This technique was introduced in [31] and reports an input impedance corresponding to 0.01 pF//1 TΩ. An example schematic to illustrate the increase in input impedance due to supply bootstrapping is shown in Fig. 12, right. The ExG input signal is bootstrapped to the supply rails; effectively no current can flow through C_{par} as the supply nodes also move along with the input signal. This effectively guards the input terminals resulting in a very large input impedance. An active electrode with power supply bootstrapping [32] to be used with ECG/EEG systems demonstrated an input impedance corresponding to a 71 fF capacitor. Further examples of power supply bootstrapping to enhance the input impedance of the readout chain are discussed in [19, 33, 34].

2.3.3 Summary

A summary of different methods to improve the input impedance of the ExG system to enable interfacing with dry electrodes is shown in Table 3.

Table 3 Summary of techniques to boost the input impedance of the readout chain

Ref.	Solution for increasing the readout input impedance	Achieved input impedance
[13]	Active electrode with buffer/amplifier placed close to the skin-electrode contact	1.2 GΩ at 20 Hz
[30]	Calibrated positive feedback loop to compensate the input current of the amplifier	50 GΩ at 50 Hz
[31]	Power supply bootstrapping to increase the input impedance	0.01 pF//1 TΩ [31]
[32]		71 fF [32]

2.4 *Cabling Complexity and Patient Mobility*

2.4.1 Problem Description

Long-term monitoring of ExG signals in a nonclinical setup requires progress and development in the readout systems on two major fronts. Measurements acquired from such systems (a) should preferably use dry electrodes and be robust in the presence of EMI and motion artifacts, and (b) the cabling complexity should not inhibit the patient mobility. State-of-the-art approaches for interfacing with dry electrodes have been discussed earlier in Sect. 2.3. This section discusses system solutions that do not inhibit patient mobility by reducing the cabling complexity for long-term monitoring in nonclinical environments.

2.4.2 Existing Solutions

This section discusses a cooperative sensor [33, 35]-based solution for ECG measurement that lends its self for an easy integration into a wearable vest and ExG measurement systems that have an integrated RF transceiver [3, 36] for wireless transfer of data from the patient to a recording station:

- **Cooperative sensing for long-term ambulatory ECG measurements**: The long-term ambulatory monitoring system, based on CSEM's patented cooperative sensor approach [33, 35], operates in master-slave configuration [38] and is shown in Fig. 13. The master node simultaneously powers the bus-connected slave nodes and then measures the ECG signal with each connected slave node in turns. The two main advantages of the system are (a) only a single battery is required on the master node while the slave nodes are powered over the single wire bus connection and (b) the single wire bus connection between the master and the slave nodes enables easy integration of the ECG readout system in garments [33] and is particularly suitable to long-term monitoring. The readout chain uses power supply bootstrapping at the input to enable interfacing with a dry electrode and is detailed in [38]. Example integration of such a cooperative sensor system into a wearable garment is illustrated in Fig. 14.

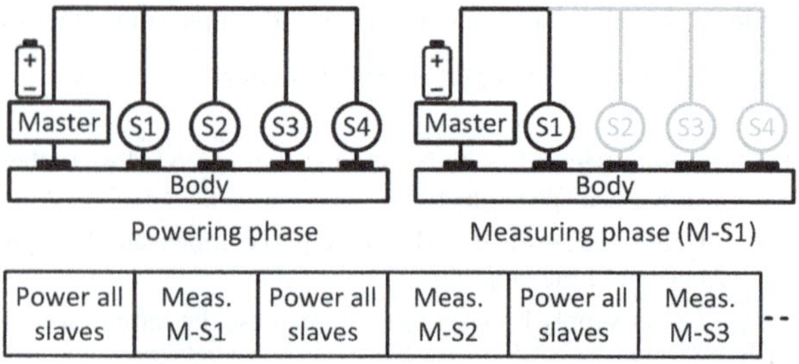

Fig. 13 Master slave configuration for a cooperative sensor-based ECG readout system

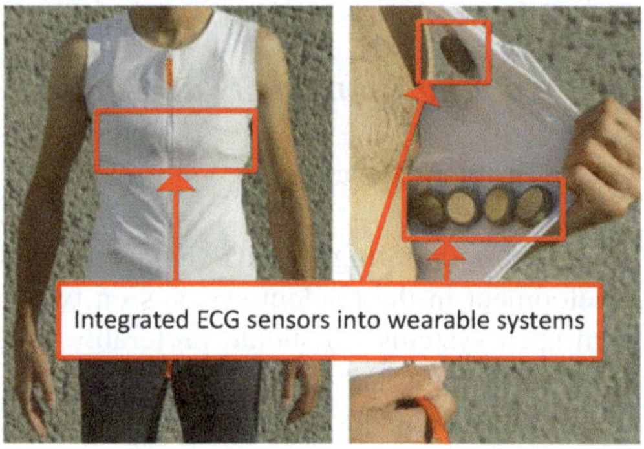

Fig. 14 Example ECG readout system integration into garment for long-term ambulatory recording [37]

- **Integrated wireless communication**: Battery-operated ExG systems with an integrated wireless transceiver significantly improve the patient comfort and mobility, thus enabling long-term monitoring. A three-channel ECG measurement system [3] has an integrated RF transceiver enabling a cable-free ECG measurement system. The die micrograph of the SoC is shown in Fig. 15. As another example, a vital sign monitoring system with an integrated BLE transceiver and compatible with dry electrodes is discussed in [36]. Such integrated systems that enable operation within the home environment have a strong potential to reduce the healthcare costs by enabling on-demand and timely diagnostics.

2.4.3 Summary

A summary of different methods to reduce the cabling complexity of an ExG system is shown in Table 4.

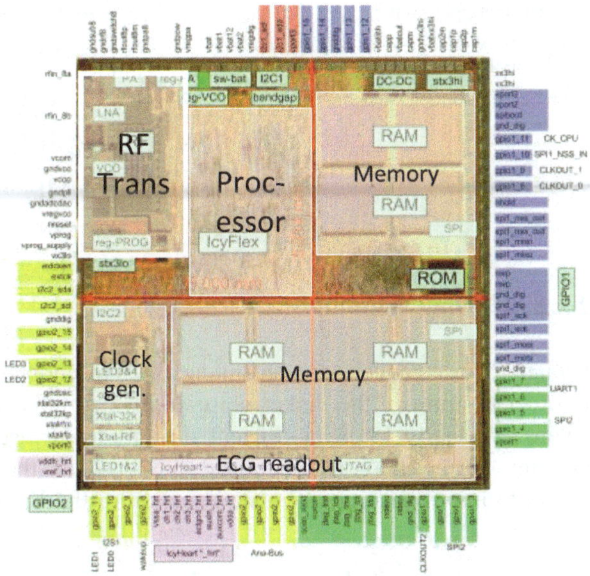

Fig. 15 Example of a highly integrated SOC containing an RF transceiver for ECG readout [3]

Table 4 Summary of solutions to reduce the cabling complexity of the ExG readout system

Ref.	Solution for reducing cabling complexity	Number of cables
[38]	Cooperative sensing approach based on master-slave configuration and exploiting the return current path through the body	1 + return path through the body
[37]	Cooperative sensing approach with bus-connected sensors + wearable integration	2 cables for 12-lead ECG
[3, 36]	EXG data transfer through an integrated RF transceiver	Enables a cable-free system

2.5 *Low-Noise Design for Power- and Area-Efficient Systems*

2.5.1 Problem Description

The requirement of longer battery life for wearable ExG systems has increased the focus on the power efficiency of the readout chain. Power consumption of the ExG sensing system mainly depends on the acceptable input-referred noise of the amplifier. As ExG signals occupy a bandwidth from the sub-1 Hz frequency range to a maximum of a few kHz, the noise spectral density is dominated by flicker noise. Area increase to reduce the flicker noise is often not the most optimal solution since a 4X area increase is required to halve the flicker noise. While the area increases, the parasitics at the amplifier input worsen, which then degrades the input impedance and the noise performance of the feedback amplifier. Furthermore, newer biomedical applications demand an increasingly higher number of readout channels and hence require an area-efficient solution.

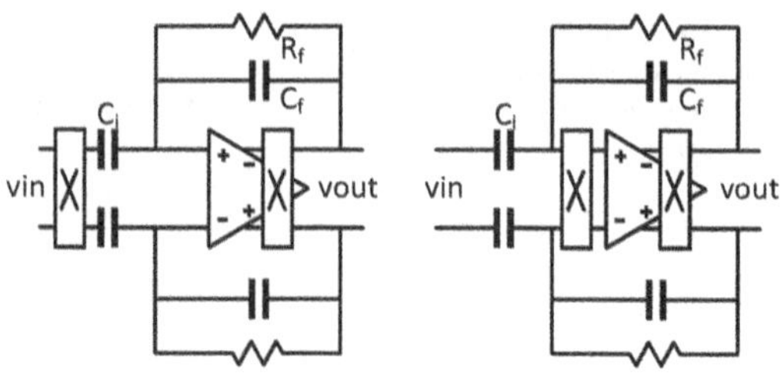

Fig. 16 Chopper switch placement in AC-coupled amplifiers for CMRR enhancement

2.5.2 Existing Solutions

Dynamic noise cancellation techniques such as chopping are the de facto standard in order to improve the flicker noise-dominated noise performance of the readout chains especially for area-efficient designs to be used in multichannel readout systems. Furthermore, as mentioned in Sect. 2.2.2, the chopping technique is also useful to improve common-mode rejection of the ExG amplifier and is discussed further:

- **Chopping for flicker noise reduction**: To improve the flicker noise with chopping, two possibilities for the placement of chopper switch exist, before the AC-coupled capacitors (Fig. 16 left) and after the AC-coupled capacitors within the feedback loop (Fig. 16 right) as discussed in [14, 39, 40]. When the chopper switches are placed before the capacitors, it lowers the input impedance of the amplifier and also limits its electrode-offset handling capacity as the amplifier appears as DC-coupled [14] to the signal source. Chopping at the opamp input increases the input current noise spectral density of the opamp due to switching currents through the chopper switch. A detailed analysis can be found in [40]. Dynamic offset cancellation methods can push the noise floor to well below $100\,\mathrm{nV}/\sqrt{\mathrm{Hz}}$ and have achievable CMRR values well above the 100 dB range as discussed in [11, 14].

3 Conclusions

This chapter discussed the main challenges in the design of a biomedical readout circuit/system and provided a brief overview of state-of-the-art solutions from the literature. The remarkable progress in the miniaturization and power efficiency of ExG readout systems has moved them from bulky clinical setups to untethered wearable devices. This will enable timely and on-demand diagnostics with a potential to reduce the healthcare costs and improve the quality of life.

References

1. ADS1293 Low-Power, 3-Channel, 24-Bit Analog Front-End for Biopotential Measurements, Texas Instruments, 12 2014, rev. C
2. ADS1299-x Low-Noise, 4-, 6-, 8-Channel, 24-Bit, Analog-to-Digital Converter for EEG and Biopotential Measurements, Texas Instruments, 01 2017, rev. C
3. F. Giroud, P. Persechini, P.-F. Ruedi, F. Viemon, F. Masson, ECG sensor interface in a low power SoC for wireless portable ECG monitoring, in *2014 IEEE Biomedical Circuits and Systems Conference (BioCAS) Proceedings* (IEEE, Piscataway, 2014), pp. 149–152
4. M.J. Burke, D.T. Gleeson, A micropower dry-electrode ECG preamplifier. IEEE Trans. Biomed. Eng. **47**(2), 155–162 (2000)
5. M.P. Sole, K. Badami, J. Deng, T. Mavrogordatos, E. Azarkhish, L. Zahnd, C. Cosentino, M. Augustyniak, Y. Zha, A. Bischof, L. Bergamini, P. Dallemagne, S. Emery, A 20 channel EMG SoC with an integrated 32b RISC core for real-time wireless prosthetic control, in *ESSCIRC 2019 – IEEE 45th European Solid State Circuits Conference (ESSCIRC)* (2019), pp. 73–76
6. C.M. Lopez, J. Putzeys, B.C. Raducanu, M. Ballini, S. Wang, A. Andrei, V. Rochus, R. Vandebriel, S. Severi, C. Van Hoof, et al., A neural probe with up to 966 electrodes and up to 384 configurable channels in 0.13μm SOI CMOS. IEEE Trans. Biomed. Circuits Syst. **11**(3), 510–522 (2017)
7. R. Harrison, C. Charles, A low-power low-noise CMOS amplifier for neural recording applications. IEEE J. Solid State Circuits **38**(6), 958–965 (2003)
8. E. Guglielmi, F. Toso, F. Zanetto, G. Sciortino, A. Mesri, M. Sampietro, G. Ferrari, High-value tunable pseudo-resistors design. IEEE J. Solid State Circuits **55**(8), 2094–2105 (2020)
9. P. Harpe, H. Gao, R. van Dommele, E. Cantatore, A.H. van Roermund, A 0.20 mm^2 3nW signal acquisition IC for miniature sensor nodes in 65 nm CMOS. IEEE J. Solid State Circuits **51**(1), 240–248 (2015)
10. H. Chandrakumar, D. Marković, A 2μW 40mVpp linear-input-range chopper-stabilized bio-signal amplifier with boosted input impedance of 300MΩ and electrode-offset filtering, in *2016 IEEE International Solid-State Circuits Conference (ISSCC)* (IEEE, Piscataway, 2016), pp. 96–97
11. T. Denison, K. Consoer, W. Santa, A.-T. Avestruz, J. Cooley, A. Kelly, A 2 μW 100nV/$\sqrt{}$Hz chopper-stabilized instrumentation amplifier for chronic measurement of neural field potentials. IEEE J. Solid State Circuits **42**(12), 2934–2945 (2007)
12. N. Verma, A. Shoeb, J. Bohorquez, J. Dawson, J. Guttag, A.P. Chandrakasan, A micro-power EEG acquisition SoC with integrated feature extraction processor for a chronic seizure detection system. IEEE J. Solid State Circuits **45**(4), 804–816 (2010)
13. J. Xu, S. Mitra, A. Matsumoto, S. Patki, C. Van Hoof, K.A.A. Makinwa, R.F. Yazicioglu, A wearable 8-channel active-electrode EEG/ETI acquisition system for body area networks. IEEE J. Solid State Circuits **49**(9), 2005–2016 (2014)
14. Q. Fan, F. Sebastiano, J.H. Huijsing, K.A.A. Makinwa, A 1.8 μW 60nV/$\sqrt{}$Hz capacitively-coupled chopper instrumentation amplifier in 65 nm CMOS for wireless sensor nodes. IEEE J. Solid State Circuits **46**(7), 1534–1543 (2011)
15. N. Van Helleputte, M. Konijnenburg, J. Pettine, D.-W. Jee, H. Kim, A. Morgado, R. Van Wegberg, T. Torfs, R. Mohan, A. Breeschoten, H. de Groot, C. Van Hoof, R.F. Yazicioglu, A 345 μW multi-sensor biomedical SoC with bio-impedance, 3-channel ECG, motion artifact reduction, and integrated DSP. IEEE J. Solid State Circuits **50**(1), 230–244 (2015)
16. K. Nagaraj, A parasitic-insensitive area-efficient approach to realizing very large time constants in switched-capacitor circuits. IEEE Trans. Circuits Syst. **36**(9), 1210–1216 (1989)
17. M. Chen, H.S. Chun, I.D. Castro, T. Torfs, Q. Lin, C. van Hoof, G. Wang, Y. Lian, N. van Helleputte, A 400 GΩ input-impedance active electrode for non-contact capacitively coupled ECG acquisition with large linear-input-range and high CM-interference-tolerance. IEEE Trans. Biomed. Circuits Syst. **13**(2), 376–386 (2019)

18. B.B. Winter, J.G. Webster, Reduction of interference due to common mode voltage in biopotential amplifiers. IEEE Trans. Biomed. Eng. (1), 58–62 (1983)
19. T.W. Degen, Portable devices for mobile health monitoring, Ph.D. dissertation, ETH Zurich (2011)
20. J. Xu, R.F. Yazicioglu, B. Grundlehner, P. Harpe, K.A.A. Makinwa, C. Van Hoof, A 160 μW 8-channel active electrode system for EEG monitoring. IEEE Trans. Biomed. Circuits Syst. **5**(6), 555–567 (2011)
21. A. Dabbaghian, H. Kassiri, An active electrode IC with embedded analog CMRR enhancement for interference- and gain-mismatch-resilient EEG recording, in *2021 IEEE Biomedical Circuits and Systems Conference (BioCAS)* (2021), pp. 01–05
22. B.B. Winter, J.G. Webster, Driven-right-leg circuit design. IEEE Trans. Biomed. Eng. (1), 62–66 (1983)
23. E. Spinelli, N. Martinez, M. Mayosky, A transconductance driven-right-leg circuit. IEEE Trans. Biomed. Eng. **46**(12), 1466–1470 (1999)
24. V. Acharya, Improving Common-Mode Rejection Using the Right-Leg Drive Amplifier, Texas Instruments, Technical Report (2011)
25. J. Xu, B. Büsze, C. Van Hoof, K.A. Makinwa, R.F. Yazicioglu, A 15-channel digital active electrode system for multi-parameter biopotential measurement. IEEE J. Solid State Circuits **50**(9), 2090–2100 (2015)
26. N. Meziane, J. Webster, M. Attari, A. Nimunkar, Dry electrodes for electrocardiography. Physiol. Meas. **34**(9), R47 (2013)
27. M.A. Lopez-Gordo, D. Sanchez-Morillo, F.P. Valle, Dry EEG electrodes. Sensors **14**(7), 12 847–12 870 (2014)
28. E. Habibzadeh Tonekabony Shad, M. Molinas, T. Ytterdal, Impedance and noise of passive and active dry EEG electrodes: a review. IEEE Sensors J. **20**(24), 14 565–14 577 (2020)
29. Y.M. Chi, T.-P. Jung, G. Cauwenberghs, Dry-contact and noncontact biopotential electrodes: methodological review. IEEE Rev. Biomed. Eng. **3**, 106–119 (2010)
30. J. Lee, H. Kim, and S. Cho, A 255 nW ultra-high input impedance analog front-end for non-contact ECG monitoring, in *2017 IEEE Custom Integrated Circuits Conference (CICC)* (2017), pp. 1–4
31. J.M. Kootsey, E.A. Johnson, Buffer amplifier with femtofarad input capacity using operational amplifiers. IEEE Trans. Biomed. Eng. (5), 389–391 (1973)
32. F.N. Guerrero, E.M. Spinelli, A two-wired ultra-high input impedance active electrode. IEEE Trans. Biomed. Circuits Syst. **12**(2), 437–445 (2018)
33. M. Rapin, F. Braun, A. Adler, J. Wacker, I. Frerichs, B. Vogt, O. Chetelat, Wearable sensors for frequency-multiplexed EIT and multilead ECG data acquisition. IEEE Trans. Biomed. Eng. **66**(3), 810–820 (2018)
34. O. Chetelat, J.S.I. CAROS, Floating front-end amplifier and one-wire measuring devices, Sept. 16, 2009, US Patent 8,427,181
35. O. Chételat, J. Oster, O. Grossenbacher, A. Hutter, J. Krauss, A. Giannakis, A highly integrated wearable multi-parameter monitoring system for athletes, in *15th Nordic-Baltic Conference on Biomedical Engineering and Medical Physics (NBC 2011)* (Springer, Berlin, 2011), pp. 148–151
36. S. Song, M. Konijnenburg, R. Van Wegberg, J. Xu, H. Ha, W. Sijbers, S. Stanzione, D. Biswas, A. Breeschoten, P. Vis, et al., A 769 μW battery-powered single-chip SoC with BLE for multi-modal vital sign monitoring health patches. IEEE Trans. Biomed. Circuits Syst. **13**(6), 1506–1517 (2019)
37. M. Rapin, J. Wacker, O. Chételat, Two-wire bus combining full duplex body-sensor network and multilead biopotential measurements. IEEE Trans. Biomed. Eng. **65**(1), 113–122 (2017)
38. K. Badami, M. Pons-Sole, E. Azarkhish, A. Fivaz, M. Rapin, O. Chételat, S. Emery, Single-battery cooperative sensors for multi-lead long term ambulatory ECG measurement, in *2021 IEEE Biomedical Circuits and Systems Conference (BioCAS)* (IEEE, Piscataway, 2021), pp. 1–4

39. J. Xu, S. Mitra, C. Van Hoof, R.F. Yazicioglu, K.A.A. Makinwa, Active electrodes for wearable EEG acquisition: review and electronics design methodology. IEEE Rev. Biomed. Eng. **10**, 187–198 (2017)
40. J. Xu, Q. Fan, J.H. Huijsing, C. Van Hoof, R.F. Yazicioglu, K.A.A. Makinwa, Measurement and analysis of current noise in chopper amplifiers. IEEE J. Solid State Circuits **48**(7), 1575–1584 (2013)

VCO-Based ADCs for Direct Digitization of ExG Signals

Corentin Pochet and Drew A. Hall

Abstract Wearable health sensors can disrupt the healthcare system by providing health professionals with continuous, long-term, high-accuracy physiological data of patients, allowing for more accurate and personalized care. These wearable sensors have specific challenges regarding dynamic range, input impedance, and power consumption. The dynamic range must be wide (>90 dB) to process the small signals of interest (<1 mV) in the presence of large common-mode and differential-mode artifacts (>100 mV). The input impedance of the analog front-end must be high to interface with the high source impedance of small dry electrodes used in wearable applications, all this while consuming very low power and allowing on-chip digital processing to compress and secure patient data. This pushes the technology choice toward digital-friendly, deep submicron nodes where designing such a high-performance analog front-end is challenging. Therefore, new time-based digitization paradigms leveraging voltage-controlled oscillators (VCOs) have been explored to allow for more digital-friendly implementations of the analog-to-digital conversion.

1 Introduction

With the rise of the Internet of things (IoT), there has been mounting interest in wearable devices for long-term, continuous health and wellness monitoring [1–5]. This chapter describes the challenges associated with accurately digitizing biopotential signals in the context of wearable devices, which are prone to artifacts and must therefore be able to absorb or cancel these while maintaining low power and low noise for long-term operation. The remainder of the chapter focuses on architectures that leverage time-domain techniques to enable direct digitization in an energy-efficient manner that are scalable with technology. More specifically, a

C. Pochet · D. A. Hall (✉)
Department of Electrical and Computer Engineering, University of California San Diego (UCSD), La Jolla, CA, USA
e-mail: drewhall@ucsd.edu

P. Harpe et al. (eds.), *Biomedical Electronics, Noise Shaping ADCs, and Frequency References*, https://doi.org/10.1007/978-3-031-28912-5_2

second-order VCO-based front-end will be described, which allows for the direct digitization of biopotential signals with high accuracy and energy efficiency even in the presence of artifacts.

2 ExG Biosignals

Electrocardiography (ECG), electroencephalography (EEG), electrooculography (EOG), and electromyography (EMG) are examples of biopotential recording applications that can be used in wearable devices to provide diagnostics, long-term monitoring for post-operation recovery or chronic disease management. These signals are collectively referred to as ExG signals [6, 7]. When recorded using non-invasive electrodes, ExG signals typically have amplitudes between a few μVs and 10 s of mV over a couple of kHz bandwidth (BW), as shown in Table 1. From this table, it is possible to extract the basic requirements for a wearable analog front-end (AFE) in terms of bandwidth and input-referred noise. The ExG AFE should, at a minimum, be designed to have an input-referred noise of less than 5 μV_{rms} over a bandwidth of at least 1 kHz to accurately digitize the small biopotential signals of interest.

However, low input-referred noise with ultra-low power consumption to ensure high-quality measurements over an extended period is not the only challenge when designing a wearable biopotential AFE. Several other issues must be addressed to acquire high-fidelity, clinical-grade physiological data from a person outside a stationary, controlled environment like a hospital or clinic. First, the electrode-to-tissue interface must be considered. Electrodes introduce up to ±100 mV dc offset due to the polarization of the skin-electrode interface, which can vary with electrode material and skin condition. Acquiring high-quality signals is also challenging due to the impedance formed at the interface between the body and the electrode, which depends on the electrode material, area, and presence/absence of a conductive gel. Several studies have characterized the impedance of different electrodes and shown that the sub-kHz resistivity ranges from ~1 to 100 kΩ/cm^2 [8, 9]. Miniaturizing the

Table 1 ExG signal characteristics and example wearable applications

	Amplitude [μV]	Bandwidth [Hz]	Example applications
ECG	100–5000	0.1–200	Monitor cardiac health
			Diagnosis of arrhythmia
EEG	10–1000	0.1–100	Monitor sleep quality
			Monitor neural activity
EOG	50–2500	0.1–10	Eye direction
			Monitor attention
EMG	100–10,000	50–2000	Muscle activity monitoring
			Prosthesis control

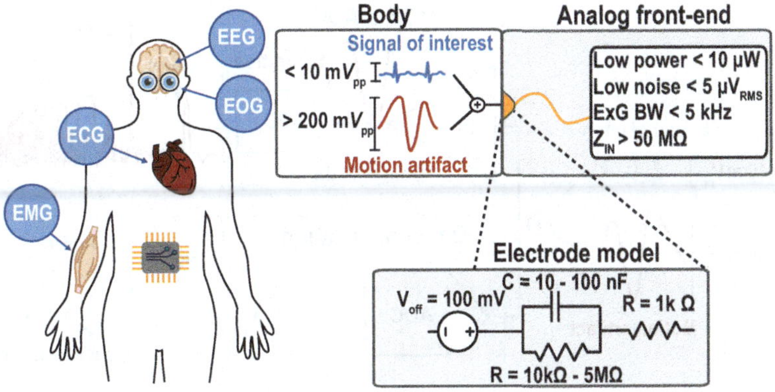

Fig. 1 ExG signal, electrode model, and analog front-end requirements. (Figure adapted from Ref. [31])

electrode to fit in a wearable form factor results in electrode impedances between 10 kΩ and 5 MΩ, requiring a front-end with a high input impedance to not attenuate the signal. More challenging, in a wearable setting, motion artifacts and 50/60 Hz power line coupling can induce large in-band signals (common- and differential-mode), requiring an input range larger than 200 mV [10]. Thus, ExG AFEs for wearable applications require a dynamic range (DR) greater than 90 dB and a high common-mode rejection ratio (CMRR) to digitize the ExG signal with electrode offset and motion artifacts without saturating. Importantly, most motion artifact removal algorithms operate on the assumption that the acquired signal is a linear combination of the artifact and ExG signal [11, 12], imposing a strict linearity requirement (>100 dB). Figure 1 illustrates the source of the various ExG signals (*i.e.*, where the electrodes need to be placed), along with a typical electrode model and a summary of the AFE requirements.

3 Acquisition Circuits

Researchers have worked on designing circuits capable of efficiently and accurately digitizing biopotential signals for the past few decades while tackling the challenges highlighted in the previous section. The conventional AFE for biopotential acquisition is illustrated in Fig. 2. It is composed of a high gain programmable gain amplifier (PGA) that amplifies and filters the ExG signal [2, 10] before digitization while also providing high input impedance (>10 MΩ). Due to the amplification and filtering, a simple Nyquist rate analog-to-digital converter (ADC), such as a low-power successive-approximation (SAR) ADC, can be used to digitize the amplifier output in a power-efficient manner. This structure works well for stationary recording, but motion artifacts can cause saturation and/or distortion due to the PGA's limited DR and linearity leading to information loss and long recovery times.

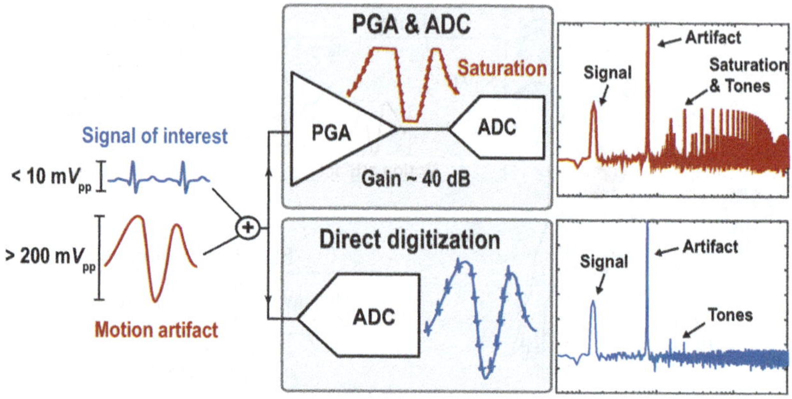

Fig. 2 Acquisition methods for biopotential acquisition and output spectra in the presence of large artifacts

As such, there has been increasing effort toward removing the PGA and directly digitizing the signal [1, 13–20] with specialized ADCs, as shown in Fig. 2.

Continuous-time (CT) delta-sigma modulators (DSM) are one candidate to replace the classic PGA and SAR-ADC front-end as they have intrinsic anti-aliasing, can achieve high DR through oversampling, and can be designed to have high input impedance [1, 15, 16, 21]. Several designs have achieved the necessary DR for wearable applications [1, 14, 17]; however, they do not have sufficient linearity, complicating downstream signal processing and analysis used to remove the artifacts. Conventionally, CT-DSMs are designed using amplifiers and comparators processing the signal in the voltage domain. While these can achieve excellent performance [21, 22], the need for wearable sensors which combine analog sensing functionality with digital intensive processing, such as edge compression or classification, is driving designers to use more advanced digital-friendly processes. Using these highly scaled technology comes at the cost of reduced supply voltage and intrinsic gain [23], which makes designing the analog-heavy CT-DSM implementations difficult. A key element to achieving high DR and linearity with a DSM relies on multi-bit internal quantizers, which require a mismatch shaping technique to remove mismatch-induced non-linearity in the feedback path, adding delay and power. Due to these drawbacks, there has been growing interest in time-domain CT-DSMs with phase-domain integration using voltage-controlled oscillators (VCO). By processing the signal in the time/phase domain, VCO-based ADCs do not suffer as much from the supply voltage reduction and benefit from the shorter transition times associated with advanced process nodes. Furthermore, in some implementations, the circular nature of the ring oscillator can be leveraged to provide intrinsic data-weighted averaging (DWA) [9, 13].

4 VCO-Based ADCs

Due to the aforementioned advantages, several architectures have been proposed in the past few years that rely on time-domain signal processing to digitize biopotential signals. This chapter presents an example of one such architecture, a high-performance second-order VCO-based ADC that directly interfaces with the electrodes. It achieves high accuracy and low noise digitization with high linearity and artifact-dependent power savings.

4.1 The VCO as an Integrator

At the heart of any CT-DSM ADC is the integrator. This block is responsible for the noise shaping, provides gain, and sets the input-referred noise. The integrators are critical to the ADC's linearity and achievable signal-to-quantization noise ratio (SQNR). In time-domain architectures, this integration function is performed by the VCO. As their name indicates, VCOs are a circuit architecture where the oscillation frequency, f_{VCO}, depends on an input voltage. The instantaneous frequency can be expressed as

$$f_{\mathrm{VCO}}(t) = K_{\mathrm{VCO}} v(t), \tag{1}$$

where K_{VCO} is the voltage-to-frequency gain in Hz/V and $v(t)$ is the instantaneous input voltage. Equation (1) indicates that there is a proportional relationship between the oscillator frequency and the input voltage, but, perhaps more interestingly, one can observe that the relationship between the phase of the VCO and the input can be expressed as

$$\phi_{\mathrm{VCO}}(t) = \int_0^t 2\pi K_{\mathrm{VCO}} v(t) dt, \tag{2}$$

where it is apparent that in the phase domain, the VCO can be used as an integrator and thus provide high gain and enable analog signal processing. As shown in Eq. (2), the integrator gain is set by the open-loop parameter K_{VCO}, which will depend on the architecture and sizing of the oscillator.

This integrator-like behavior is the key to using a VCO to replace analog-heavy amplifiers in many applications. In the past several decades, VCO-based integrators have been used to perform functions such as amplification [25, 26], voltage regulation [27], filtering [28], and analog-to-digital conversion [29]. The latter application is the focus of this chapter, where the properties of the VCO will be leveraged to implement delta-sigma ADC architectures.

A typical VCO-based integrator implementation is shown in Fig. 3. The integrator is composed of a transconductor (G_{m}) that converts the input voltage to

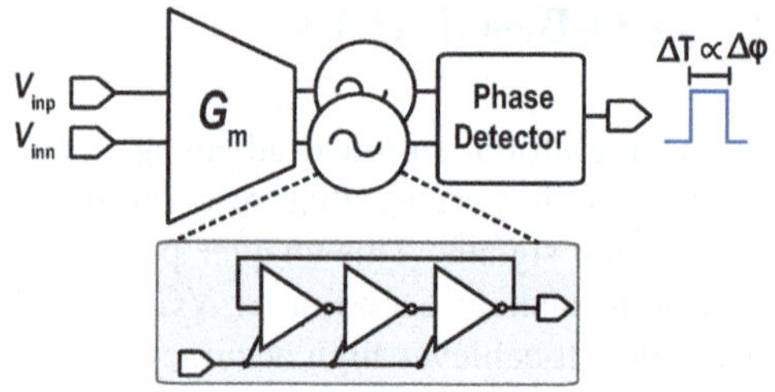

Fig. 3 Block diagram of a VCO-based integrator

an output current, which is then used to drive two pseudo-differential current-controlled oscillators (CCO) composed of several inverter-like cells. The phase difference between the two oscillators, $\Delta\varphi$, can be extracted with a phase detector (PD), leading to a time-domain representation of the integrator's output value with longer pulses corresponding to a larger integration value. K_{VCO} can be expressed as

$$K_{VCO} = G_m K_{CCO} G_{PD} \tag{3}$$

where G_m is the transconductance, K_{CCO} is the CCO gain in Hz/A, and G_{PD} is the phase detector gain.

4.2 *Quantizer*

As discussed in the previous section, the VCO converts an input voltage to a frequency with a linear gain. Therefore, the quantizer of the VCO often operates as a frequency-to-digital converter (FDC) to extract the input component. Two solutions have been popularized over the years with different complexity and properties, but they operate on the same principle: the phase difference of the VCO is sampled and differentiated to obtain the frequency and thus resolve the input signal. As explained later, this is also key to providing the noise-shaping property in a VCO-based ADC. The two main techniques to digitize the phase of the VCO are (1) an XOR-based FDC, where each node of the VCO is tapped, and (2) a counter-based FDC, where a counter is connected to a single phase of the VCO. These techniques are illustrated in Fig. 4, and the pros and cons are discussed next.

4.2.1 XOR-Based FDC

The XOR-based FDC takes advantage of all of the phases of the VCO. As shown in Fig. 4a, each node of the VCO is sampled and compared to the previous sample using an XOR gate. This enables the comparison of the phase between two samples and allows for the VCO frequency to be extracted. To avoid overflow, the designer must ensure that the VCO frequency, f_{VCO}, is such that $0.5f_s < f_{VCO} < 1.5f_s$. If either

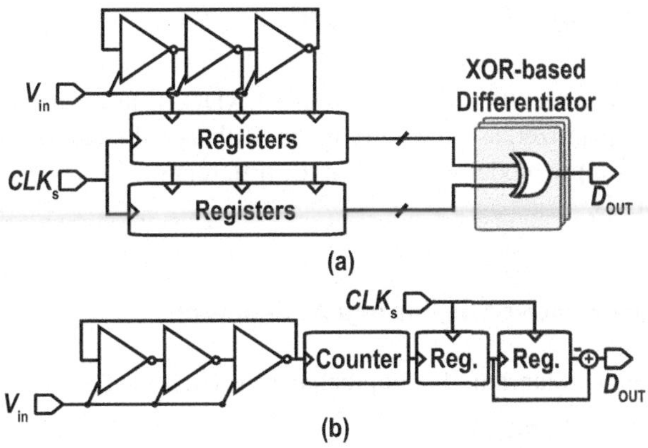

Fig. 4 Schematics of an (**a**) XOR-based FDC and (**b**) counter-based FDC

of these bounds is exceeded, it will cause phase wrapping and an incorrect output. An advantage of the XOR-based FDC is that it naturally provides DWA of the quantizer output such that any mismatch between the inverter stages is shaped and that if the output is used to drive a feedback DAC, it also shapes the DAC mismatch. The XOR-based FDC is very compact and efficient; however, it is complicated to implement more than a 4-bit quantizer due to the thermometer encoding leading to a rapidly increasing number of wires complicating the layout.

4.2.2 Counter-Based FDC

The counter-based FDC, illustrated in Fig. 4b, operates by counting the number of VCO edges in a sampling period. To achieve high accuracy, the sampling clock is typically much slower than f_{CCO} such that multiple edges are counted each period. Using the difference between the counter's end value after two sampling instances, the frequency is extracted, and the input signal can be extracted. The counter does not need to be reset, and 2's complement arithmetic can be used to calculate the result of the subtraction between the two results if the system is designed such that only one wrap-around can occur by limiting the maximum VCO frequency. This counter-based technique is popular for large quantizers (>4-bit) and an area-efficient architecture as it allows for a small VCO due to the relaxed requirements on f_{CCO} and the need for high-frequency operation. It should be noted that the counter is typically implemented with a gray counter to avoid causing a large quantization error when the sampling edge is triggered close to the asynchronous VCO edge and, thus, during a counter transition. One drawback of the counter-based technique is that the quantizer output must go through a mismatch-shaping algorithm to guarantee high linearity if it is to be used in a closed-loop design with a multi-bit feedback DAC.

4.3 Properties of VCO-Based ADCs

The simplest implementation of a VCO-based ADC is an open-loop VCO-based integrator and a quantizer, as shown in Fig. 5a, with an *s*-domain model of the ADC in Fig. 5b. This simple model will explain the different properties and challenges of VCO-based ADCs.

4.3.1 First-Order Noise Shaping and Anti-aliasing

Analyzing the block diagram, the loop behavior is expressed as

$$D_{\text{out}}\left[n\right] = \int_{(n-1)T_s}^{nT_s} K_{\text{VCO}} v(t) dt + \left(E_{\text{Q}}\left[n\right] - E_{\text{Q}}\left[n-1\right]\right) \quad (4)$$

where D_{out} is the digital representation of the input signal and E_{Q} is the quantization error. The second term highlights that the differentiation at the output effectively high-pass filters the quantization error, thus providing first-order noise shaping to the system. The first term, on the other hand, can also be expanded and shown to be equivalent to passing the input signal $v(t)$ through a sinc filter with a frequency response, H_{F}, of

$$H_{\text{F}}(f) = \frac{K_{\text{VCO}} e^{-j\pi T_s f} \sin\left(\pi T_s f\right)}{\pi f}. \quad (5)$$

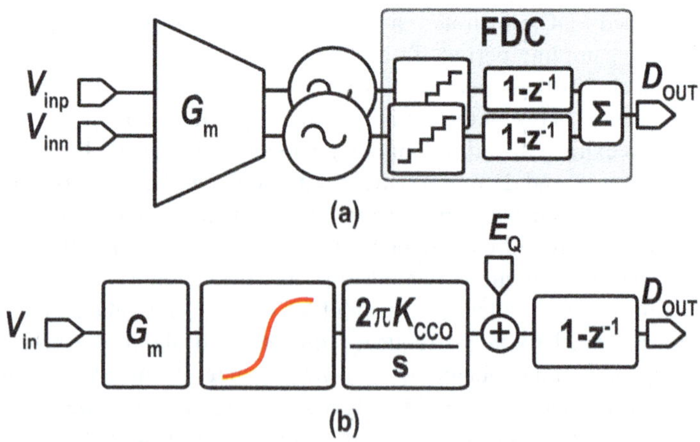

Fig. 5 (**a**) Open-loop VCO-based ADC; (**b**) *s*-domain block diagram

At low frequencies, the filter response is ~1, and thus the output is a good approximation of the input voltage. It also shows that, like a standard voltage domain architecture, the VCO-based ADC provides anti-aliasing by filtering the input signal at high frequencies and has nulls around multiples of the sampling frequency, f_S. Another advantage of using VCO as integrators is that the dc gain is infinite, which is a key factor in avoiding issues such as dead-bands and SQNR degradation in CT-DSM architectures due to the limited dc gain of the opamps in the *RC* integrators.

4.3.2 Linearity

A key challenge in VCO-based ADCs is the non-linearity of the voltage-to-frequency gain. This non-linearity is caused by the non-linear transconductor and the non-linear response of the ring oscillator. Due to these limitations, the linearity of VCO-based ADCs has typically been less than 50 dB in open loop [29]. This limited linearity has been one of the main challenges in the design of high SNDR VCO-based ADCs, and, as such, many techniques have been explored to enhance the voltage-to-frequency conversion linearity using techniques such as feedback or new oscillators that have better linearity than the classic inverter-based ring oscillator.

4.3.3 Resilience to Metastability

Metastability is a major concern when implementing multi-bit quantizers, especially with low supply voltages. A quantizer with a large number of bits, *N*, reduces the quantization step, V_{DD}/N, which is challenging when decreasing the supply voltage as it increases the power and area of the comparator. However, the inner stages of the VCO-based integrators are mostly either at ground or V_{DD}, except for the state that is currently transitioning. This significantly reduces the likelihood of a metastable state.

4.4 ExG VCO-Based AFE Literature Survey

Several architectures have been proposed over the past several decades to leverage the valuable properties of VCO-based ADCs (supply insensitivity, anti-aliasing, compactness) while addressing the linearity issue and the challenges associated with wearable ExG AFEs, namely, high input impedance, low power, and low noise. Several of these architectures are illustrated in Fig. 6. In [15], Jiang et al. proposed using an open-loop VCO and linearizing it with digital non-linearity correction (Fig. 6a). While this non-linearity correction algorithm increases the SFDR to ~70 dB, the input amplitude is limited to 50 mV_{pp}. Implementing the large digital block required for the non-linearity correction also consumed significant

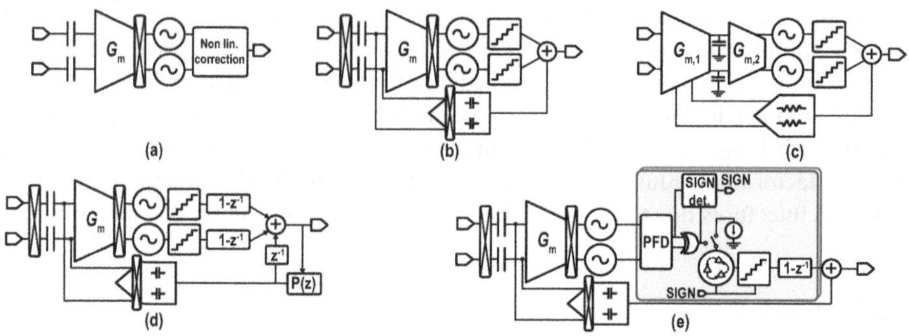

Fig. 6 Direct digitization VCO-based sensor front-end architectures: (**a**) open-loop with digital correction, (**b**) closed-loop ac-coupled, (**c**) hybrid dc-coupled, (**d**) DPCM-based, and (**e**) second-order multi-phase quantizer. (Figure adapted from Ref. [31])

power due to leakage (20× higher than the switching power), leading to high power consumption. To further improve the ADC's linearity and increase the input range, a closed-loop architecture, shown in Fig. 6b, was reported in [16]. This design achieved 90 dB SFDR without requiring an explicit dynamic element matching (DEM) algorithm by leveraging the intrinsic data shuffling of VCO-based quantizers discussed earlier. However, the input range was still limited to 100 mV_{pp}, and the first-order noise shaping required a high oversampling ratio (OSR). When using chopping to mitigate $1/f$ noise, the OSR requires a high chopping frequency leading to a low input impedance (220 kΩ). In [24], a dc-coupled architecture combines a G_m-C integrator with a VCO-based quantizer, as shown in Fig. 6c. The dc-coupled architecture ensures a high input impedance, but it comes at the cost of sensitivity to input common-mode and an analog-heavy first stage. A differential pulse-code modulation (DPCM)-based predictor is used in [17], as shown in Fig. 6d, to maintain first-order quantization noise shaping while allowing for second-order shaping of the input. However, this requires a large 9-bit capacitive DAC (CDAC) with a large input capacitor, resulting in a low input impedance.

5 A Second-Order VCO-Based ExG Interface

This section focuses on designing a single-loop, second-order VCO-based ADC that combines two VCO-based integrators and leverages a novel multi-phase, multi-quantizer technique to achieve high DR and SNDR in a power-efficient fashion, as shown in Fig. 6e. The approach uses an auxiliary input impedance booster for high input impedance and offers an elegant solution for wearable ExG front-ends.

5.1 ADC Architecture

5.1.1 Basic Operation

The proposed architecture is inspired by [30], a high-speed ADC with second-order noise shaping using only VCO-based integrators. The architecture, shown in Fig. 7a, comprises two main blocks to achieve second-order noise shaping. The first stage is a standard VCO-based integrator, as discussed in the previous section. It contains a G_m cell and a current-controlled oscillator followed by a phase detector. The second stage is a noise-shaped dual-mode ring oscillator (DMRO) time-to-digital converter (TDC), essentially a VCO-based integrator that operates between two levels. The TDC quantizes the pulse at the PD output and feeds it back to the input through a multi-bit DAC guaranteeing a small input swing at the G_m-CCO integrator input and, thus, high linearity. The time-based encoding, which is used inside the loop, has three notable benefits: (1) It enables the use of a low supply voltage after the first stage since the information is encoded in the pulse width, not the amplitude, (2) the second stage's linearity is guaranteed since a two-level signal drives it, and (3) the DMRO power is duty-cycled by the output of the PD as it is proportional to the input voltage. As shown in Fig. 7b, using the models developed in the previous section, an s-domain representation can be derived and shows that the loop contains two integrators and has second-order noise shaping. The loop is stable for a wide range of coefficients and can tolerate significant

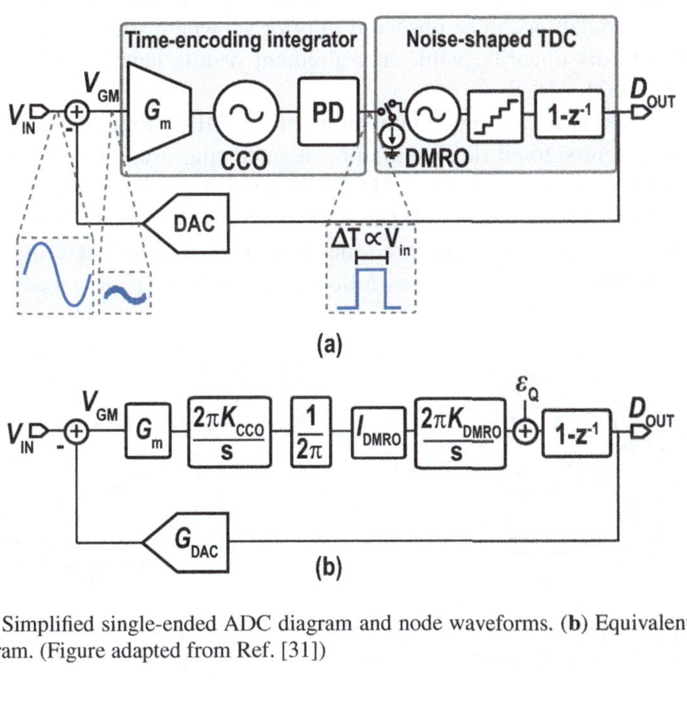

Fig. 7 (**a**) Simplified single-ended ADC diagram and node waveforms. (**b**) Equivalent s-domain block diagram. (Figure adapted from Ref. [31])

quantizer delay, guaranteeing reliable and stable operation, even if the coefficients shift due to process, voltage, and temperature (PVT) variation [32]. The following sections discuss the implementation innovations that allowed for improvements in power efficiency, reliability, and linearity compared to the original design.

5.2 *Design of a Mimatch Resilient Quantizer*

One of the first challenges associated with this architecture is its sensitivity to quantizer path mismatch. As shown in Fig. 7b, the TDC is modeled as a CT integrator followed by a digital differentiator, which is equivalent to having a block with $2\pi K_{\text{DMRO}}$ gain [33]. To cancel the ADC input, the pulse width at the output of the first integrator must be proportional to the input amplitude. Therefore, the non-saturating range of the PD directly defines the ADC's input range, as an overflow would cause the loop to be unstable. Typically, the PD is implemented using a phase-frequency detector (PFD), which has a 2× larger non-saturating input range than an XOR-based PD [28]. The PFD-based TDC operation is shown in Fig. 8, where the positive or negative path is activated depending on the signal polarity. Despite the high loop gain, this pseudo-differential operation leads to significant SFDR degradation with path mismatch. Figure 8b shows the simulated linearity where the path mismatch was varied from 0% to 5%. To achieve better than 100 dB SFDR, the mismatch between the two paths must be less than 0.5% and degrades rapidly, reaching 80 dB with a 5% mismatch due to even-order harmonic distortion. This SFDR degradation can be observed in prior art where simulation results show more than 100 dB linearity, while measurement results have an SFDR less than 80 dB [30, 32, 33].

There are two common implementations of a DMRO noise-shaped TDC, both based on a current-starved ring oscillator – a gated ring oscillator (GRO) [35] and a switched ring oscillator (SRO) [33]. While a GRO has lower power and better noise performance due to the on/off behavior, the SRO reduces the timing skew and thus improves the linearity. The three major sources of frequency mismatch in a current-starved ring oscillator are as follows: (1) the total node capacitance, C_{tot},

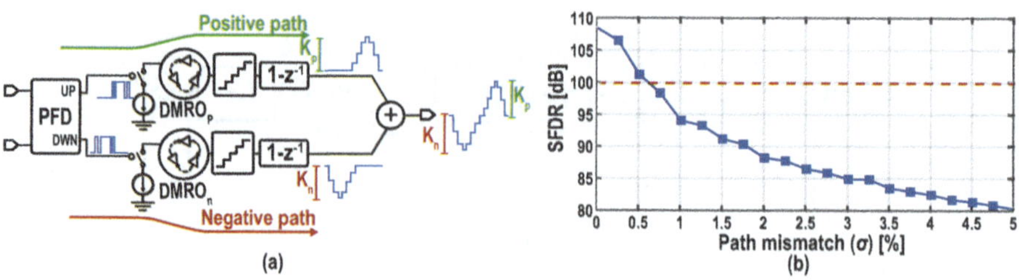

Fig. 8 (**a**) Block diagram of mismatch-induced non-linearity in a dual-path TDC. (**b**) Simulated SFDR as a function of path mismatch. (Figure adapted from Ref. [31])

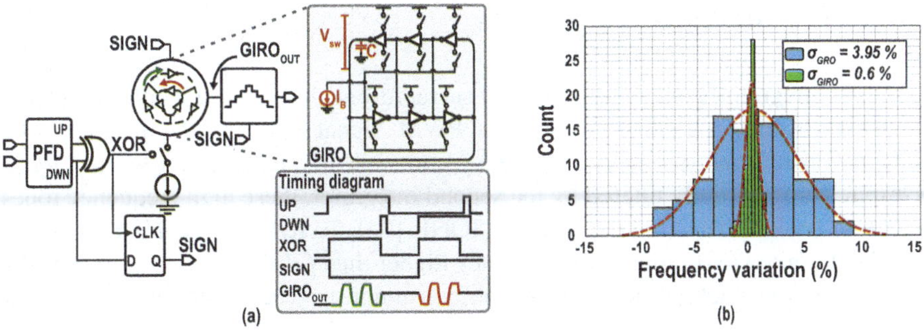

Fig. 9 (**a**) GIRO and sign detection circuit with timing diagram and (**b**) Monte Carlo simulation results comparing a GRO and GIRO. (Figure adapted from Ref. [31])

which is influenced by variation in sizing and the oxide capacitance; (2) the voltage swing, V_{SW}, due to threshold voltage variation; and (3) bias current variation, I_B. To improve the path matching of a GRO, we reported the GIRO structure shown in Fig. 9. The GIRO combines an inverted ring oscillator-based TDC [36] and a sign detection circuit, similar to a digital PLL [37]. The sign detection is implemented with an XOR gate and a DFF. The XOR takes the "absolute value," outputting only the pulse width, and clocks the DFF, which samples the DWN path, thus extracting the "sign" information. This is slightly different from [37], where the other PFD output clocked the sign detection circuit. Clocking with the XOR gate ensures that the SIGN bit updates at the start of the XOR pulse instead of flipping between the start and end, as in the original implementation. With this sign detection circuit, the two TDC paths can be merged into a single path, and the polarity correction is pushed to the digital domain.

The noise shaping is maintained by using an inverted ring oscillator structure, which merges two ring oscillators such that the TDC oscillates in a clockwise or counter-clockwise direction depending on the SIGN bit polarity. This allows the nodes to share the same capacitance and bias current, significantly improving the matching. The path merging also reduces the leakage by ensuring that the TDC only holds the charge during the short time between pulses instead of between polarity flips, as required in a dual-path GRO.

Figure 9b shows Monte Carlo simulation results ($n = 100$) of the frequency variation of a GRO and GIRO. For the GRO, the mismatch is zero mean with a 4% variance, limiting the SFDR to ~82 dB, per Fig. 8b. The GIRO also has a zero-mean mismatch, but the variation is reduced to 0.6%. For similar matching, the size of the dual-path SRO/GRO would need to be increased by ~36× (based on Pelgrom's law), which would require significantly increasing the quantizer's power consumption. Thus, the GIRO structure improves the matching by 6.5× over a GRO and enables the system to achieve 100 dB SFDR much more efficiently than the two-path approach.

5.3 Multi-quantizer Structure

The first integrator's output after the PFD is represented by a pulse width-modulated (PWM) encoded signal at f_{CCO}. This PWM signal contains the result of the integration information at low-frequency and high-frequency tones around f_{CCO}. Despite being partially filtered by the second integrator, these high-frequency tones can fold back in-band and degrade the SQNR [32]. To reduce this effect, the CCO can be designed to oscillate at a frequency higher than f_{s} [32]; however, this has the significant drawback of increasing power consumption. The integrator gain, K_{int}, is

$$K_{\text{int}} = G_{\text{m}} K_{\text{CCO}} \propto \frac{G_{\text{m}}}{I_{\text{B}}} f_{\text{CCO}}. \tag{6}$$

This shows that for a target K_{int}, a higher f_{CCO} reduces the achievable G_{m} cell's efficiency. Since the noise target often sets G_{m}, it is critical to minimize f_{CCO} to minimize the integrator's power.

In [30], a multi-phase approach was proposed to break this tradeoff, where f_{CCO} is virtually increased by tapping and combining multiple phases of the ring oscillator in the current domain. While this technique lowers f_{CCO}, it comes at the cost of losing the intrinsic linearity of the TDC-based quantizer. Instead, one can use multiple noise-shaped TDCs in parallel, which maintains the intrinsic linearity of each channel and allows a lower f_{CCO}. This approach is illustrated in Fig. 10a using three phases. Similar to [30], multiple out-of-phase components of the ring oscillator are tapped, but instead of being combined in an analog fashion using a current summer, each of the individual channel's SIGN bit is detected, and then the output of the XOR gate is fed to the GIRO-based TDC. The results are then summed digitally. This ensures that each channel only operates between two levels by pushing this summation to the digital domain, thus preserving the inherent linearity.

Figure 10b illustrates how the digital tone cancellation works. Each XOR output pulse contains the integration information and high-frequency tones quantized by a separate TDC. Each TDC output comprises input tones scaled by 1/*N*, where *N*

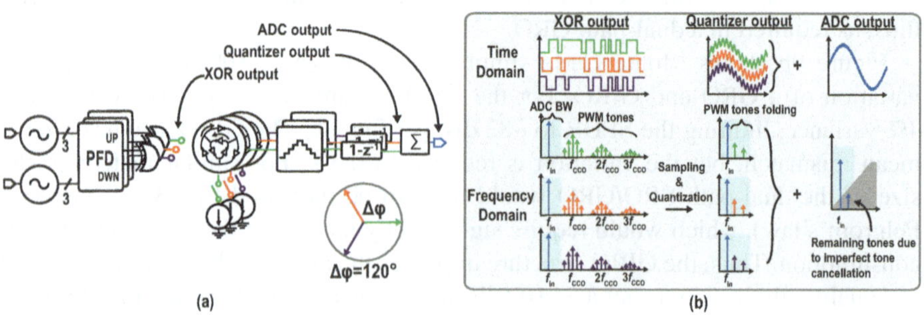

Fig. 10 (**a**) Multi-quantizer-based TDC ($N = 3$). (**b**) Time- and frequency-domain operation of the constructive and destructive summing. (Figure adapted from Ref. [31])

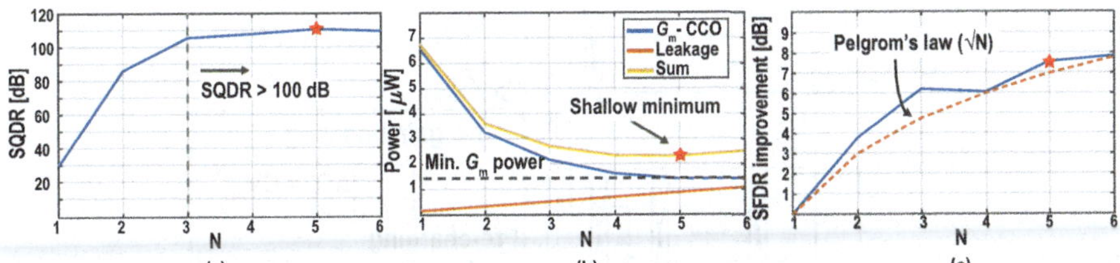

Fig. 11 (**a**) SQDR vs. *N*, (**b**) power vs. *N*, and (**c**) SFDR improvement vs. *N*. (Figure adapted from Ref. [31])

is the number of channels, and folded back tones. The phase and frequency of the input tone are matched across the channels, while the folded tones generated by the PWM encoding are both out-of-phase *and* at the same time folded frequency. These conditions enable the signal tones to add constructively during digital summation, while the folded tones add destructively. This leads to a high SQNR and is reasonably insensitive to phase and gain mismatch between the channels. Selecting N is a tradeoff between area, power, and the target signal-to-quantization and distortion ratio (SQDR). As illustrated in Fig. 11a, b, as the number of phases increases, the SQDR increases, while the power required by the G_m-CCO decreases. However, the digital cells' leakage power increases, leading to a shallow power optimum. Balancing these tradeoffs, $N = 5$ was selected for this design.

Another advantage of the multi-quantizer approach is that it improves the linearity due to averaging by reducing the distortion and the quantization noise, as observed in [38]. Here, the multiple TDCs act as staggered quantizers improving the SQNR and linearity. This was simulated and is shown in Fig. 11c, where the matching increases with N and follows the well-known Pelgrom's law [39, 40], which trades off area (the number of quantizers) for improved matching. For $N = 5$, this improves the SFDR by 7 dB. This multi-quantizer averaging also reduces the effect of inter-channel gain and phase mismatch. Simulations showed no SQNR degradation occurs for gain and phase mismatch <15%, which is easily achieved with appropriate sizing and layout.

5.4 *Chopping and Impedance Boosting*

Chopping is a well-known technique to allow high precision and low noise in sensor applications with low bandwidth. The technique relies on a set of switches driven by a clock to upmodulate the offset and the flicker noise of the AFE. One drawback of chopping is that it reduces the impedance seen at the input of the chopping switch. The input capacitor must be charged at each chopping edge, drawing current from the input. This input impedance depends on the input capacitor and the chopping frequency. In a CT-DSM, the chopping frequency must be at a multiple of $f_s/2$ such

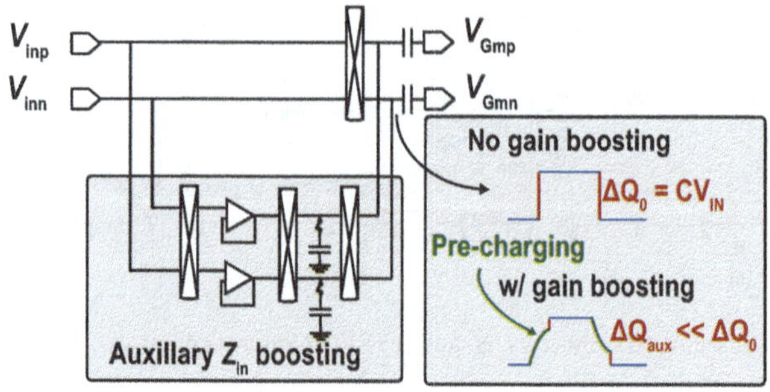

Fig. 12 Auxiliary impedance boosting concept

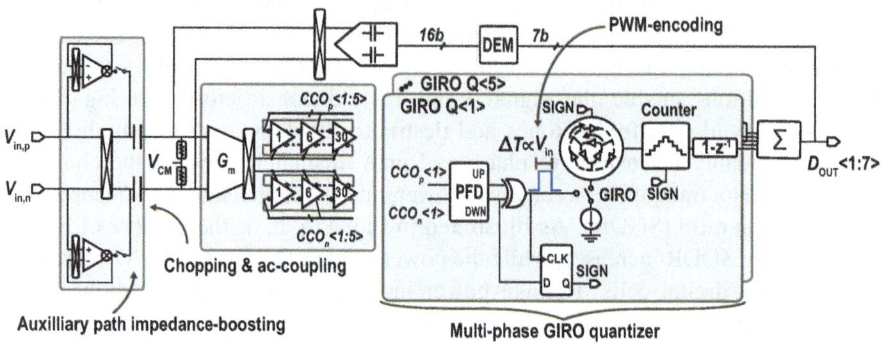

Fig. 13 ExG front-end architecture. (Figure adapted from Ref. [31])

that the chopping artifact lands in the nulls of the sinc filter response of the ADC, avoiding quantization noise folding.

To reduce this impedance degradation, it was proposed to precharge the ac-coupling capacitors with auxiliary buffers such that less charge needs to be supplied by the source, as shown in Fig. 12. The auxiliary buffers can be duty-cycled to reduce the power of the auxiliary buffers. It should be noted that the buffers themselves need to be chopped to reduce their flicker noise and offset, which creates a parasitic impedance that limits the impedance boosting.

5.5 *Architecture Summary*

Putting these techniques together, the proposed ADC is shown in Fig. 13. The auxiliary impedance booster increases the AFE input impedance to a high enough value to interface with the electrodes. The multi-quantizer and GIRO quantizer enable high linearity and high power efficiency.

5.6 Measurement Results

The VCO-based ExG AFE was fabricated in a 65 nm TSMC LP process and occupies an active area of 0.075 mm^2. An annotated die micrograph is shown in Fig. 14. The first integrator, impedance booster, and CDAC are powered from a 1.2 V supply, while the multi-phase TDC quantizer and the digital blocks containing the summer and DEM logic operate at 0.8 V.

5.6.1 Electrical Characterization

The ADC's SNDR, SFDR, and DR were characterized with an ultra-low distortion signal generator (APx555) that generated a full-scale, 400 mV_{pp} sinusoid. The measured output spectrum is shown in Fig. 15. The ADC achieved an SNDR of 92.3 dB in 1 kHz BW and a 110.3 dB SFDR, limited by the third harmonic. The 40 dB/decade noise shaping expected from the second-order noise transfer function (NTF) is evident. Figure 16 shows the measured 92.3 dB DR and the power as a function of the input amplitude. The pie charts show that the system power reduced from 5.8 to 4.25 μW (35%) with low input swing due to a lower GIRO and digital power, as expected from the input-dependent time-encoding, which duty-cycles the back-end. A 30% reduction in the power consumption is observed when the input amplitude was below 10 mV, which is within the typical range of ExG signal values, enabling lower power in the absence of motion artifacts and significantly increasing the lifetime of the wearable device.

The AFE input-referred noise was measured with shorted inputs. The PSD is shown in Fig. 17. The integrated noise from 1 Hz to 1 kHz is 3.5 μV_{rms}, and the average spot noise over the bandwidth is 110 nV/$\sqrt{Hz}$. The $1/f$ noise in the spectrum is mainly due to the inverters in the first CCO, which are not chopped. However, this is less than 15% of the integrated noise.

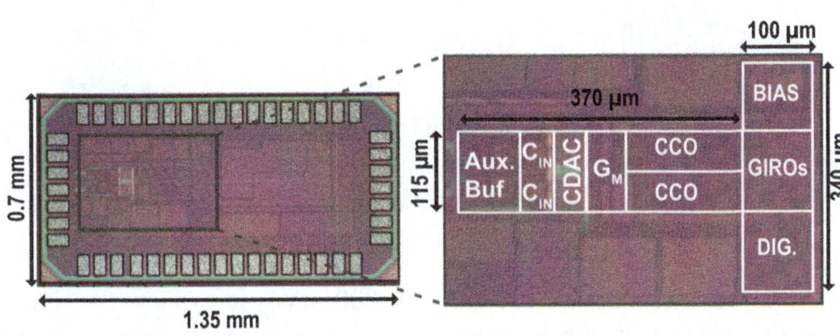

Fig. 14 Die micrograph. (Figure adapted from Ref. [31])

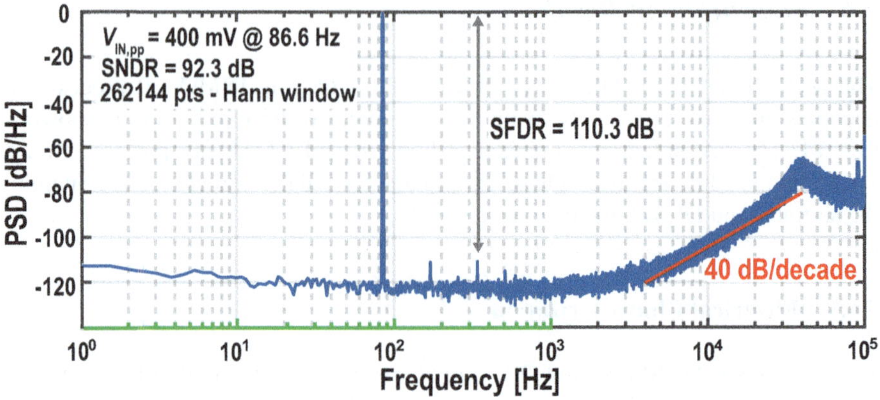

Fig. 15 Measured ADC output spectrum. (Figure adapted from Ref. [31])

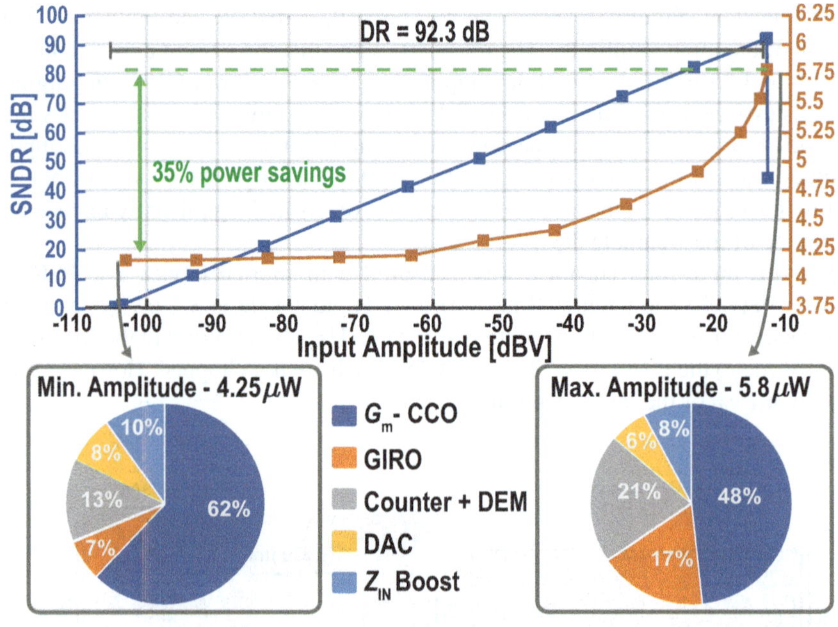

Fig. 16 Measured SNDR and power vs. input amplitude. (Figure adapted from Ref. [31])

The ADC was excited with two tones at 200 mV_{pp} to evaluate the impact of large motion artifacts on the linearity. The measured intermodulation distortion (IMD) was 99.8 dBc with the intermodulation products ~1 μV, which is below the 3.5 μV_{rms} integrated noise of the AFE. This demonstrates that even in the presence of significant motion artifacts/interference, the distortion would be below the noise floor and thus not introduce any unwanted signals. The common-mode

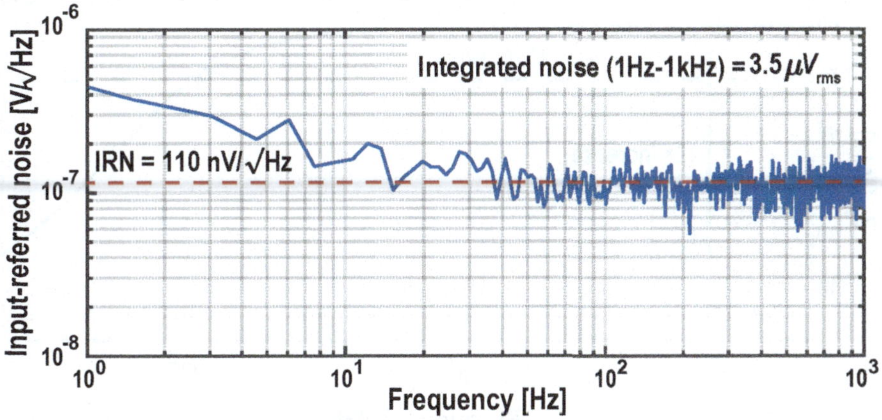

Fig. 17 Measured input-referred noise. (Figure adapted from Ref. [31])

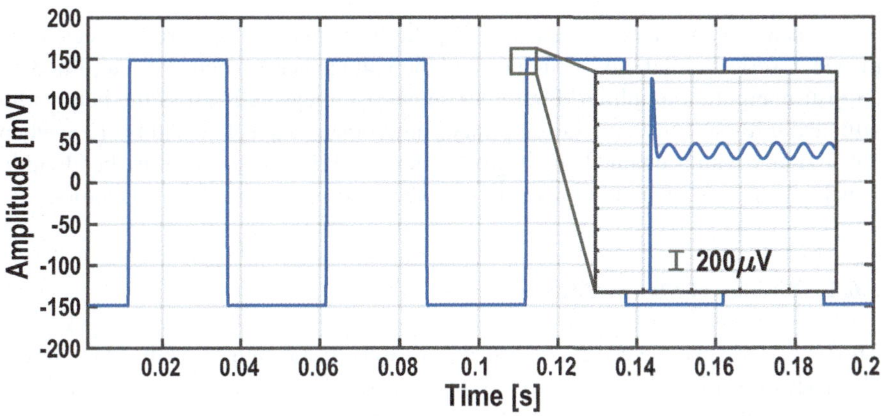

Fig. 18 Measured square wave and sinusoid combined. (Figure adapted from Ref. [31])

rejection ratio (CMRR) was measured across the bandwidth with a full-scale ADC input (200 mV_{pp} single-ended) and was above 95 dB over the entire frequency range. Since a larger common-mode artifact is expected in a wearable device (up to 400 mV_{pp}), the ADC's CMRR was measured with this larger input to be 82 dB. Common-mode artifacts larger than 400 mV_{pp} cause the ADC's loop stability to degrade, leading to a loss in performance. The insensitivity to large and rapid artifacts, as expected from sudden movement or stimulation artifacts, was tested by superimposing a 150 μV_{pp} sinusoid signal on a 300 mV_{pp} square wave. The resulting decimated time-domain waveform is illustrated in Fig. 18, demonstrating that the ADC quickly recovers from the artifact, allowing the small sinusoidal to be clearly observed.

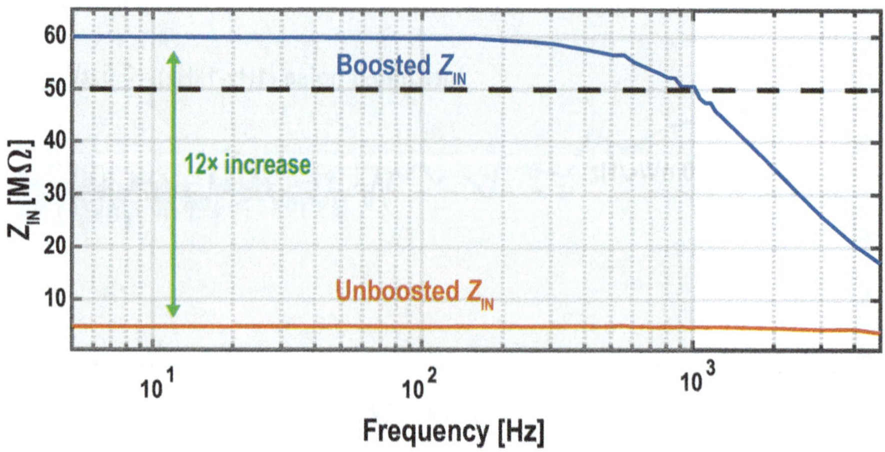

Fig. 19 Measured input impedance with and without boosting. (Figure adapted from Ref. [31])

To measure the input impedance, Z_{in}, high-precision 1 MΩ resistances were placed in series with the ADC inputs, and the voltage across them was amplified by an instrumentation amplifier. Figure 19 shows Z_{in} measured across the bandwidth with and without activating the auxiliary path booster. Z_{in} is boosted by 12× when the auxiliary path is enabled and is greater than 50 MΩ across the entire bandwidth, which allows it to operate with small, dry electrodes on wearable form factors.

5.6.2 Biological Measurements

This AFE was validated by measuring various ExG signals with 3M red dot electrodes on a healthy volunteer. Figure 20a shows the waveform from a three-lead chest ECG recording. Around 8 s into the recording, motion artifacts were purposely generated to showcase the ability of the system to correctly digitize the ECG waveform in the presence of large motion artifacts like one would encounter with a wearable in a non-stationary environment. The system does not saturate, and the ECG waveform can be observed and extracted despite the large motion artifact. The system was also used to measure EOG (Fig. 20b) and EMG (Fig. 20c) showing that the system can measure standard ExG waveforms.

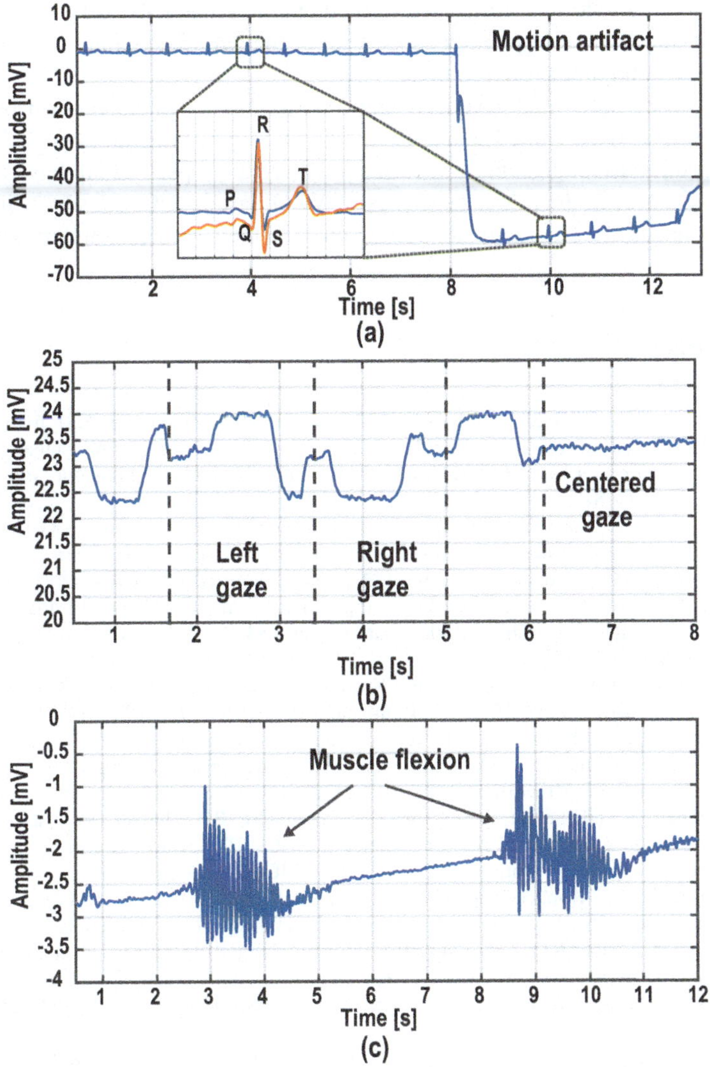

Fig. 20 ExG measurements: (**a**) ECG, (**b**) EOG, and (**c**) EMG. (Figure adapted from Ref. [31])

6 Conclusions

This chapter presented an ADC architecture that leverages VCO-based integrators to achieve higher-order noise shaping with time-only signal processing. This allows a very efficient AFE for use in direct digitization applications. The time-domain processing allows for a low supply voltage and input-dependent power duty cycling, which enables significant power savings in the absence of artifacts. The proposed architecture was verified in measurements to have all of the requirements for continuous wearable monitoring – a wide dynamic range (>90 dB), a large SFDR (>100 dB), and a high input impedance – all of that while maintaining a low power consumption and input-referred noise.

It should be noted that several recent architectures have also shown promising results by leveraging mixed-domain (voltage and time) integrators or feedforward of the pseudo-virtual ground to linearize the integrators [41]. There have been significant advances in the power efficiency of VCO-based ADCs for sensor interfaces over the past decade, and they are getting close to their voltage counterparts.

References

1. X. Yang et al., A 108dB DR hybrid-CTDT direct-digitalization $\Delta\Sigma$-ΣM front-end with 720mVpp input range and >300mV offset removal for wearable bio-signal recording, in *2019 Symposium on VLSI Circuits*, (2019), pp. C296–C297. https://doi.org/10.23919/VLSIC.2019.8778185
2. S. Mondal, C.-L. Hsu, R. Jafari, D. Hall, A dynamically reconfigurable ECG analog front-end with a 2.5 × data-dependent power reduction, in *2017 IEEE Custom Integrated Circuits Conference (CICC)*, (Austin, 2017), pp. 1–4. https://doi.org/10.1109/CICC.2017.7993705
3. S. Mondal, D.A. Hall, A 13.9-nA ECG amplifier achieving 0.86/0.99 NEF/PEF using AC-coupled OTA-stacking. IEEE J. Solid State Circuits **55**(2), 414–425 (2020). https://doi.org/10.1109/JSSC.2019.2957193
4. Y. Zheng et al., Unobtrusive sensing and wearable devices for health informatics. IEEE Trans. Biomed. Eng. **61**(5), 1538–1554 (2014). https://doi.org/10.1109/TBME.2014.2309951
5. J. Andreu-Perez, D.R. Leff, H.M.D. Ip, G. Yang, From wearable sensors to smart implants–toward pervasive and personalized healthcare. IEEE Trans. Biomed. Eng. **62**(12), 2750–2762 (2015). https://doi.org/10.1109/TBME.2015.2422751
6. J. Warchall, P. Theilmann, Y. Ouyang, H. Garudadri, P.P. Mercier, Robust biopotential acquisition via a distributed multi-channel FM-ADC. IEEE Trans. Biomed. Circuits Syst. **13**(6), 1229–1242 (2019). https://doi.org/10.1109/TBCAS.2019.2941846
7. S. Lee, L. Yan, T. Roh, S. Hong, H. Yoo, A 75 μ W real-time scalable body area network controller and a 25 μW ExG sensor IC for compact sleep monitoring applications. IEEE J. Solid State Circuits **47**(1), 323–334 (2012). https://doi.org/10.1109/JSSC.2011.2170636
8. A. Searle, L. Kirkup, A direct comparison of wet, dry and insulating bioelectric recording electrodes. Physiol. Meas. **21**(2), 271–283 (2000). https://doi.org/10.1088/0967-3334/21/2/307
9. G. Li, S. Wang, Y.Y. Duan, Towards gel-free electrodes: A systematic study of electrode-skin impedance. Sensors Actuators B Chem. **241**, 1244–1255 (2017). https://doi.org/10.1016/j.snb.2016.10.005
10. H. Chandrakumar, D. Markovic, An 80-mVpp linear-input range, 1.6-GΩ input impedance, low-power chopper amplifier for closed-loop neural recording that is tolerant to 650-mVpp common-mode interference. IEEE J. Solid State Circuits, 1–18 (2017). https://doi.org/10.1109/JSSC.2017.2753824
11. D. Seok, S. Lee, M. Kim, J. Cho, C. Kim, Motion artifact removal techniques for wearable EEG and PPG sensor systems. Front. Electron. **2**, 4 (2021). https://doi.org/10.3389/felec.2021.685513
12. X. Jiang, G.-B. Bian, Z. Tian, Removal of artifacts from EEG signals: A review. Sensors (Basel) **19**(5), 987 (2019). https://doi.org/10.3390/s19050987
13. H. Kassiri et al., Rail-to-rail-input dual-radio 64-channel closed-loop neurostimulator. IEEE J. Solid State Circuits **52**(11), 2793–2810 (2017). https://doi.org/10.1109/JSSC.2017.2749426

14. C. Kim, S. Joshi, H. Courellis, J. Wang, C. Miller, G. Cauwenberghs, Sub- μ Vrms-noise sub- μ W/channel ADC-direct neural recording with 200-mV/ms transient recovery through predictive digital autoranging. IEEE J. Solid State Circuits **53**(11), 3101–3110 (2018). https://doi.org/10.1109/JSSC.2018.2870555
15. W. Jiang, V. Hokhikyan, H. Chandrakumar, V. Karkare, D. Markovic, A $\pm$50-mV linear-input-range VCO-based neural-recording front-end with digital nonlinearity correction. IEEE J. Solid State Circuits **52**(1), 173–184 (2017). https://doi.org/10.1109/JSSC.2016.2624989
16. S. Li et al., A 0.025-mm^2 0.8-V 78.5dB-SNDR VCO-based sensor readout circuit in a hybrid PLL-$\Delta\Sigma$M structure, in *2019 IEEE Custom Integrated Circuits Conference (CICC)*, (2019), pp. 1–3. https://doi.org/10.1109/CICC.2019.8780175
17. M. Reza Pazhouhandeh, M. Chang, T.A. Valiante, R. Genov, Track-and-zoom neural analog-to-digital converter with blind stimulation artifact rejection. IEEE J. Solid State Circuits **55**(7), 1984–1997 (2020). https://doi.org/10.1109/JSSC.2020.2991526
18. J. Huang, P.P. Mercier, A 112-dB SFDR 89-dB SNDR VCO-based sensor front-end enabled by background-calibrated differential pulse code modulation. IEEE J. Solid State Circuits **56**(4), 1046–1057 (2021). https://doi.org/10.1109/JSSC.2020.3037833
19. J. Huang, P.P. Mercier, 28.1 a distortion-free VCO-based sensor-to-digital front-end achieving 178.9dB FoM and 128dB SFDR with a calibration-free differential pulse-code modulation technique, in *2021 IEEE International Solid- State Circuits Conference (ISSCC)*, vol. 64, (2021), pp. 386–388. https://doi.org/10.1109/ISSCC42613.2021.9365950
20. H. Jeon, J. Bang, Y. Jung, I. Choi, M. Je, A high DR, DC-coupled, time-based neural-recording IC with degeneration R-DAC for bidirectional neural interface. IEEE J. Solid State Circuits **54**(10), 2658–2670 (2019). https://doi.org/10.1109/JSSC.2019.2930903
21. H. Chandrakumar, D. Marković, A 15.2-ENOB 5-kHz BW 4.5-μW chopped CT $\Delta\Sigma$-ADC for artifact-tolerant neural recording front ends. IEEE J. Solid State Circuits **53**(12), 3470–3483 (2018). https://doi.org/10.1109/JSSC.2018.2876468
22. J.-S. Bang, H. Jeon, M. Je, G.-H. Cho, A 6.5μW 92.3dB-DR biopotential-recording front-end with 360mVpp linear input range, in *Symposium on VLSI Circuits Digest of Technical Papers*, (2018), p. 2
23. P.R. Kinget, Scaling analog circuits into deep nanoscale CMOS: Obstacles and ways to overcome them, in *2015 IEEE Custom Integrated Circuits Conference (CICC)*, (2015), pp. 1–8. https://doi.org/10.1109/CICC.2015.7338394
24. C. Lee et al., A 6.5-μW 10-kHz BW 80.4-dB SNDR G_m-C-based CT $\Delta\Sigma$ modulator with a feedback-assisted G_m linearization for artifact-tolerant neural recording. IEEE J. Solid State Circuits **60**(3), 1–1 (2020). https://doi.org/10.1109/JSSC.2020.3018478
25. S. Kalani, T. Haque, R. Gupta, P.R. Kinget, Benefits of using VCO-OTAs to construct TIAs in wideband current-mode receivers over inverter-based OTAs. IEEE Trans. Circuits Syst. I: Regul. Pap. **66**(5), 1681–1691 (2019). https://doi.org/10.1109/TCSI.2018.2883711
26. S. Kalani, P.R. Kinget, Zero-crossing-time-difference model for stability analysis of VCO-based OTAs. IEEE Trans. Circuits Syst. I: Regul. Pap. **67**(3), 839–851 (2020). https://doi.org/10.1109/TCSI.2019.2954833
27. Z. Dai, S. Gupta, S. Kalani, P.R. Kinget, 3.7μW 0.8V VCO-integrator-based high-efficiency capacitor-free low-dropout voltage regulator, in *2016 IEEE International Symposium on Circuits and Systems (ISCAS)*, (2016), pp. 498–501. https://doi.org/10.1109/ISCAS.2016.7527286
28. B. Drost, M. Talegaonkar, P.K. Hanumolu, Analog filter design using ring oscillator integrators. IEEE J. Solid State Circuits **47**(12), 3120–3129 (2012). https://doi.org/10.1109/JSSC.2012.2225738
29. G. Taylor, I. Galton, A reconfigurable mostly-digital delta-sigma ADC with a worst-case FOM of 160 dB. IEEE J. Solid State Circuits **48**(4), 983–995 (2013). https://doi.org/10.1109/JSSC.2013.2239113
30. Y. Zhong et al., A second-order purely VCO-based CT $\Delta\Sigma$ ADC using a modified DPLL structure in 40-nm CMOS. IEEE J. Solid State Circuits **55**(2), 356–368 (2020). https://doi.org/10.1109/JSSC.2019.2948008

31. C. Pochet, J. Huang, P. Mercier, D. Hall, A 174.7-dB FoM, 2nd-order VCO-based ExG-to-digital front-end using a multi-phase gated-inverted-ring oscillator quantizer. IEEE Trans. Biomed. Circuits Syst. **15**(6), 1283–1294 (2021). https://doi.org/10.1109/TBCAS.2021.3133531
32. A. Jayaraj, M. Danesh, S.T. Chandrasekaran, A. Sanyal, Highly digital second-order $\Delta\Sigma$ VCO ADC. IEEE Trans. Circuits Syst. I **66**(7), 2415–2425 (2019). https://doi.org/10.1109/TCSI.2019.2898415
33. A. Elshazly, S. Rao, B. Young, P.K. Hanumolu, A noise-shaping time-to-digital converter using switched-ring oscillators – Analysis, design, and measurement techniques. IEEE J. Solid State Circuits **49**(5), 1184–1197 (2014). https://doi.org/10.1109/JSSC.2014.2305651
34. A. Jayaraj, M. Danesh, S.T. Chandrasekaran, A. Sanyal, 76-dB DR, 48 fJ/step second-order VCO-based current-to-digital converter. IEEE Trans. Circuits Syst. I: Regul. Pap. **67**(4), 1149–1157 (2020). https://doi.org/10.1109/TCSI.2019.2941956
35. M.Z. Straayer, M.H. Perrott, A multi-path gated ring oscillator TDC with first-order noise shaping. IEEE J. Solid State Circuits **44**(4), 1089–1098 (2009). https://doi.org/10.1109/JSSC.2009.2014709
36. X. Yi, Z. Liang, C.C. Boon, G. Feng, F. Meng, K. Yang, An inverted ring oscillator noise-shaping time-to-digital converter with in-band noise reduction and coherent noise cancellation. IEEE Trans. Circuits Syst. I: Regul. Pap. **67**(2), 686–698 (2020). https://doi.org/10.1109/TCSI.2019.2949732
37. C.-W. Yao et al., A 14-nm 0.14-psrms fractional-N digital PLL with a 0.2-ps resolution ADC-assisted coarse/fine-conversion chopping TDC and TDC nonlinearity calibration. IEEE J. Solid State Circuits **52**(12), 3446–3457 (2017). https://doi.org/10.1109/JSSC.2017.2742518
38. S.T. Chandrasekaran, A. Sanyal, Stochastic $\Delta\Sigma$ VCO-ADC utilizing 4$\times$ staggered averaging, in *2020 IEEE International Symposium on Circuits and Systems (ISCAS)*, (2020), pp. 1–5. https://doi.org/10.1109/ISCAS45731.2020.9180835
39. M.J.M. Pelgrom, A.C.J. Duinmaijer, A.P.G. Welbers, Matching properties of MOS transistors. IEEE J. Solid State Circuits **24**(5), 1433–1439 (1989). https://doi.org/10.1109/JSSC.1989.572629
40. A. Sheikholeslami, Process variation and Pelgrom's law [circuit intuitions]. IEEE Solid-State Circuits Mag. **7**(1), 8–9 (2015). https://doi.org/10.1109/MSSC.2014.2369331
41. C. Pochet, D.A. Hall, A 4.4μW 2.5kHz-BW 92.1dB-SNDR 3rd-order VCO-based ADC with pseudo virtual ground feedforward linearization, in *2022 IEEE International Solid-State Circuits Conference (ISSCC)*, vol. 65, (2022), pp. 408–410. https://doi.org/10.1109/ISSCC42614.2022.9731549

Circuits and Architectures for Neural Recording Interfaces

Carolina Mora Lopez and Xiaohua Huang

Abstract Large-scale neural interfacing is needed to provide better understanding of the brain at the cellular level and to develop more advanced prosthetic devices and brain-machine interfaces. However, interfacing with biological tissues poses many challenges such as biocompatibility, the size scale of cellular features, the stability and longevity of the materials in a moist environment at body temperature, and the tiny signal amplitudes to be captured. Therefore, the devices and circuits designed for these interfaces need to fulfil a series of requirements to tackle these challenges. This chapter provides a general overview of existing techniques and requirements to perform neural recording, as well as a review of the state-of-the-art neural readout chips and neural probes.

1 Introduction

Implantable neural probes are the most widely used tool to monitor electrical neural activity at single-cell level [1, 2]. Although they have been fabricated using diverse techniques and materials, silicon neural probes have become popular because they offer important advantages such as precise definition of shank shapes and recording sites, accurate fabrication processes, and automation capabilities to produce low-cost microprobes in large volumes. Additionally, silicon probes can integrate CMOS circuits in the same silicon substrate [3–8], thus enabling the implementation of a large number of electrodes with a reduced number of connecting wires.

In order to design effective and robust tools for neural recording, the following challenges need to be tackled:

- *Neural signal amplitude and frequency:* With implantable devices, it is possible to sense two types of neural signals [9]. The action potentials (APs) or spikes are fast transients that represent single-neuron activity and have amplitudes from

C. M. Lopez (✉) · X. Huang
imec, Leuven, Belgium
e-mail: carolina.moralopez@imec.be

P. Harpe et al. (eds.), *Biomedical Electronics, Noise Shaping ADCs, and Frequency References*, https://doi.org/10.1007/978-3-031-28912-5_3

hundreds to thousands of μV. Their signal bandwidth is from ~300 Hz to 10 kHz. The local field potentials (LFPs) in the lower frequency (<1 Hz to ~1 kHz) represent the combined activity of many neurons in a volume of tissue and have amplitudes from hundreds of μV to a few mV.

A neural readout device must be able to capture both signals with sufficient signal-to-noise ratio (SNR >10 bits) in the full signal bandwidth. However, such signal bandwidth gets affected by both flicker noise and thermal noise, which makes the trade-off among area, power, and noise very difficult to solve.

- *High-density neural interfaces:* In order to increase the recording yield and cover larger brain volumes, high-density neural probes connected to high-channel-count readout chips are required [3–6]. In such tools, the readout channels must achieve very low power consumption and very small area so that they can easily be scaled. Furthermore, the channel electrical performance must be robust over process, voltage, and temperature (PVT) variations so that a high channel-to-channel uniformity can be achieved.
- *Electrode DC offset (EDO):* When the recording electrodes are in contact with the tissue, a DC potential develops at the electrode-electrolyte surface due to the flow of electrons (i.e., half-cell potential) [10]. The DC voltage between the recording and reference electrode will mostly depend on the materials of both electrodes, and it can reach hundreds of mV. However, the total EDO that is seen by the readout channel will depend on the attenuation caused by the channel's DC input impedance. Typically, the channel must be able to tolerate EDOs in the range of tens of mV after attenuation, which is still much larger than the neural signals, without saturation.
- *Electrode impedance:* Very small (e.g., <30-μm-diameter) electrodes are desirable for the recording of APs and LFPs with implantable probes. Therefore, depending on the material, these electrodes can exhibit very large impedances and high thermal noise. Although many new materials and fabrication techniques have been proposed in the last decades to maximize the surface area and minimize the impedance of an electrode [11], the electrode impedance is still in the range of hundreds of kΩ to a few MΩ at 1 kHz. To avoid signal attenuation, the input impedance of the readout channel must be at least ten times higher than the electrode impedance.
- *Biocompatibility and stability:* All the materials in contact with the brain or living tissue must be biocompatible to prevent tissue toxicity and cell death. This poses important limitations on the technologies and materials of the implantable device. For example, CMOS devices with copper metallization may require encapsulation techniques to prevent the diffusion of copper into the biological tissue.

The shapes and dimensions of the implantable devices must be carefully designed to minimize the foreign body reaction. Moreover, for long-term chronic implants, it is also important that the electrode material is stable when exposed to the chemical substances present in the tissue. This is crucial to maintain a stable impedance and signal quality.

2 Neural Readout Architectures

Two very important trends in neural readout architectures can be found in literature to tackle the abovementioned challenges: (i) a conventional architecture consisting of one or more amplification stages followed by an analog-to-digital converter (ADC) and (ii) a trendy direct digitization approach where the input transconductance stage is merged within the ADC loop. The characteristics, advantages, and disadvantages of these architectures are discussed in this section.

2.1 IA-ADC Readout Architectures

The most widespread architecture for neural signal readout consists of an AC-coupled instrumentation amplifier (IA) followed by a time-multiplexed moderate-resolution (8–12-bit) ADC [3–5, 12–21] (see Fig. 1). AC coupling is the most effective way to deal with EDOs, since it is possible to achieve rail-to-rail EDO rejection without complex feedback loops. In order to bias the IA input nodes at DC, pseudo-resistor elements that can achieve up to TΩ resistances are utilized [22]. Although these pseudo-resistor elements have some important disadvantages, namely, high sensitivity to process variation and light as well as strong nonlinearity with voltage, they are still the most power- and area-effective way of implementing the ultra-low cutoff frequencies required to record LFPs (<1 Hz). The AC coupling capacitor must be large enough to provide sufficient gain (e.g., 50–100 V/V) in a closed-loop IA, but not too large to avoid the excessive deterioration of the AC input impedance. To solve this trade-off, the minimum feedback capacitor size, the electrode impedance, the supply voltage, and the area constraints must be taken into account. Contrary to the common belief that AC coupling requires large area, the area of this input capacitor is normally not a big concern since it can be placed on top of the active circuits when it is implemented using metal-insulator-metal (MIM) or metal-oxide-metal (MOM) devices. Therefore, EDO rejection using AC coupling and pseudo-resistors can be both area and power efficient.

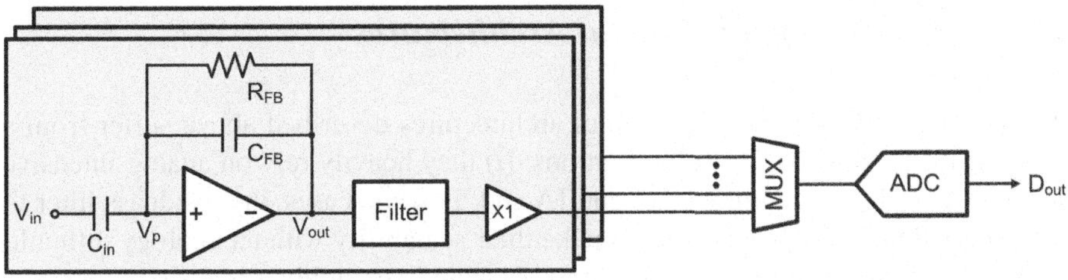

Fig. 1 Diagram of a conventional IA-ADC readout

Diverse operational transconductance amplifier (OTA) architectures can be used to achieve a good trade-off among power, area, and noise in the IA. The most popular choices are folded cascode [3–5, 13–15, 21] and inverter-based [12, 17, 20] OTAs. The latter can achieve impressive area and power performances while maintaining a low noise. However, it is very difficult to keep this OTA well biased over PVT variations, making it less robust for high-density recording architectures where a uniform channel-to-channel performance is crucial to guarantee signal quality.

In order to solve the trade-off between the power and area required for the ADC and its preceding driver, the multiplexing ratio needs to be optimized as proposed in [23]. A SAR ADC is normally used in this application due to its good power efficiency in the moderate-resolution and low-frequency ranges (see Fig. 2) [24]. However, hybrid ADCs combining properties of different traditional architectures are also popular since they can be customized to achieve the required power, area, resolution, and speed.

An example of a conventional IA-ADC readout architecture can be found in the 384-channel CMOS neural probe reported in [4]. The high-level architecture of this probe is shown in Fig. 3. In this design, each channel is split in two signal paths to filter and amplify the APs and LFPs separately. In this way, the resolution of the SAR ADC can be kept lower (10 bits) to reduce its area contribution. A shared fixed (50 V/V) amplification and an independent variable (1–50 V/V) amplification are included in the signal paths. A 12:1 multiplexing ratio is used here to optimize the channel area and power. The channel design was optimized to achieve PVT robustness and a very good channel-to-channel uniformity. Today, this neural probe is widely used by the neuroscience community, becoming the new gold standard in *in vivo* electrophysiology.

To significantly reduce the area in [4], the CMOS neural probe in [5] implements a full-band readout channel followed by a higher-resolution (14-bit) hybrid SAR-assisted pipeline ADC (Fig. 4). In this way, the area-consuming band-splitting filters could be eliminated. The channel includes only one fixed amplification stage (80 V/V), while a multiplexing ratio of 16:1 is used. This design achieved an area reduction of >3× for the same number of channels.

2.2 *Direct-to-Digital Readout Architectures*

The conventional AC-coupled readout architectures described above suffer from a few significant drawbacks and limitations: (i) they heavily rely on analog-intensive techniques to implement the front-end IA and, in some cases, the bandpass filter to separate AP and LFP bands, which make their scalability with technology difficult; (ii) the pseudo-resistors used for input biasing are highly sensitive to process variation and light, nonlinear, and very prone to large offsets caused by leakage currents; and (iii) due to the front-end high-gain amplification, such designs have a limited input dynamic range (typically <20 mV_{pp}), which makes them prone

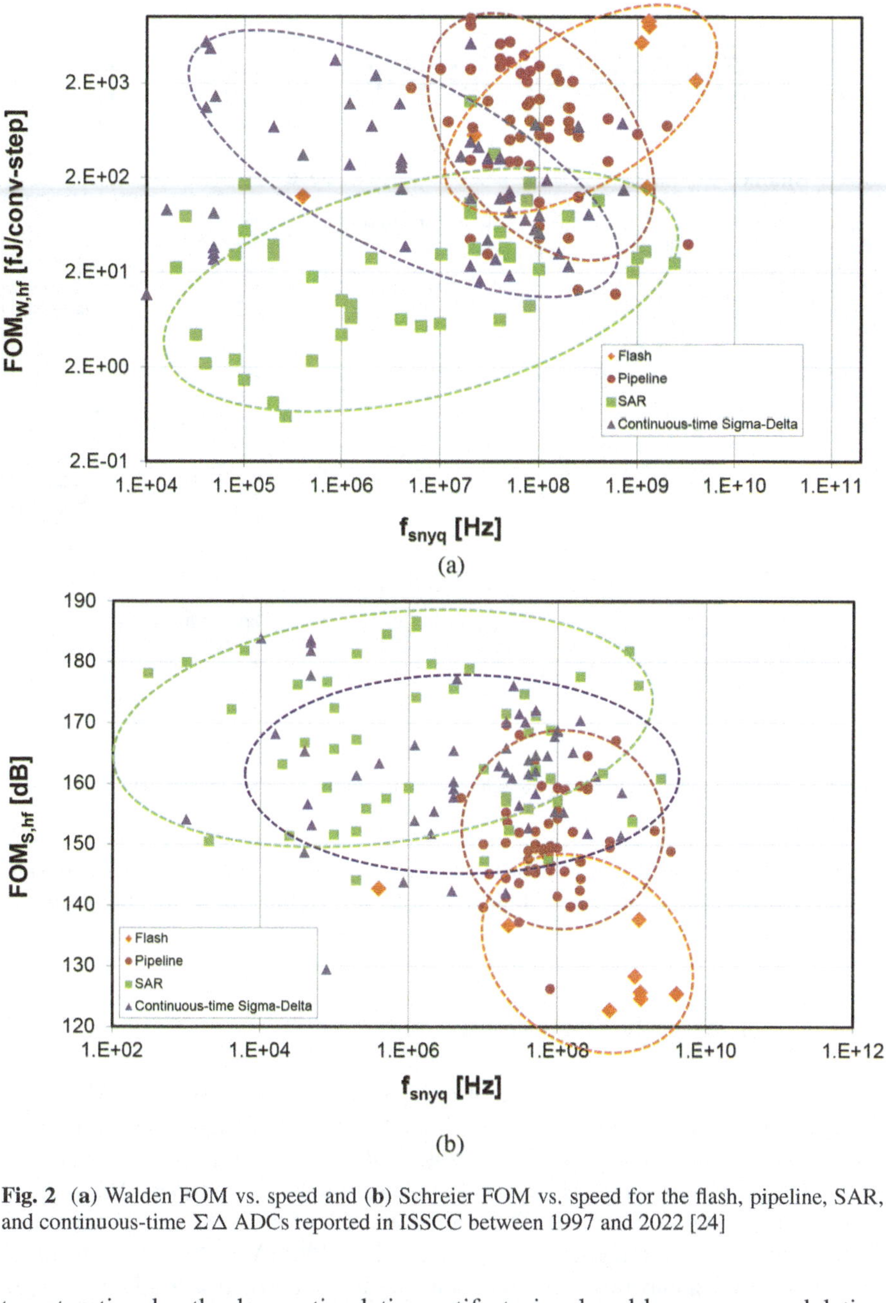

Fig. 2 (**a**) Walden FOM vs. speed and (**b**) Schreier FOM vs. speed for the flash, pipeline, SAR, and continuous-time $\Sigma\Delta$ ADCs reported in ISSCC between 1997 and 2022 [24]

to saturation by the large stimulation artifacts in closed-loop neuromodulation applications. To address these challenges, direct-to-digital readout architectures (i.e., without front-end high-gain amplification) that use oversampling ADCs have been explored extensively in the last years [8, 25–47]. Depending on the required

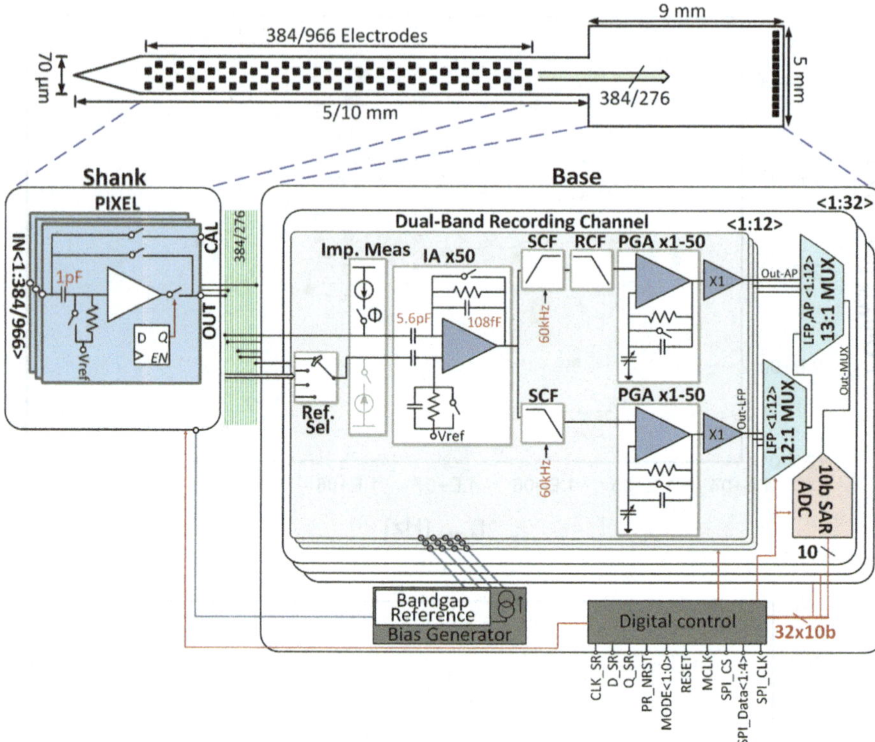

Fig. 3 High-level architecture of the 384-channel neural probe reported in Ref. [4]

input range, these designs can generally be classified into two different categories: recording-only architectures and artifact-tolerant architectures for bidirectional neural interfaces.

2.2.1 Recording-Only Architectures

For neural recording in the absence of artifacts, a moderate-resolution ADC (~8–11 bits) can be employed to directly digitize the raw neural signals in a very area- and power-efficient manner, while the large EDOs can be either compensated by a mixed-signal DC-servo loop (DSL) [25] or filtered out by conventional AC coupling [29, 32], as shown in Fig. 5a, b. The main drawback of the DSL-based approaches is the limited EDO compensation range, typically ~100 mV_{pp}. Although this range can be extended, it may result in significant area and/or power overhead for the high-resolution and wide-dynamic-range feedback DACs. For this reason, significant area reduction has been obtained in [8] by removing the EDO compensation loop (see Fig. 5c). Nevertheless, such approach suffers from either degraded noise (20.19 μV_{rms}) or limited EDO tolerance (22.5 mV_{pp}). While adding a coarse offset

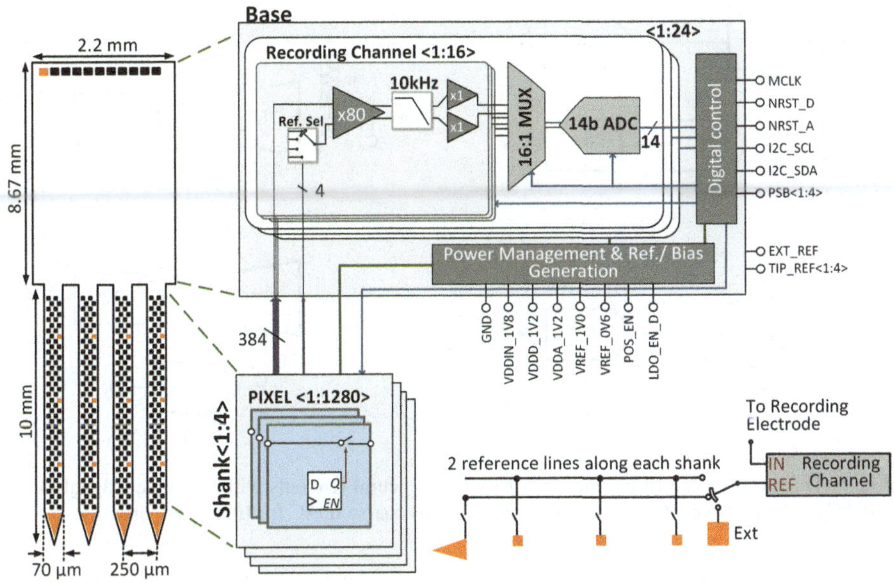

Fig. 4 High-level architecture of the 384-channel neural probe reported in Ref. [5]

IDAC (Fig. 5d) extends the EDO cancellation range up to 120 mV_{pp} [30], it results in significant noise degradation due to the unbalanced input differential pair when large EDOs are present. In contrast, rail-to-rail EDOs can be tolerated by using AC coupling. The penalty is the required large area coupling capacitors. This claim, however, has been challenged in some recent designs [29, 32]. This is because the capacitor can be designed reasonably small via proper engineering, and its density increases with technology scaling as more metal layers are available and the spacing between two adjacent metal tracks becomes narrower. Additionally, the AC coupling capacitor can be stacked above the active circuits for further area saving. At the same time, the drawbacks associated with the pseudo-resistor biasing can be avoided by periodically resetting the OTA input to its common mode [44, 48], applying a reset when the OTA input drifts outside the linear input range of the ADC [29], or resetting via the incremental operation [36]. The reset-induced kT/C noise is usually not a big concern since it gets averaged and becomes negligible compared to the overall ADC input-referred noise due to the very infrequent reset operation [29] or converted into out-of-band chopper ripple and then filtered out by the subsequent integrator [36].

2.2.2 Artifact-Tolerant Architectures for Bidirectional Neural Interfaces

For artifact-tolerant recording, many different readout architectures based on delta-sigma ($\Delta\Sigma$) modulation, Δ modulation, and a combination of both (i.e., Δ-$\Delta\Sigma$

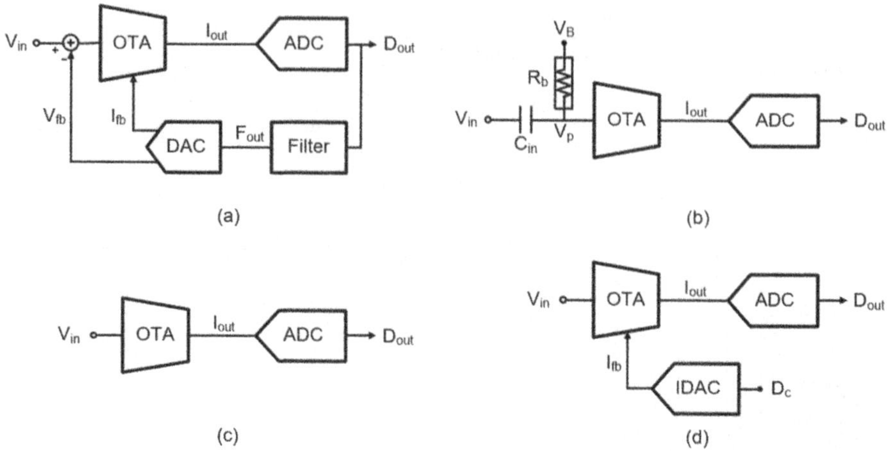

Fig. 5 Diagrams of recording-only direct-to-digital neural readouts with (**a**) mixed-signal DSL, (**b**) AC coupling, (**c**) ADC linear input range, and (**d**) coarse IDAC for EDO compensation

or $\Delta^2\Sigma$) have been proposed. On the one hand, to accommodate both the large EDOs and stimulation artifacts on top of the weak neural signals, the linear input range of the $\Delta\Sigma$ ADC can be extended to tens or hundreds of mV [33, 36, 42, 43, 45, 46]. This comes at the expense of substantially larger area and higher power compared to those designs with a small linear input range (typically around a few mV) mentioned above. While the power consumption can be reduced by employing high-order modulators with reduced oversampling ratios, the power and area overhead of the back-end decimation filters grow rapidly, which unfortunately are rarely mentioned and implemented in most designs. For this reason, the reported channel counts are typically very limited (one or a few). On the other hand, the large EDOs and stimulation artifacts can be compensated by the feedback loop in Δ modulators. Nevertheless, due to the limited tracking speed of the Δ modulation, this usually requires some recovery time for the ADC to settle after a stimulation event. Additionally, very large oversampling ratios are required in order to keep the quantization noise sufficiently low. As a result, these architectures have mostly been reported in low-bandwidth readouts used for electrocorticography (ECoG) [26, 27, 47] and cannot easily be applied to high-frequency AP recording.

To reduce the oversampling ratio while maintaining the tracking for artifacts and/or EDO compensation, Δ-$\Delta\Sigma$ or $\Delta^2\Sigma$ topologies have been reported widely [28, 41, 44]. Good area and power efficiencies have already been achieved for several low-bandwidth designs [41, 44], while the number of high-bandwidth designs is still limited. In [28], a high-bandwidth readout architecture has been proposed, which achieves a promising channel area of 0.0077 mm^2. In this example, a 16-channel readout IC was implemented, featuring a DC-coupled second-order Δ-$\Delta\Sigma$ architecture as shown in Fig. 6. Combining Δ-$\Delta\Sigma$ modulation with new bootstrapping and chopping schemes, this design achieves large dynamic range

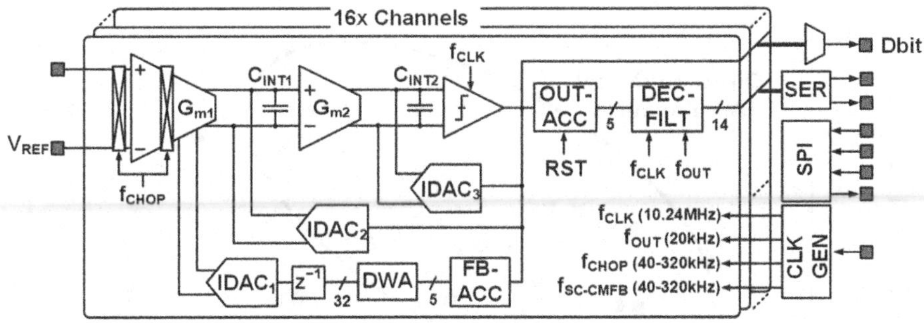

Fig. 6 High-level architecture of the 16-channel Δ-ΔΣ neural readout reported in Ref. [28]

while maintaining high input impedance and low noise. However, the reported power efficiency is not sufficient for high-channel-count implementations. It is worth mentioning that while almost rail-to-rail artifact and/or EDO tolerance can be achieved when the feedback integrator is realized with a charge pump, such scheme is prone to saturation in the final output reconstruction step. This is because the digital accumulation in the back-end reconstruction circuit does not match the leaky integration performed by the charge pump.

Finally, several mixed-signal techniques have also been proposed to reduce the transient recovery time after a stimulation event. In [44], adaptive step size (or digital auto-ranging) is utilized in the Δ-modulation path, achieving a recovery from >200 mV_{pp} artifacts within <1 ms. The transient recovery time can further be reduced by boosting the ADC sampling clock when an artifact is detected [47]. Recently, adaptive clock boosting combined with SAR-assisted artifact cancellation has been employed, resulting in a maximum recovery time of 20 μs for ±175 mV artifacts [40].

2.3 Performance Benchmarking

The different state-of-the-art IA-ADC and direct-to-digital designs discussed above are compared in terms of power, area, and noise performance in Fig. 7. As shown in the figure, smaller areas can be achieved with direct-to-digital architectures. This is mostly because their digitally intensive designs can scale with technology. However, the power and noise performances of direct-to-digital architectures are not yet as good as IA-ADC architectures. The larger power can be attributed to the high oversampling ratios and the additional digital circuits required, e.g., for decimation, while the larger noise is mainly due to the lack of high-gain preamplification.

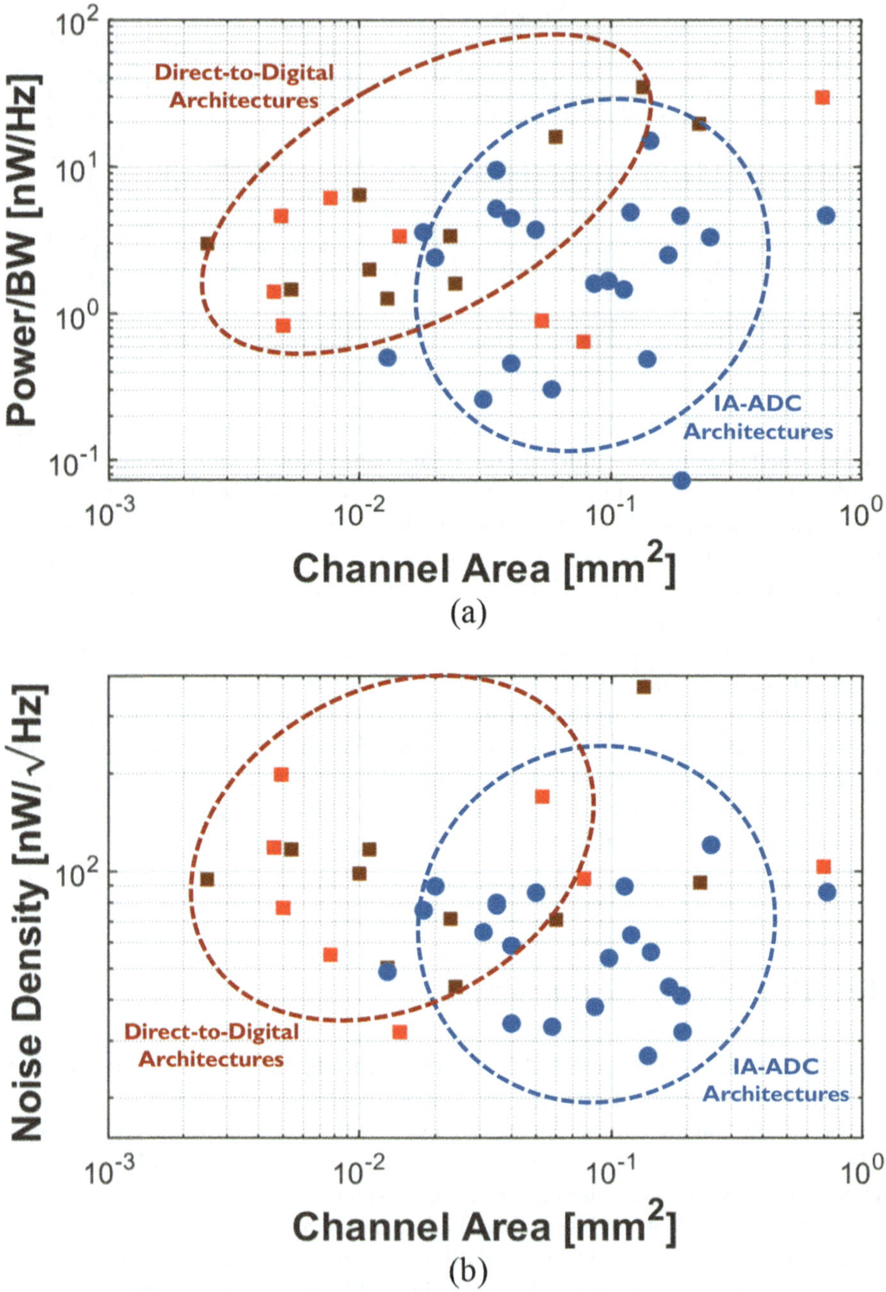

Fig. 7 Performance comparison of two types of neural recording architectures: IA-ADC (shown as blue circles) and direct-to-digital (shown as red squares: in light red are the designs that cover the 5~10 kHz bandwidth for AP recording, while in dark red are those covering the ≤1 kHz ECoG bandwidth). (Data obtained from Refs. [3–5, 8, 12–21, 25–47])

3 Conclusions

The development of high-density neural interfaces requires area- and power-efficient neural readout circuits that can also address other application challenges such as low noise, wide input range, sufficient EDO rejection/tolerance, and high input impedance. This chapter reviews the state-of-the-art neural readout chips reported in the last years, classifying them in two categories: conventional architectures consisting of an instrumentation amplifier (IA) followed by an analog-to-digital converter (ADC) and trendy direct digitization approaches where the input transconductance stage is merged within the ADC loop. The performance benchmark of these two approaches shows that direct-to-digital architectures can achieve smaller areas, while IA-ADC architectures achieve lower noise and power. This trend can change in the future if direct-to-digital designs can further profit from technology scaling.

References

1. J.J. Jun et al., Fully integrated silicon probes for high-density recording of neural activity. Nature **551**, 232–236 (2017)
2. N.A. Steinmetz et al., Neuropixels 2.0: A miniaturized high-density probe for stable, long-term brain recordings. Science (80-) **372**(6539), eabf4588 (2021)
3. C.M. Lopez et al., An implantable 455-active-electrode 52-channel CMOS neural probe. IEEE J. Solid State Circuits **49**(1), 248–261 (2014)
4. C. Mora Lopez et al., A neural probe with up to 966 electrodes and up to 384 configurable channels in 0.13 μm SOI CMOS. IEEE Trans. Biomed. Circuits Syst. **11**(3), 510–522 (2017)
5. S. Wang et al., A compact quad-shank CMOS neural probe with 5,120 addressable recording sites and 384 fully differential parallel channels. IEEE Trans. Biomed. Circuits Syst. **13**, 1 (2019)
6. B.C. Raducanu et al., Time multiplexed active neural probe with 1356 parallel recording sites. Sensors **17**(10), 2388 (2017)
7. G.N. Angotzi et al., SiNAPS: An implantable active pixel sensor CMOS-probe for simultaneous large-scale neural recordings. Biosens. Bioelectron. **126**(2018), 355–364 (2019)
8. D. De Dorigo et al., Fully immersible subcortical neural probes with modular architecture and a delta-sigma ADC integrated under each electrode for parallel readout of 144 recording sites. IEEE J. Solid State Circuits **53**(11), 3111–3125 (2018)
9. C. Mora Lopez et al., A multichannel integrated circuit for electrical recording of neural activity, with independent channel programmability. IEEE Trans. Biomed. Circuits Syst. **6**(2), 101–110 (2012)
10. L.A. Geddes, L.E. Baker, Principles of applied biomedical instrumentation. J. Clin. Eng. **2**(1), 86 (1977)
11. G. Hong, C.M. Lieber, Novel electrode technologies for neural recordings. Nat. Rev. Neurosci. **20**(6), 330–345 (2019)
12. T. Lee et al., A multimodal multichannel neural activity readout IC with 0.7μW/channel Ca2+-probe-based fluorescence recording and electrical recording, in *2019 Symposium on VLSI Circuits*, (2019), pp. C290–C291
13. G. Gagnon-Turcotte, M.N. Khiarak, C. Ethier, Y. De Koninck, B. Gosselin, A 0.13um CMOS SoC for simultaneous multichannel optogenetics and neural recording. IEEE J. Solid State Circuits **53**(11), 3087–3100 (2018)

14. Y. Jung et al., A wide-dynamic-range neural-recording IC with automatic-gain-controlled AFE and CT dynamic-zoom ΔΣ ADC for saturation-free closed-loop neural interfaces. IEEE J. Solid State Circuits **57**(10), 1–12 (2022)
15. D.Y. Yoon, S. Pinto, S.W. Chung, P. Merolla, T.W. Koh, D. Seo, A 1024-channel simultaneous recording neural SoC with stimulation and real-time spike detection, in *IEEE Symposium on VLSI Circuits, Digest of Technical Papers*, (2021)
16. B. Gosselin et al., A mixed-signal multichip neural recording interface with bandwidth reduction. IEEE Trans. Biomed. Circuits Syst. **3**(3), 129–141 (2009)
17. D. Han, Y. Zheng, R. Rajkumar, G. Dawe, M. Je, A 0.45V 100-channel neural-recording IC with sub-μW/channel consumption in 0.18μm CMOS. IEEE Trans. Biomed. Circuits Syst. **7**(6), 735–746 (2014)
18. S.-Y. Park, J. Cho, K. Na, E. Yoon, Modular 128-channel Δ-ΔΣ analog front-end architecture using spectrum equalization scheme for 1024-channel 3-D neural recording microsystems. IEEE J. Solid State Circuits **53**(2), 501–514 (2018)
19. H. Gao et al., HermesE: A 96-channel full data rate direct neural interface in 0.13 μm CMOS. IEEE J. Solid State Circuits **47**(4), 1043–1055 (2012)
20. X. Liu et al., A fully integrated wireless compressed sensing neural signal acquisition system for chronic recording and brain machine interface. IEEE Trans. Biomed. Circuits Syst. **10**(4), 874–883 (2016)
21. J. Xu, T. Wu, W. Liu, Z. Yang, A frequency shaping neural recorder with 3 pF input capacitance and 11 plus 4.5 bits dynamic range. IEEE Trans. Biomed. Circuits Syst. **8**(4), 510–527 (2014)
22. R.R. Harrison, C. Charles, A low-power low-noise CMOS amplifier for neural recording applications. Solid-State Circuits IEEE J. **38**(6), 958–965 (2003)
23. M.S. Chae, W. Liu, M. Sivaprakasam, Design optimization for integrated neural recording systems. IEEE J. Solid State Circuits **43**(9), 1931–1939 (2008)
24. B. Murmann, ADC Performance Survey 1997–2022, [Online] (2022). Available http://web.stanford.edu/~murmann/adcsurvey.html
25. E. Greenwald et al., A bidirectional neural interface IC with chopper stabilized BioADC array and charge balanced stimulator. IEEE Trans. Biomed. Circuits Syst. **10**(5), 990–1002 (2016)
26. M. Reza Pazhouhandeh, M. Chang, T.A. Valiante, R. Genov, Track-and-zoom neural analog-to-digital converter with blind stimulation artifact rejection. IEEE J. Solid State Circuits **55**(7), 1984–1997 (2020)
27. M.R. Pazhouhandeh, H. Kassiri, A. Shoukry, I. Weisspapir, P.L. Carlen, R. Genov, Opamp-less sub-μW/channel Δ-modulated neural-ADC with super-GΩ input impedance. IEEE J. Solid State Circuits **56**(5), 1565–1575 (2021)
28. S. Wang et al., A compact chopper stabilized Δ-ΔΣ neural readout IC with input impedance boosting. IEEE Open J. Solid-State Circuits Soc. **1**(June), 67–78 (2021)
29. X. Huang et al., A compact, low-power analog front-end with event-driven input biasing for high-density neural recording in 22-nm FDSOI. IEEE Trans. Circuits Syst. II Express Briefs **69**(3), 804–808 (2022)
30. D. Wendler et al., A 0.0046-mm^2 two-step incremental delta-sigma analog-to-digital converter neuronal recording front end with 120-mVpp offset compensation. IEEE J. Solid State Circuits **58**, 1–12 (2022)
31. X. Huang et al., A 256-channel actively-multiplexed μECoG implant with column-parallel incremental ΔΣ ADCs employing bulk-DACs in 22-nm FDSOI technology, in *2022 IEEE International Solid-State Circuits Conference (ISSCC)*, vol. 65, (2022), pp. 200–202
32. X. Yang et al., A 128-channel AC-coupled 1st-order Δ-ΔΣ IC for neural signal acquisition, in *2022 IEEE Symposium on VLSI Technology and Circuits (VLSI Technology and Circuits)*, (2022), pp. 60–61
33. W. Jiang, V. Hokhikyan, H. Chandrakumar, V. Karkare, D. Markovic, A ±50-mV linear-input-range VCO-based neural-recording front-end with digital nonlinearity correction. IEEE J. Solid State Circuits **52**(1), 173–184 (2017)

34. A. Nikas, S. Jambunathan, L. Klein, M. Voelker, M. Ortmanns, A continuous-time delta-sigma modulator using a modified instrumentation amplifier and current reuse DAC for neural recording. IEEE J. Solid State Circuits **54**(10), 2879–2891 (2019)
35. J.P. Uehlin et al., A single-chip bidirectional neural interface with high-voltage stimulation and adaptive artifact cancellation in standard CMOS. IEEE J. Solid State Circuits **55**(7), 1–1 (2020)
36. B.C. Johnson et al., An implantable 700μW 64-channel neuromodulation IC for simultaneous recording and stimulation with rapid artifact recovery, in *2017 Symposium on VLSI Circuits*, (2017), pp. C48–C49
37. W.A. Smith, J.P. Uehlin, S.I. Perlmutter, J.C. Rudell, V.S. Sathe, A scalable, highly-multiplexed delta-encoded digital feedback ECoG recording amplifier with common and differential-mode artifact suppression, in *2017 Symposium on VLSI Circuits*, (2017), pp. C172–C173
38. J. Huang et al., A 0.01-mm^2 mostly digital capacitor-less AFE for distributed autonomous neural sensor nodes. IEEE Solid-State Circuits Lett. **1**(7), 162–165 (2018)
39. C. Tu, Y. Wang, T. Lin, A low-noise area-efficient chopped VCO-based CTDSM for sensor applications in 40-nm CMOS. IEEE J. Solid State Circuits **52**(10), 2523–2532 (2017)
40. T. Moeinfard, G. Zoidl, H. Kassiri, A SAR-assisted DC-coupled chopper-stabilized 20μs-artifact-recovery $\Delta\Sigma$ ADC for simultaneous neural recording and stimulation, in *2022 IEEE Custom Integrated Circuits Conference (CICC)*, (2022), pp. 1–2
41. H. Kassiri et al., Rail-to-rail-input dual-radio 64-channel closed-loop neurostimulator. IEEE J. Solid State Circuits **52**(11), 1–18 (2017)
42. J. Bang, H. Jeon, M. Je, G. Cho, A 6.5mμW 92.3dB-DR biopotential-recording front-end with 360mVpp linear input range, in *VLSI Circuits (VLSIC), 2018 Symposium on*, (2018), pp. 239–240
43. H. Chandrakumar, D. Markovic, A 15.2-ENOB 5-kHz BW 4.5-μW chopped CT $\Delta\Sigma$-ADC for artifact-tolerant neural recording front ends. IEEE J. Solid State Circuits **53**(12), 3470–3483 (2018)
44. C. Kim, S. Joshi, H. Courellis, J. Wang, C. Miller, G. Cauwenberghs, Sub-μVrms-noise Sub-μW/channel ADC-direct neural recording with 200-mV/ms transient recovery through predictive digital autoranging. IEEE J. Solid State Circuits **53**(11), 3101–3110 (2018)
45. H. Jeon, J.S. Bang, Y. Jung, I. Choi, M. Je, A high DR, DC-coupled, time-based neural-recording IC with degeneration R-DAC for bidirectional neural interface. IEEE J. Solid State Circuits **54**(10), 2658–2670 (2019)
46. C. Lee et al., A 6.5-μW 10-kHz BW 80.4-dB SNDR G_m-C-based CT $\Delta\Sigma$ modulator with a feedback-assisted G_m linearization for artifact-tolerant neural recording. IEEE J. Solid State Circuits **60**(3), 1–1 (2020)
47. M.R. Pazhouhandeh et al., 22.8 adaptively clock-boosted auto-ranging responsive neurostimulator for emerging neuromodulation applications, in *2019 IEEE International Solid-State Circuits Conference – (ISSCC)*, vol. 2, (2019), pp. 374–376
48. M. Sharma et al., Acquisition of neural action potentials using rapid multiplexing directly at the electrodes. Micromachines **9**(10), 477 (2018)

Chip-Integrated Spin Detection for Biomedical Applications

Jens Anders, Daniel Krüger, Frederik Dreyer, Qing Yang, and Michal Kern

Abstract In this chapter, we discuss recent developments in the field of chip-integrated electronics for nuclear magnetic resonance (NMR) and electron paramagnetic resonance (EPR)–NMR- and EPR-on-a-chip—for biomedical applications. More specifically, starting from a brief review of the underlying physical working principles, we discuss potential applications of NMR and EPR in the biomedical context together with an overview of the state of the art of (chip-integrated) systems in these fields. This is followed by a detailed discussion of the architectures and the main building blocks of NMR- and EPR-on-a-chip detectors. Finally, we discuss chip-integrated hyperpolarization as a potential path to solving the sensitivity problem of MR-based methods before we conclude the chapter with a summary and an outlook on future research directions of NMR- and EPR-on-a-chip detectors in the biomedical context.

1 Motivation

With recent advances in microelectronics, micro- and nano-fabrication technologies, and artificial intelligence (AI), the field of biomedical sensors is booming. As a result, the monitoring of well-being, comfort, and health, e.g., via smartwatches,

J. Anders (✉)
Institute of Smart Sensors, University of Stuttgart, Stuttgart, Germany

Center for Integrated Quantum Science and Technology (IQST), Stuttgart and Ulm, Germany

Institute for Microelectronics Stuttgart, IMS CHIPS, Stuttgart, Germany
e-mail: jens.anders@iis.uni-stuttgart.de

D. Krüger
Institute of Smart Sensors, University of Stuttgart, Stuttgart, Germany

John A. Paulson School of Engineering and Applied Sciences, Harvard University, Cambridge, MA, USA

F. Dreyer · Q. Yang · M. Kern
Institute of Smart Sensors, University of Stuttgart, Stuttgart, Germany

P. Harpe et al. (eds.), *Biomedical Electronics, Noise Shaping ADCs, and Frequency References*, https://doi.org/10.1007/978-3-031-28912-5_4

smart patches, or other wearables, has become an integral part of our everyday lives. Similarly, the newly available technological possibilities have enabled benchtop and even portable biomedical sensing platforms of unprecedented sensitivity and specificity in a smaller form factor and with reduced power consumption. While, in principle, any sensor technology can benefit from microelectronic integration in the required interface electronics, the miniaturization and cost reduction potential associated with CMOS integration varies largely depending on the used sensing modality. More specifically, the degree to which CMOS integration can contribute to the overall size and cost reduction largely depends on the required additional periphery, which varies significantly across different sensor types.

While the possibility of using a certain sensor type in a portable or even wearable device can open up many new application scenarios, especially in the biomedical domain, this reduction in form factor and weight usually must not come with a degradation in performance. Here, apart from displaying an excellent limit of detection (LOD), which is typically the most important specification of a certain sensor type, especially biomedical sensors also have to provide sufficient specificity for the target biomarker. Here, spin-based sensors have an intrinsic advantage over many other sensor types since they use the spin of a certain particle as a nanoscopic probe that monitors its intramolecular and extramolecular surroundings. In this way, spin-based detection methods can provide a broad range of information, including details of the chemical structure of a molecule, the composition of an analyte, as well as the chemical and physical bulk properties of a material such as pore spaces of porous media or the fat content of tissue.

Interestingly, a net spin, as it is required for spin-based sensing, can be found in certain nuclei[1] as well as unpaired electrons. The resonant readout of the spins' magnetic moment using a combination of static and (resonant) AC magnetic fields is referred to as "magnetic resonance" (MR). If the utilized spin is associated with a nucleus, the corresponding technique is called "nuclear magnetic resonance" (NMR), cf. Sect. 2. If the spin under investigation originates from an unpaired electron, one refers to the corresponding method as "electron spin resonance" (ESR) or "electron paramagnetic resonance" (EPR), cf. Sect. 3.

Interestingly, NMR and EPR are complementary in nature—the former being broadly applicable due to the abundance of NMR active nuclei in almost any chemical substance and the latter being restricted to samples with unpaired electrons, i.e., paramagnetic samples, such as free radicals in the biomedical context. Although NMR on paramagnetic samples is, in principle, possible, the presence of the spin of the unpaired electrons diminishes the resolution of an NMR spectrum to the extent that spin couplings are rarely resolved. Therefore, in the biomedical context, NMR is an ideal tool for the study of all diamagnetic samples, whereas EPR can be considered as the gold standard for the measurement of free radicals.

As will be discussed in detail in Sects. 2 and 3, an important distinction between NMR and EPR for the circuit designer is the widely different operating

[1] All nuclei with an odd number of nucleons.

frequencies of the two methods, which is a direct consequence of the largely different gyromagnetic ratios of nuclei—displaying gyromagnetic ratios between approximately 42 MHz/T for protons and approximately 6 MHz/T for oxygen—which are around three orders of magnitude smaller compared to the gyromagnetic ratio of a free electron of 28 GHz/T. As a consequence, proton NMR at a static field of 1 T requires RF electronics with an operating frequency around 42 MHz, while EPR equipment for the same target field operates at millimeter-wave frequencies around 28 GHz.

A similarity of both NMR and EPR is that the two methods require relatively large static magnetic fields ranging from several hundred millitesla up to more than 20 T for NMR. The large static magnetic fields—commonly referred to as B_0 field—are mostly required to provide sufficient signal strength, i.e., sensitivity. This is because the interaction energy between the spin and the external magnetic field is tiny compared to the thermal energy kT, requiring large B_0 fields to produce reasonable polarization levels inside the sample. We will come back to this challenge in Sect. 5, where we will discuss means of overcoming the small thermal polarization levels using so-called hyperpolarization techniques.

Fortunately, there has been great progress in the design and manufacturing of permanent magnets up to around 3 T [1–3], which achieve high spectral resolutions in the range of a few ppm, required for magnetic resonance applications, resulting in the availability of commercial benchtop NMR and EPR spectrometers with a price range between $ 20k and $ 100k as well as research prototypes as small as a ping-pong ball [4]. This relatively recent availability of such small, lightweight, and relatively cheap magnets presents a paradigm shift in magnetic resonance because—with the previous main bottleneck toward miniaturization and cost reduction being removed—today, magnetic resonance detection platforms can fully benefit from the power of chip integration for the design of next-generation portable and wearable NMR and EPR devices with price tags below 1000 USD.

This chapter is organized as follows: In Sects. 2 and 3, we will briefly recap the relevant basics of NMR and EPR for biomedical applications, including a short review of state-of-the-art (chip-integrated) systems for NMR and EPR. This is followed by an overview of the required NMR and EPR electronics, including the commonly used transceiver architectures and the performance critical building blocks in Sect. 4. In Sect. 5, we will then take a look at one of the latest research trends in MR applications, hyperpolarization, before we conclude the chapter with a brief summary and an outlook on the next generation of chip-integrated MR detection platforms in Sect. 6.

2 Introduction to NMR

Nuclei with an odd number of nucleons, e.g., ^{1}H, ^{13}C, and ^{19}F, possess a net (nuclear) spin. In the presence of a static magnetic field (B_0), the energy levels associated with the different spin states become distinguishable (Zeeman splitting), resulting in different populations of the different energy levels, producing a net magnetization (M_0) in the direction of the B_0 field. The net magnetization can be manipulated by an applied RF magnetic field (B_1), whose frequency matches the condition $\omega_{RF} \approx \omega_L = \gamma B_0$, γ being the gyromagnetic ratio, and whose direction is perpendicular to B_0. More specifically, by applying a pulsed B_1 field, the net magnetization can be rotated around the axis defined by the B_1 field, resulting in a magnetization component perpendicular to B_0, which precesses around the B_0 axis at the Larmor frequency ω_L. This precessing transversal magnetization can be picked up as an electromotive force (emf) by a simple coil. Since relaxation processes lead to an (approximately) exponential decay of the transversal magnetization, with a characteristic time constant called T_2, and a restoration of the original (longitudinal) magnetization along B_0, with a characteristic time constant called T_1, the signature NMR signal picked up by the detection coil after a pulsed excitation (the so-called free induction decay (FID)) is an exponentially –with time constant T_2– decaying sine wave at the Larmor frequency ω_L, cf. Fig. 1.

Depending on the target application, modern NMR has developed in three directions: relaxometry, spectroscopy, and imaging. NMR relaxometry is useful for obtaining information about bulk properties of the sample of interest. It started with spin echoes [5] and has developed into a variety of more sophisticated pulse sequences, e.g., Carr-Purcell-Meiboom-Gill (CPMG) [6, 7], T_1–T_2 correlation, and T_2–T_2 exchange, that are capable of measuring various kinds of information, including, for example, the distinction of bound and unbound water in porous media or the quantification of the fat content in biological tissue. NMR spectroscopy is focused on extracting molecular structure information, e.g., the identification of functional groups or the structure of an entire molecule, from the so-called chemical shifts and J-couplings [8], both of which are due to the different microenvironments in which the nuclear spins are located. Here, it should be noted that, today, virtually all NMR experiments are carried out in the pulsed mode described above, which is, in general, more powerful than the continuous-wave (cw) measurements that are still commonly employed in most EPR experiments, cf. Sect. 3.

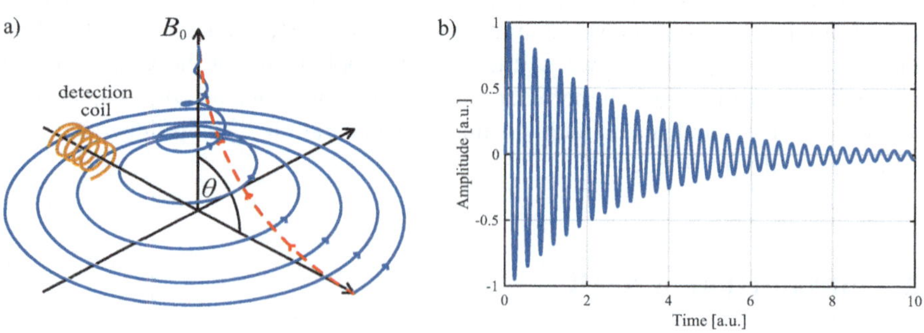

Fig. 1 (**a**) Detection of the transversal magnetization after a pulsed excitation using a pickup coil in the xy-plane and (**b**) the corresponding free induction decay (FID)

Due to its noninvasive and nondestructive nature, NMR has been widely used in biomedical applications. The probably best-known example is magnetic resonance imaging (MRI), which is used in almost every larger hospital. At its basis, MRI uses gradients in the B_0 field along the x-, y-, and z-axis to spatially encode the NMR signal from different voxels (volume pixels) in the sample. Depending on the applied pulse sequence, MRI can produce several contrasts (spin/proton density, T_1, T_2, and diffusion weighted) in the tissue with the same detection hardware, resulting in an outstanding diagnostic power of MRI. Very recently, CMOS integration has been used to design very compact MRI scanners [9].

In addition to MRI, various NMR biomedical detection platforms based on CPMG measurements have been developed over the past decade [10–13], e.g., to identify circulating tumor cells (CTCs) in the blood for early cancer diagnosis. In drug discovery, NMR spectroscopy is known for its unmatched power in identifying/characterizing molecular structures, dynamics, and interactions [14, 15].

3 EPR for Biomedical Applications

The fundamental physics underlying both NMR and EPR are very similar. However, the consequence of the abovementioned three-orders-of-magnitude difference between the nuclear and electron gyromagnetic ratios has significant implications for practical applications. While the first EPR experiments were performed only a few years after the first NMR experiments [16], their development has differed significantly since.

NMR requires RF fields in the range of several MHz to hundreds of MHz, which can be relatively simply generated and guided to the sample using simple (coaxial) cables. By contrast, conventional EPR, even at low B_0 fields of a few hundred mT, requires GHz radiation that necessitates bulky waveguides for its transmission and microwave resonators to excite and read out the spin magnetization with high efficiency. Moreover, the timescales of the spin dynamics are largely different for NMR and EPR. Here, transversal relaxation times (T_2), which are decisive for the observation of a time domain signal after a pulsed excitation, of nuclei at room temperature are typically in the order of milliseconds, whereas electrons relax in a few hundred nanoseconds. As a consequence, as already mentioned above, today, NMR is performed almost exclusively in the pulsed mode, whereas, until today, EPR is predominantly a cw technique. In a conventional cw experiment, the EPR spectrum is recorded by irradiating the sample with a B_1 field at a fixed frequency, and the B_0 field is swept through the resonance condition, in general, resulting in longer acquisition times and less experimental freedom. The predominance of cw EPR is mainly due to the prohibitive costs of pulsed EPR spectrometers between one million and two million USD. Therefore, chip-based pulsed EPR spectrometers have a potential for true technological disruption in this area, potentially leading to a much wider adoption and faster development of pulsed EPR techniques [17].

However, even in cw mode, EPR can provide an unrivaled amount of information on paramagnetic species, which, in a biomedical context, are mostly free radicals and reactive oxygen species (ROS). Free radicals are intimately related to human physiology and can have both positive and negative effects; as an example, their imbalance can cause cancer but can also be used for therapy [18]. An excess of free radicals in cells or tissue is a consequence of oxidative stress and can, for example, be used to assess the antioxidative capacity of the human skin [19]. These radical species are usually too short-lived to be captured by cw methods directly, and most current detection methods therefore rely on using so-called spin traps or radical scavengers [20]. These are molecules that are initially diamagnetic but can easily react with free radicals in their vicinity, resulting in a paramagnetic species with a corresponding EPR signature. By monitoring the number of spin traps in the sample, i.e., the EPR signal intensity, it is therefore possible to monitor the evolution of free radical species. A second widespread application of EPR in the biomedical context is tissue oximetry, which is required for the diagnosis and treatment of many clinical problems, such as chronic wounds and diabetic ulcers [21]. The detection of oxygen levels is typically also performed by the use of an additional spin probe with a material whose T_2 times sensitively react to changes in oxygen concentration. While these measurements can, in principle, be performed with cw spectrometers as well, pulsed EPR measurements are much faster and more reliable for T_2 extraction. The first pulsed EPR spectrometers for this application are currently being developed and even clinically tested [21]. However, these instruments still require a rather large microwave bridge with a significant power consumption.

4 Chip-Integrated NMR and EPR Electronics

Spin detection, especially in low-field biomedical applications, suffers from a low signal-to-noise ratio (SNR) due to the low polarization levels and the low operating frequency. Therefore, the receiver electronics should preserve the small intrinsic SNR of the MR detector as faithfully as possible, leading to stringent requirements on the receiver noise figure. Here, typical noise figures of conventional, discrete MR LNAs are in the range of 0.4 dB to 0.6 dB. Since performance is very critical in both NMR and EPR detection systems, any chip-integrated NMR/EPR detection electronics have to achieve a similar (ideally even better) noise performance. An important advantage of monolithically integrated NMR and EPR transceivers lies in the fact that the entire transceiver can be integrated on a tiny footprint, greatly facilitating the design of large-scale arrays of detectors. Moreover, the CMOS realization of the transceiver electronics allows for a fully differential realization of all building blocks, greatly increasing the electronics' immunity against the Hall effect and magnetoresistive effects, thereby enabling the placement of the transceiver electronics inside the B_0 field in close proximity to the MR detection coil. This configuration removes the strict need for an impedance-matched connection between the tuned NMR/EPR resonator and the RF frontend, leading to

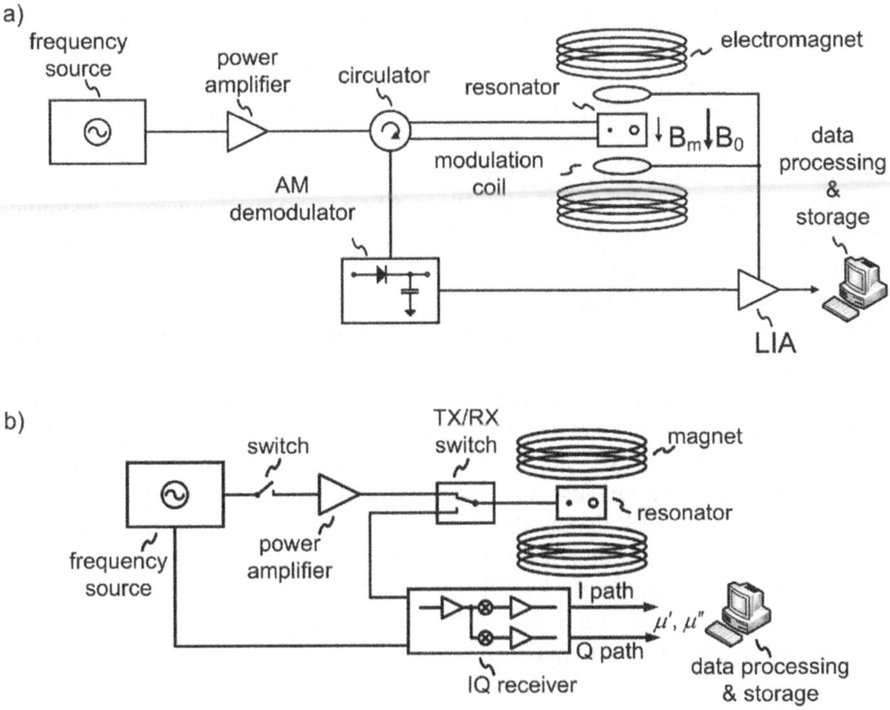

Fig. 2 Conventional setups for (**a**) continuous-wave and (**b**) pulsed magnetic resonance experiments

extra degrees of freedom in the design of the low-noise amplifier (LNA) and power amplifier (PA) that can be exploited to improve the system performance.

Typical setups for cw and pulsed magnetic resonance experiments are shown in Fig. 2. According to Fig. 2a, the cw setup requires a circulator (or a similar device), which cannot be easily integrated in standard integrated circuit technologies. As an alternative to this conventional architecture, in Sect. 4.2, we will discuss the oscillator-based detection approach, which provides a detection architecture that is much more compatible with CMOS integration. Here, it should be pointed out that cw NMR experiments are very uncommon these days and cw detection methods are used almost exclusively in EPR.

Figure 2b shows the conventional experimental setup for a pulsed magnetic resonance experiment. Typically, the same coil/resonator is used both for exciting the spin ensemble and for detecting the resulting change in spin magnetization. Due to the nature of the pulsed MR signal (see Sect. 2), excitation and detection can be performed in a time-multiplexed fashion, requiring a simple TX/RX switch to toggle between the transmit and receive modes of operation. Therefore, this architecture can be easily integrated into standard CMOS and is the starting point for virtually all NMR-on-a-chip realizations presented until today. By contrast, the significantly higher operating frequencies of EPR render the design of very-low-noise LNAs and CMOS PAs with sufficient driving strengths difficult, resulting in only a few publications that integrate the architecture of Fig. 2b.

Since the TX/RX switch is placed in the receive path prior to the LNA, its noise performance is very critical. Moreover, during TX, it has to provide sufficient isolation between the PA and the sensitive LNA. Conventional, discrete, impedance-matched MR systems often implement the well-known $\lambda/4$ arrangement, a simple, transmitter-receiver separation based on a $\lambda/4$ transformer and switching diodes to protect the LNA input [22].

For chip-integrated MR transceivers without the need for impedance matching, typically (on- or off-chip) high-speed, low-loss protection switches are used to provide sufficient isolation for TX and a good noise figure during RX.

An important problem associated with the pulsed detection mode is the energy stored inside the coil/resonator during TX. After the pulse, this energy needs to be removed before the LNA can be safely connected to the coil/resonator, causing a certain dead-time after the pulse. For NMR, where the signal typically lasts for a few up to a few hundred milliseconds, this dead-time problem is less critical compared to EPR, where typical relaxation times are in the range of a few tens of nanoseconds up to microseconds. This issue for EPR detection can be solved by the oscillator-based detection approach discussed in Sect. 4.2, which provides an intrinsically dead-time free detection after the excitation pulse.

4.1 Transceiver Architectures for NMR-on-a-Chip Detectors

As mentioned above, most NMR-on-a-chip realizations integrate the architecture of Fig. 2b. The block diagram of a typical implementation is shown in Fig. 3. The transceiver can be split into three main building blocks: the quadrature (I/Q) receiver (blue), the transmitter (orange), and the frequency synthesizer (green). These three main building blocks will be described in detail in the following:

Quadrature Receiver

The quadrature receiver typically consists of an LNA, followed by a quadrature down-conversion mixer, an optional automatic offset compensation, and intermediate frequency (IF) variable gain amplifiers (VGAs). As mentioned above, the SNR of the detected MR signal is intrinsically weak, mandating a very low overall receiver noise figure (NF) to preserve this weak intrinsic SNR as faithfully as possible. According to the Friis formula, for a well-designed LNA with sufficient gain, the overall receiver NF is determined by the NF of the LNA, requiring an LNA with low input-referred noise and sufficient gain to overcome the noise of the subsequent mixer stage. An example LNA topology that has been used in several NMR-on-a-chip publications from the open literature is shown in Fig. 4. The LNA features a differential input stage, which is ideally suited to process the emf induced in the NMR coil, with a current reuse technique to provide the highest possible

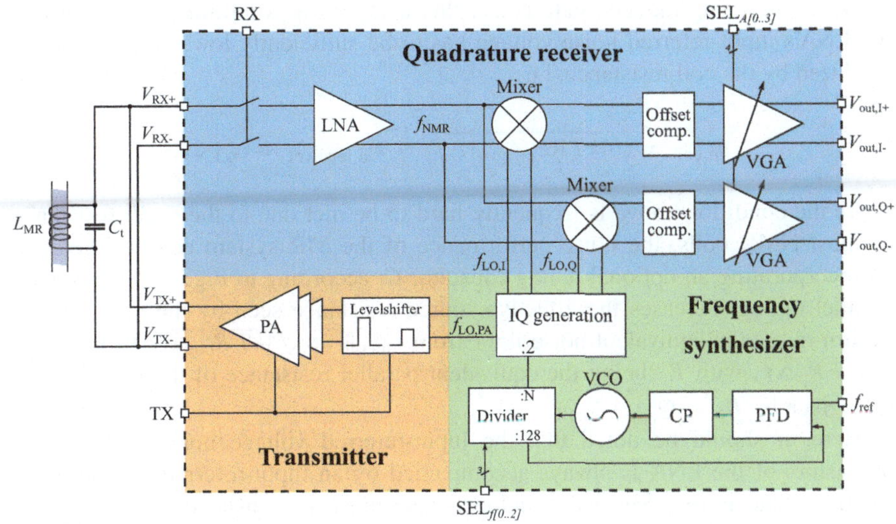

Fig. 3 Block diagram of a typical NMR-on-a-chip transceiver, cf. [23]

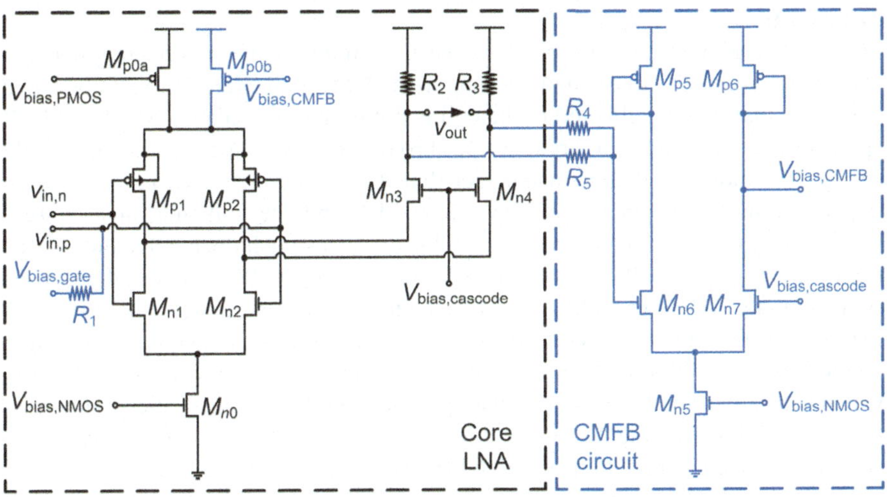

Fig. 4 Typical LNA topology for NMR-on-a-chip transceivers

g_m/I_{BIAS} to minimize the input-referred voltage noise for a given bias current. A common-mode feedback (CMFB) circuit is used to ensure that the (folded) cascode branch is biased with sufficient current by monitoring the common-mode voltage at the output.

An important difference between conventional discrete LNAs for NMR and NMR-on-a-chip implementations is the close proximity of the MR sensor coil and the ASIC, removing the need for a 50 Ω impedance matching between the coil and the LNA. Ignoring induced gate noise, this leads to the simplified condition that the LNA's input-referred noise voltage has to be sufficiently lower than the noise produced by the coil resistance, i.e.:

$$v_{\mathrm{n,coil}} = \sqrt{4kT\mathrm{Re}\{Z_{\mathrm{coil}}\}\Delta f} \overset{!}{>} v_{\mathrm{n,receiver}} \approx v_{\mathrm{n,LNA}}. \quad (1)$$

Since the condition above is frequently hard to be met due to the small resistance of most NMR coils, the noise performance of the MR system can be improved by incorporating an optional tuning capacitor C_t according to Fig. 3. This resonant parallel tuning increases the effective source resistance seen by the LNA, which in turn raises the equivalent noise floor from $v_{\mathrm{n,coil}} = \sqrt{4kTR_{\mathrm{coil}}\Delta f}$ to $v_{\mathrm{n,tank}} = \sqrt{4kTR_{\mathrm{p}}\Delta f}$, with R_{p} being the equivalent parallel resistance of the resonant LC tank given by $R_{\mathrm{p}} = Q_{\mathrm{coil}}^2 \cdot R_{\mathrm{coil}}$.

Here, it should be noted that the input-referred voltage noise of the input transistors of the LNA is always accompanied by an input-referred current noise (induced gate noise). Since this induced gate noise is translated into equivalent voltage noise by the impedance of the coil/the resonant tank, induced gate noise is particularly important when the NMR coil is resonantly tuned using C_t.

Transmitter

The most important building block in the transmitter is the power amplifier that delivers the required large driving current to the NMR coil. Here, it should be noted that the actual power delivered to the coil is not important because NMR and EPR applications only require a large current through the coil to produce a sufficiently strong B_1 field. Together with the fact that the in-field PA does not require an impedance-matched connection between the PA and the coil, the most energy-efficient way of delivering current to a coil is to use a switch operated at the desired frequency to periodically connect and disconnect the NMR coil to and from the supper supply. While providing the highest energy efficiency, such a nonlinear PA requires more sophisticated concepts to provide the amplitude modulation required by most modern NMR applications. The most simple way of driving the NMR coil, which naturally provides the required amplitude modulation capabilities, is the class A amplifier shown in Fig. 5a, where a MOSFET operated in saturation is used as a controlled current source to drive the NMR coil.

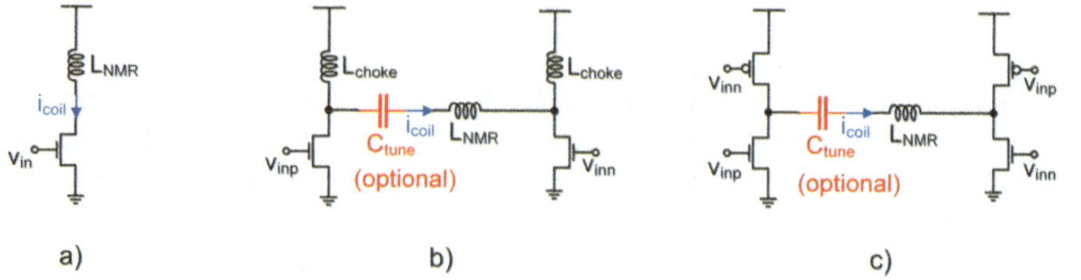

Fig. 5 Typical power amplifier topologies for NMR-on-a-chip transmitters

For coil-frequency combinations, for which the effective coil impedance is compatible with the voltage compliance of the PA, this structure provides a broadband, linear excitation with a reasonable efficiency. Here, efficiency refers to the signal current in the coil divided by the required DC bias current. While the single-ended structure of Fig. 5a is good for illustration purposes, a more realistic implementation of a CMOS or BiCMOS in-field PA would resort to the differential version shown in Fig. 5b, which is more immune against interferences due to the large B_0 field, which can, for example, be caused by the Hall effect. Here, in order to maximize the achievable output swing, chokes are used. However, due to their large inductance values, the chokes frequently have to be placed off-chip, significantly increasing the overall footprint of the electronics. When integrated on-chip, the large required circuit area can significantly increase the overall cost of the in-field PA. In the differential structure of Fig. 5b, the two MOSFETs are again operated in saturation, resulting in a linear amplifier. If the voltage compliance level does not allow for sufficient current driving capabilities, one can series-tune the NMR coil to remove its reactance (see Fig. 5b). If a higher efficiency and/or higher output current levels are required, one has to resort to the switching PAs mentioned above. One typical implementation using an H-bridge is shown in Fig. 5c. Being a nonlinear switching PA, the basic H-bridge configuration only allows for phase modulation. If amplitude modulation is required as well, either pulse-width modulation (PWM); pulse density modulation (PD), e.g., produced by a $\Sigma\Delta$ modulator); or a polar modulation concept has to be used with the associated difficulties of, for example, switching tones in the PWM case. To implement a simple polar PA, one can replace the PMOS transistors in Fig. 5c by NMOS high-side switches with appropriate drivers [24]. A practical realization of such a configuration is shown in Fig. 6a. The required amplitude modulation can be performed by modulating the supply voltage (polar PA). Using high-voltage transistors, the circuit of Fig. 6a can deliver currents up to 8.3 A_{pp} to standard 3-mm NMR coils, allowing for larger sensitive volumes. These large currents in turn allow for short, microsecond excitation pulses (see Fig. 6b indicating a 90° pulse length of only 4 μs [24]), which in turn allow for the excitation of very broad spectral ranges.

Frequency Synthesis

A faithful extraction of the MR information from the sample requires not only an LNA implementation with an excellent noise performance but also the possibility to excite the sample with an arbitrarily adjustable phase and to detect the resulting spin magnetization with a coherent local oscillator signal. Typical MR architectures utilize the quadrature detection scheme, cf. Fig. 3, to down-convert the high-frequency NMR/EPR signal to an intermediate frequency (IF) signal, which in turn relaxes the requirements on the subsequent amplifier and/or A/D converter. Furthermore, with the quadrature detection scheme, both homodyne (the receiver LO frequency equals the excitation frequency) and low-IF detection (the LO frequency and the excitation frequency display an offset) can be performed. The former benefits from the intrinsically coherent output phase across different scans but introduces a trade-off between excitation efficiency (on-resonance vs. off-resonance excitation) and

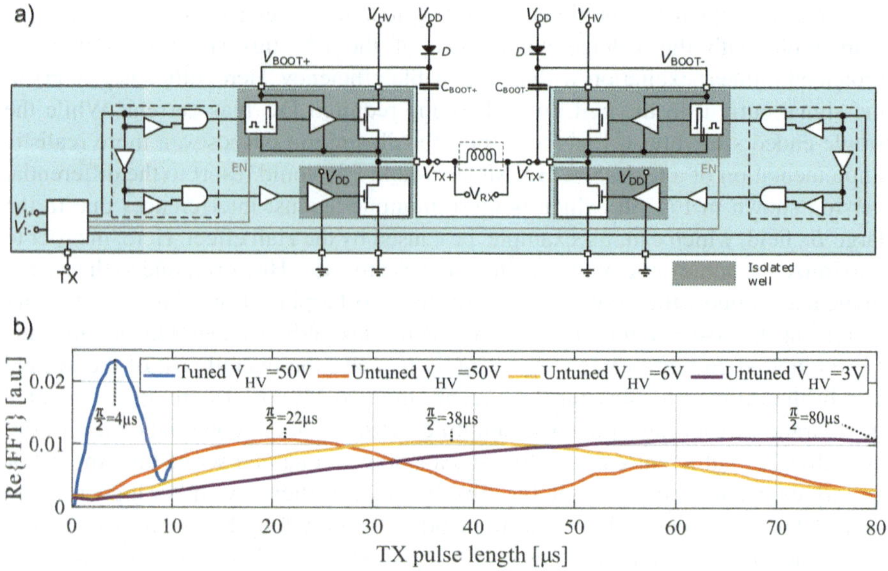

Fig. 6 (**a**) A practical realization of an H-bridge switching PA achieving up to 8.3 A_{pp} current in the NMR coil according to [24]. (**b**) Nutation plot measured with the chip from [24]

1/f noise (for a single spectral line, on-resonance excitation results in zero IF, while off-resonance excitation produces an intrinsic nonzero IF [25]). The latter allows for both on-resonance excitation and a nonzero IF to alleviate the influence of 1/f noise on the overall SNR. However, the low-IF approach also comes at the expense of an uncertain receiver phase if no dedicated countermeasures are taken [25]. To address the phase coherence issue, two different approaches are commonly used. The first one, which works very well for standard pulse sequences with fixed inter-pulse spacings, delays the transmitter signal until the LO and excitation frequency are in phase [26]. However, it is not suitable for sequences requiring arbitrarily and precisely definable time intervals between consecutive excitation pulses. In order to achieve a broader compatibility for both standard and more complex pulse sequences, it must be possible to not only synchronize the receiver LO signal to the transmitter waveform without introducing nonconstant delays in the transmitter timing, but also the phase of the excitation pulse has to be arbitrarily adjustable. This leads to the second approach, which makes use of direct digital synthesis (DDS), which is explained in detail in [25]. Figure 7 shows a typical DDS-based frequency synthesis system for MR applications.

Moreover, the DDS approach also allows for so-called phase cycling, which is a very useful method to mitigate the effect of gain and phase mismatches between the in-phase and quadrature outputs which in turn, without proper countermeasures, would cause an imperfect cancellation of the image frequency, i.e., distortion of the MR spectrum.

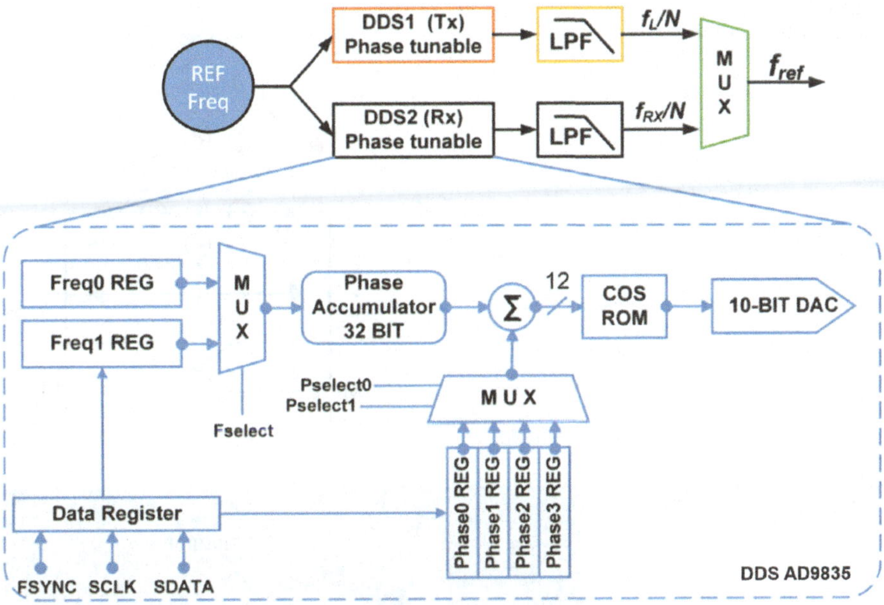

Fig. 7 Typical DDS-based frequency synthesis

4.2 Oscillator-Based EPR

Over the last decade, oscillator-based EPR has emerged as an alternative to the conventional transceiver-based architecture discussed in the previous section for conducting EPR experiments. In the oscillator-based approach, according to Fig. 8, the sample is placed in the sensitive volume of the inductor of an LC oscillator. Therefore, the current running through the inductor produces the microwave B_1 field required to excite the electron spins. A cw experiment is then performed by sweeping the B_0 field through the resonance condition $\omega_{osc} = \gamma B_0$, where ω_{osc} is the free-running oscillation frequency of the LC oscillator and γ is the gyromagnetic ratio. Around resonance, the EPR effect produces an effective change of the sample susceptibility, which, in turn, changes the net inductance and resistance of the tank inductor, and, thereby, the oscillation frequency and amplitude [27]. As shown in [28], under certain mild assumptions, the frequency- and amplitude-sensitive detection modes have the same limit of detection, which is equal to that of conventional coil-/resonator-based EPR.

Since its introduction, the basic oscillator-based sensing principle has been improved in several aspects. In [29], the use of voltage-controlled oscillators (VCOs) was proposed to allow for frequency-sweep EPR instead of the conventional field-sweep EPR experiments. Here, the main advantage of VCO-based EPR is the fact that the frequency sweeps can be carried out with near-constant sensitivity since a change in the VCO tuning voltage simultaneously changes the resonance

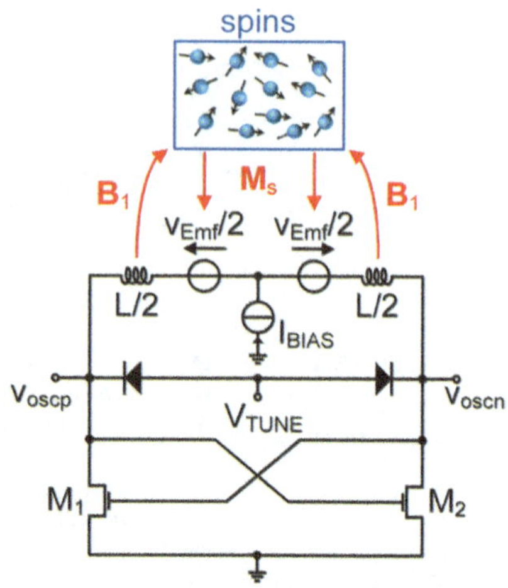

Fig. 8 Principle of oscillator-based EPR

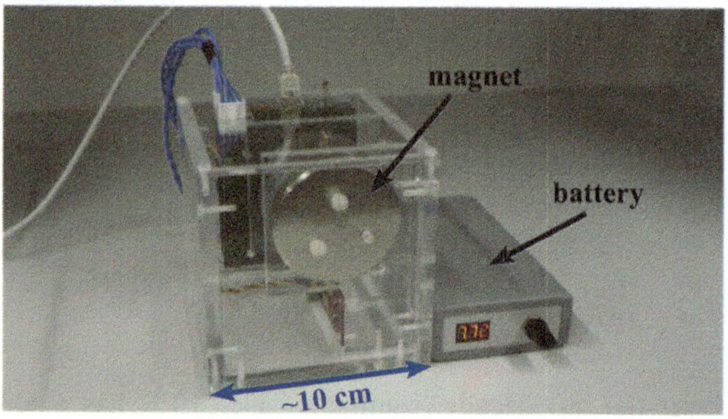

Fig. 9 The world's first portable EPR spectrometer

frequency of the LC resonator inside the VCO and the VCO's oscillation frequency. The possibility for frequency-sweep EPR allowed for the use of permanent magnets instead of the bulky electromagnets required for field-sweep EPR, enabling the design of the world's first portable EPR spectrometer shown in Fig. 9, cf. [29].

In [30], the sensitive volume of the VCO-based detectors was increased by injection locking several VCOs in a two-dimensional array. This approach not only increases the detector's sensitive volume but, at the same time, also lowers the phase noise of the array frequency compared to that of a single VCO. In this way, the injection-locked array approach improves the concentration sensitivity by a factor of $N_{\mathrm{VCO}}{}^{(3/2)}$, where N_{VCO} is the number of VCOs in the array. All early oscillator-based EPR experiments were carried out in the open-loop mode, i.e., the oscillator was not embedded in a phase-locked loop (PLL), and the EPR signal was extracted from the oscillator's amplitude and frequency using dedicated AM and FM demodulation blocks. While allowing for very simple experimental setups, the open-loop approach suffered from limited control on the precise value of the oscillation amplitude and frequency, as it is required for, for example, quantitative EPR. To improve this, in [31], the VCO-based EPR detector was embedded into a

PLL both to make the demodulated (frequency) EPR signal available at the VCO tuning voltage and to allow for a precise definition of the VCO frequency and phase from an external, stable reference.

Very recently, the VCO-based approach has been extended beyond conventional cw EPR by the use of high-bandwidth PLLs, cf. [17, 32, 33]. Here, in [32], a VCO-based detector embedded into a high-bandwidth PLL was used for rapid scan EPR experiments to improve the limit of detection by avoiding sample saturation. In [17], it was, for the first time, experimentally shown that VCO-based EPR detectors can be used to perform dead-time free pulsed EPR experiments. Practical demonstrations of these EPR modalities enabled by fast on-chip PLLs can be seen in Fig. 10 [33]. In Fig. 10a, the topology of the employed semi-digital PLL is shown together with the measurement setup. Even without a direct access to the VCO array's tuning voltage, the EPR data can be extracted through FM demodulation

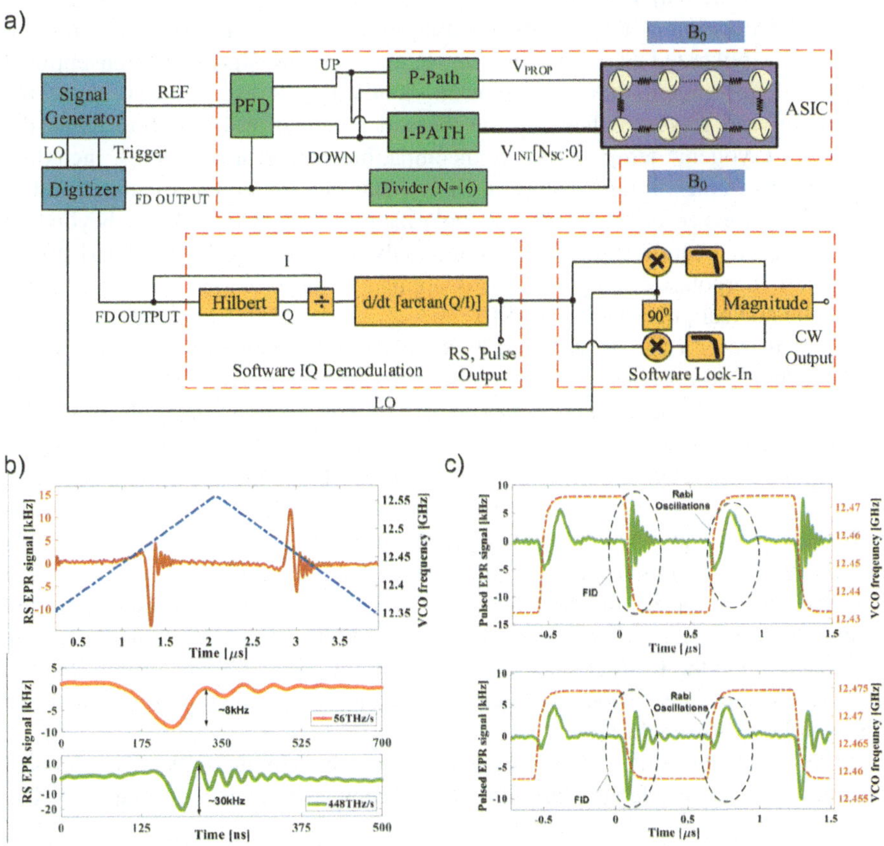

Fig. 10 A practical realization of a high bandwidth semi-digital PLL for EPR applications. (**a**) ASIC and experimental setup topology. (**b**) Rapid scan and (**c**) pulsed EPR measurements performed using the above setup

of the divided VCO output in post-processing. Figure 10b shows data from rapid scan EPR experiments, clearly indicating an increase in signal intensity with increased scan rates. Dead-time free pulsed EPR measurements, showing both Rabi oscillations and FIDs, measured with the same setup, are shown in Fig. 10c.

Overall, VCO-based EPR presents a promising path for the realization of future point-of-care EPR detection platforms for biomedical applications since it provides a high level of integration (sensor and readout electronics co-integrated in CMOS) together with a state-of-the-art sensing performance.

5 Chip-Integrated Dynamic Nuclear Polarization

As briefly mentioned in the introduction and explained in more details in Sects. 2 and 3, one of the major limitations of both NMR and EPR is the small degree of thermal polarization of the sample, resulting in approximately one in one million spins being polarized –i.e., contributing to the net MR signal– in conventional NMR and one in ten thousand spins in standard EPR experiments. Since the straightforward way of enhancing the thermal polarization of cooling down the sample to cryogenic temperatures is incompatible with many samples, including biomedical samples in their natural environment, one important research trend in magnetic resonance is to increase the polarization level of the sample beyond its value in thermal equilibrium. This is commonly referred to as "hyperpolarization" or if the hyperpolarization technique is applied to an ensemble of nuclear spins "dynamic nuclear polarization" (DNP).[2]

While there is a zoo of different hyperpolarization techniques currently being studied, they more or less all follow the same basic principle: The (for a given temperature and B_0 field) much larger polarization of an ensemble of electron spins is transferred to a nearby ensemble of nuclear spins. If the electron and nuclear spin ensemble are always kept at the same effective temperature and magnetic field, the achievable enhancement factor for the polarization level of the nuclear spins is bound by the ratio of the gyromagnetic ratios of the electron spin ensemble and the nuclear spin under investigation. As an example, this leads to a maximum enhancement factor around 660 for protons. This basic principle can be varied in numerous ways, e.g., by using different temperatures and/or B_0 fields for the polarization and the readout step. Due to the circuit-oriented focus of this chapter, we refer the interested reader to a recent book that provides a good overview of the different DNP approaches [34].

In this chapter, we will discuss two approaches in more details: classical Overhauser DNP and DNP using nitrogen vacancy (NV) centers in diamond. The

[2] DNP refers to the transfer of the intrinsically higher electron polarization to nuclei. There are two other widespread nuclear hyperpolarization methods, namely, para-hydrogen-induced polarization (PHIP) and chemically induced dynamic nuclear polarization (CIDNP).

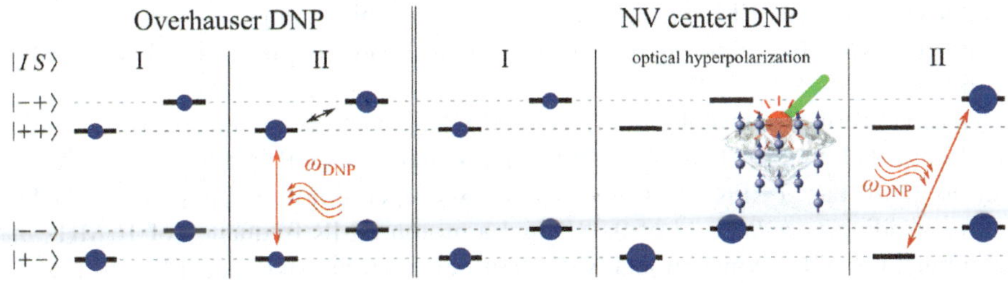

Fig. 11 Basic working principle of Overhauser DNP and NV center DNP with an intermediate optical hyperpolarization step

former is important because it has been shown to work well at lower B_0 fields, which are the most relevant for future portable and wearable NMR systems, experimentally achieving enhancement factors $\gg 100$. The latter is explicitly mentioned here, because it uses an optical hyperpolarization step, which can provide even larger enhancement factors $\gg 1000$ without the need for cryogenic temperatures and/or elevated field strengths. The basic working principle of both Overhauser DNP and NV center DNP is illustrated in Fig. 11. According to the figure, both methods rely on a transfer of the higher polarization level (step II) of the lower-energy electron spin state $|+-\rangle$ to the spin state $|-+\rangle$, which can then be read out using conventional NMR pulse sequences. The NV-based approach involves an intermediate optical hyperpolarization step, in which the lower-level electron spin state is almost fully polarized, potentially leading to much larger enhancement factors compared to Overhauser DNP, which relies on the higher thermal polarization levels of the electron spins in the sample.

Importantly, in both cases, the transfer of polarization to the nuclear spin ensemble requires irradiating the sample with a microwave magnetic field, whose frequency ω_{DNP} matches the difference of the energy levels. For Overhauser DNP, it is directly the EPR resonance frequency of the electron spins of the DNP agent in the absence of the nuclear spins, whereas for NV center DNP it is different to that only by a few MHz. Therefore, any NMR spectrometer that wants to benefit from the signal enhancement due to DNP requires a microwave transmitter with sufficient output power to saturate the transition shown in Fig. 11. Since it is, in general, advisable to be able to verify the efficiency of the polarization transfer prior to the actual DNP experiment, it is very useful to also integrate an EPR receiver in addition to the EPR transmitter, i.e., an entire EPR spectrometer, into a DNP-enhanced NMR spectrometer.

As discussed in Sect. 3, a very elegant way of realizing an EPR spectrometer in CMOS makes use of an array of injection-locked VCOs [30]. Here, an important additional constraint for a DNP-enhanced NMR spectrometer compared to a simple EPR-on-a-chip spectrometer results from the need to irradiate the sample volume with both the microwave field for the polarization transfer B_{1e} and the radio frequency field for the subsequent NMR experiment, B_1. This leads to an intrinsic trade-off between homogeneity of the NMR-B_1 field, which is best if the NMR coil is placed inside the EPR resonator, and the achievable strength of the microwave magnetic field B_{1e}, which is best if the EPR resonator is placed inside the NMR coil.

A first proof-of-concept, chip-integrated DNP spectrometer has been presented by Boero et al. in [35]. This single-chip DNP-enhanced NMR spectrometer includes an oscillator-based EPR spectrometer at a fixed operating frequency of 10.7 GHz, corresponding to a B_0 field of 0.38 T and a proton NMR frequency of 16 MHz in combination with a conventional NMR transceiver architecture according to Fig. 3. Both the EPR and the NMR spectrometer make use of chip-integrated planar spiral coils. The results presented in [35] clearly demonstrate that CMOS integration presents a very promising route toward the miniaturization and cost reduction of DNP-enhanced NMR spectrometers, thereby providing the possibility of making the power of DNP-enhanced NMR available in future generations of point-of-care NMR devices.

6 Summary and Conclusion

In this chapter, we have discussed NMR and EPR as two complementary spin-based detection methods, whose analytical powers render them prime candidates for point-of-care biomedical devices. However, as explained above, for a long time, the size and cost of conventional MR equipment prevented its widespread use. With the recent availability of high-performance permanent magnets, chip-integrated realizations of the required MR electronics (NMR-on-a-chip and EPR-on-a-chip) proved to be the key enabler for highly miniaturized and scalable MR detection platforms. Here, several academic research prototypes of such miniaturized MR detection systems have demonstrated their excellent achievable performance in a number of biomedical target applications. However, due to the relatively low B_0 field strengths of the portable MR platforms, they typically suffer from a degraded sensitivity compared to state-of-the-art MR lab equipment. As a potential solution to the sensitivity problem, we have discussed (chip-integrated) hyperpolarization techniques, which can provide sensitivity increases by several orders of magnitude. Overall, the NMR- and EPR-on-a-chip approaches have the potential to turn NMR and EPR from niche applications into widely used methods that can have a great impact on personalized medicine and the health sector in general.

References

1. E. Danieli, J. Perlo, B. Blümich, F. Casanova, Small magnets for portable NMR spectrometers. Angew. Chem. Int. Ed. **49**(24), 4133–4135 (2010)
2. B. Blümich, K. Singh, Desktop NMR and its applications from materials science to organic chemistry. Angew. Chem. Int. Ed. **57**(24), 6996–7010 (2018)

3. B.M. Alnajjar, A. Buchau, L. Baumgärtner, J. Anders, NMR magnets for portable applications using 3D printed materials. J. Magn. Reson. **326**, 106934 (2021)
4. N. Sun, T.-J. Yoon, H. Lee, W. Andress, R. Weissleder, D. Ham, Palm NMR and 1-chip NMR. IEEE J. Solid State Circuits **46**(1), 342–352 (2010)
5. E.L. Hahn, Spin echoes. Phys. Rev. **80**(4), 580 (1950)
6. H.Y. Carr, E.M. Purcell, Effects of diffusion on free precession in nuclear magnetic resonance experiments. Phys. Rev. **94**(3), 630 (1954)
7. S. Meiboom, D. Gill, Modified spin-echo method for measuring nuclear relaxation times. Rev. Sci. Instrum. **29**(8), 688–691 (1958)
8. M. Balci, *Basic 1H-and 13C-NMR Spectroscopy* (Elsevier, Amsterdam, 2005)
9. D. Krüger, A. Zhang, H. Hinton, V.M. Arnal, Y.-Q. Song, Y. Tang, K.-M. Lei, J. Anders, D. Ham, "A portable CMOS-based MRI system with 67× 67× 83 μm^3 image resolution, in *ESSCIRC 2022-IEEE 48th European Solid State Circuits Conference (ESSCIRC)* (IEEE, Piscataway, 2022), pp. 225–228
10. A.A. Ghazani, C.M. Castro, R. Gorbatov, H. Lee, R. Weissleder, Sensitive and direct detection of circulating tumor cells by multimarker μ-nuclear magnetic resonance. Neoplasia **14**(5), 388–IN2 (2012)
11. D. Issadore, Y. Park, H. Shao, C. Min, K. Lee, M. Liong, R. Weissleder, H. Lee, Magnetic sensing technology for molecular analyses. Lab Chip **14**(14), 2385–2397 (2014)
12. C.M. Castro, A.A. Ghazani, J. Chung, H. Shao, D. Issadore, T.-J. Yoon, R. Weissleder, H. Lee, Miniaturized nuclear magnetic resonance platform for detection and profiling of circulating tumor cells. Lab Chip **14**(1), 14–23 (2014)
13. F. Dreyer, Q. Yang, D. Krüger, J. Anders, A chip-based NMR relaxometry system for point-of-care analysis, in *2022 IEEE Biomedical Circuits and Systems Conference (BioCAS)* (2022), pp. 183–187
14. M. Pellecchia, I. Bertini, D. Cowburn, C. Dalvit, E. Giralt, W. Jahnke, T.L. James, S.W. Homans, H. Kessler, C. Luchinat, et al., Perspectives on NMR in drug discovery: a technique comes of age. Nat. Rev. Drug Discov. **7**(9), 738–745 (2008)
15. T. Sugiki, K. Furuita, T. Fujiwara, C. Kojima, Current NMR techniques for structure-based drug discovery. Molecules **23**(1), 148 (2018)
16. E. Zavoisky, Paramagnetic relaxation of liquid solution for perpendicular fields. J. Phys. U. S. S. R. **9**, 211 (1945)
17. M.A. Hassan, M. Kern, A. Chu, G. Kalra, E. Shabratova, A. Tsarapkin, N. MacKinnon, K. Lips, C. Teutloff, R. Bittl, J.G. Korvink, J. Anders, Towards single-cell pulsed EPR using VCO-based EPR-on-a-chip detectors. Frequenz (2022). https://doi.org/10.1515/freq-2022-0096
18. S. Prasad, S.C. Gupta, M.K. Pandey, A.K. Tyagi, L. Deb, Oxidative stress and cancer: advances and challenges. Oxidative Med. Cell. Longev. **2016**, 5010423 (2016). https://doi.org/10.1155/2016/5010423
19. S.F. Haag, B. Taskoparan, M.E. Darvin, N. Groth, J. Lademann, W. Sterry, M.C. Meinke, Determination of the antioxidative capacity of the skin in vivo using resonance Raman and electron paramagnetic resonance spectroscopy. Exp. Dermatol. **20**(6), 483–487 (2011). https://onlinelibrary.wiley.com/doi/abs/10.1111/j.1600-0625.2010.01246.x
20. M. Kopáni, P. Celec, L. Danišovič, P. Michalka, C. Biró, Oxidative stress and electron spin resonance. Clin. Chim. Acta **364**(1), 61–66 (2006). https://www.sciencedirect.com/science/article/pii/S0009898105004869
21. D. Cristea, H. Wolfson, R. Ahmad, Y. Twig, P. Kuppusamy, A. Blank, Compact electron spin resonance skin oximeter: properties and initial clinical results. Magn. Reson. Med. **85**(5), 2915–2925 (2021). https://doi.org/10.1002/mrm.28595
22. J. Anders, F. Dreyer, D. Krüger, On-chip nuclear magnetic resonance, in *Handbook of Biochips* (Springer, Berlin, 2022), pp. 667–698
23. F. Dreyer, D. Krüger, S. Baas, A. Velders, J. Anders, A broadband transceiver ASIC for X-nuclei NMR spectroscopy in 0.13 μm BiCMOS, in *2022 20th IEEE Interregional NEWCAS Conference (NEWCAS)* (2022), pp. 65–69

24. H. Bürkle, T. Klotz, R. Krapf, J. Anders, A 0.1 MHz to 200 MHz high-voltage CMOS transceiver for portable NMR systems with a maximum output current of 2.0 App, in *ESSCIRC 2021 – IEEE 47th European Solid State Circuits Conference (ESSCIRC)* (2021), pp. 327–330
25. Q. Yang, J. Zhao, F. Dreyer, D. Krüger, J. Anders, A portable NMR platform with arbitrary phase control and temperature compensation. Magn. Reson. **3**(1), 77–90 (2022). https://mr.copernicus.org/articles/3/77/2022/
26. S. Hong, N. Sun, Portable CMOS NMR system with 50-kHz IF, 10-us dead time, and frequency tracking. IEEE Trans. Circuits Syst. Regul. Pap. **68**(11), 4576–4588 (2021)
27. J. Anders, *Nonlinear Modeling of Continuous-Wave Spin Detection Using Oscillator-Based ESR-on-a-Chip Sensors* (Springer International Publishing, Cham, 2018), pp. 57–87. https://doi.org/10.1007/978-3-319-58996-1_4
28. A. Chu, B. Schlecker, M. Kern, J.L. Goodsell, A. Angerhofer, K. Lips, J. Anders, On the modeling of amplitude-sensitive electron spin resonance (ESR) detection using voltage-controlled oscillator (VCO)-based esr-on-a-chip detectors. Magn. Reson. **2**(2), 699–713 (2021). https://mr.copernicus.org/articles/2/699/2021/
29. J. Handwerker, B. Schlecker, U. Wachter, P. Radermacher, M. Ortmanns, J. Anders, A 14GHz battery-operated point-of-care ESR spectrometer based on a 0.13 µm CMOS ASIC, in *2016 IEEE International Solid-State Circuits Conference (ISSCC)* (2016), pp. 476–477
30. A. Chu, B. Schlecker, K. Lips, M. Ortmanns, J. Anders, An 8-channel 13GHz ESR-on-a-Chip injection-locked vco-array achieving 200μM-concentration sensitivity, in *2018 IEEE International Solid-State Circuits Conference-(ISSCC)* (IEEE, Piscataway, 2018), pp. 354–356
31. B. Schlecker, A. Hoffmann, A. Chu, M. Ortmanns, K. Lips, J. Anders, Towards low-cost, high-sensitivity point-of-care diagnostics using VCO-based ESR-on-a-chip detectors. IEEE Sensors J. **19**(20), 8995–9003 (2019)
32. M.A. Hassan, T. Elrifai, A. Sakr, M. Kern, K. Lips, J. Anders, A 14-channel 7 GHz VCO-based EPR-on-a-chip sensor with rapid scan capabilities, in *2021 IEEE Sensors* (2021), pp. 1–4
33. K. Khan, M.A. Hassan, M. Kern, K. Lips, I. Schwartz, M. Plenio, F. Jelezko, J. Anders, A 12.2 to 14.9 GHz injection-locked VCO array with an on-chip 50 MHz BW semi-digital PLL for transient spin manipulation and detection, in *2022 IEEE 65th International Midwest Symposium on Circuits and Systems (MWSCAS)* (2022), pp. 1–4
34. V.K. Michaelis, R.G. Griffin, B. Corzilius, S. Vega, *Handbook of High Field Dynamic Nuclear Polarization* (John Wiley & Sons, Hoboken, 2019)
35. N. Sahin Solmaz, M. Grisi, A.V. Matheoud, G. Gualco, G. Boero, Single-chip dynamic nuclear polarization microsystem. Anal. Chem. **92**(14), 9782–9789 (2020)

Models and Interfaces for Electrochemical Sensors: Architectures and Implementations

Zhongzheng Wang, Anthony Wall, Alan O'Riordan, Daniel O'Hare, Gerardo Molina Salgado, and Ivan O'Connell

Abstract To determine and optimize the electrochemical sensor interface circuit specifications, as well as to propose novel topologies, a comprehensive understanding of its equivalent model is necessary. This chapter provides a detailed analysis of the sensor's electrode-electrolyte interface and illustrates the electrical modeling of both two-electrode and three-electrode electrochemical systems. Furthermore, several realistic design assumptions and concerns are discussed, along with examples of typical electrochemical sensor interfaces, namely, potentiostats and current readout circuits. In conclusion, challenges and future trends are discussed for both the sensor modeling and the sensor interface, with the intention of researching and presenting new opportunities and future advances.

1 Introduction

Numerous chemical sensing systems and technologies have been developed for both offline and inline continuous monitoring, with applications in healthcare, pharmaceutical manufacturing, environmental protection, agriculture, and food production [1–6]. Electrochemical sensors are gaining popularity as sensing solutions in these areas. These sensors employ a chemical reaction at the electrode surface to produce a current when a known potential is applied to the sensor system, showing selectivity to a certain type of analytes in the testing environment. Electrochemical sensors enable a number of applications, including DNA detection, glucose-level monitoring, virus detection, and enzyme reactions, among others, and provide a higher resolution, faster response, and lower power dissipation. This enables the development of mobile or implantable platform applications.

With the advent of the Internet of Things (IoT), the global market for electrochemical sensors was valued at $2.19 billion in 2019, and it is expected to increase to

Z. Wang (✉) · A. Wall · A. O'Riordan · D. O'Hare · G. M. Salgado · I. O'Connell
MCCI, Tyndall National Institute, Cork, Ireland
e-mail: zhongzheng.wang@mcci.ie

P. Harpe et al. (eds.), *Biomedical Electronics, Noise Shaping ADCs, and Frequency References*, https://doi.org/10.1007/978-3-031-28912-5_5

$11.83 billion by 2025 [7, 8]. In the context of the global COVID-19 pandemic, the demand for rapid, stable, accurate, cost-effective, and easy-to-use electrochemical sensors and sensor interfaces has never been greater [9]. Therefore, electrochemical sensors and their interfaces have undergone significant development and show a promising future.

Thus far, electrochemical sensor models have been limited, and it has not been possible to simulate electrochemical sensors in electronic design automation (EDA) platforms alongside their readout circuits. Consequently, it is necessary to model the sensor with its lumped element circuits, reflecting the physical attributes of the sensor. A thorough comprehension of the sensor's equivalent model and modeling process is crucial when proposing new readout circuit topologies. This chapter aims to fill the electrochemical sensor knowledge gap and explores their application in a range of typical sensor interfaces. The chapter begins with an in-depth illustration of the electrode-electrolyte interface, then a two-electrode electrochemical system is explored, and finally the three-electrode electrochemical system is detailed. Then, varieties of sensor potentiostats and current readout circuits are mentioned, along with key design assumptions and considerations, which contribute to simplifying the design and generating new ideas. Finally, future trends for both the electrochemical sensor modeling and their interface will be shown.

2 Chemical Sensor and Electrochemical Sensor Overview

A chemical sensor is a device which transforms chemical information into electrical signals, such as voltage or current, which can be further processed, and inferences made from them. This chemical information may originate from chemical reactions of a particular analyte or caused by a physical property of the investigated system. A very common example of this is the concentration of a specific sample in a liquid or gas [10].

As shown in Fig. 1, a chemical sensor typically consists of two components: a **receptor** and a **transducer**. In the receptor section, chemical information is converted into a type of energy that the transducer is sensitive to (e.g., light

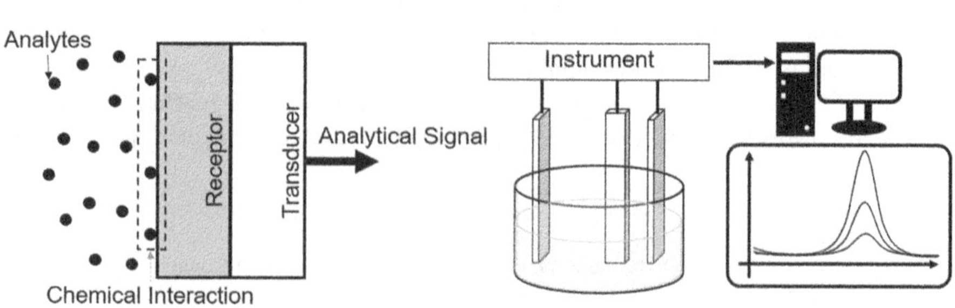

Fig. 1 The chemical sensors and the electrochemical sensor system

or charge). In the transducer section, the energy released by the receptor will be converted into an analytical signal that can be digitized. Examples of typical chemical sensor applications include the breathalyzer, carbon monoxide sensors, and gas sensors [11].

The electrochemical sensor is a subset of chemical sensors, where the transducer is an electrode. It is often submerged in a testing liquid or gas to detect a particular chemical reaction. The distinguishing factor of electrochemical sensors from chemical sensors is their ability to analyze the **quantitative relationship** between the electricity and chemical processes. The typical measurement techniques of electrochemical sensors include cyclic voltammetry (CV), amperometry, and electrochemical impedance spectroscopy (EIS). Applications of this type of sensor include environmental monitoring, agricultural production, disease diagnosis and health monitoring, etc.

This chapter focuses on electrochemical sensors and not the broader topic of chemical sensors. The goal of the chapter is to build an understanding of the three-electrode electrochemical sensor and how to best interrogate it.

3 Electrode-Electrolyte Interface and Electrochemical Sensor Systems

3.1 Electrode-Electrolyte Interface

The electrode-electrolyte interface is the interface between the electrode surface and electrolyte when an electrode is immersed in a testing electrolyte, as is shown in Fig. 2. On the left is the electrode, and its surface with the electrolyte is on the right, containing some solvated molecules, solvated cations (positively charged ions), and solvated anions [12].

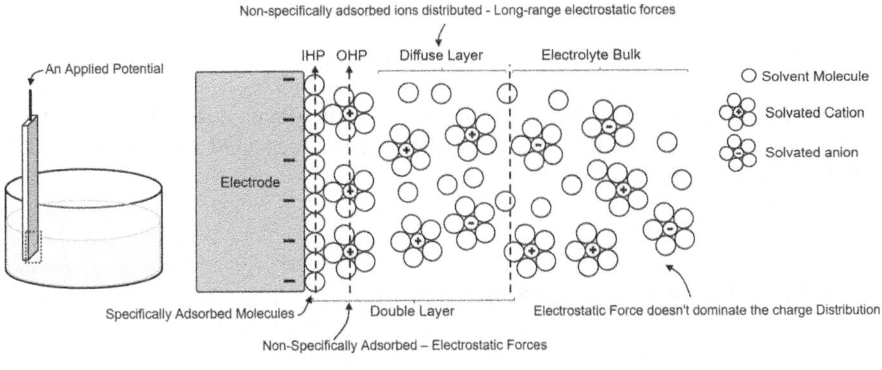

Fig. 2 The electrode-electrolyte interface for a single electrode

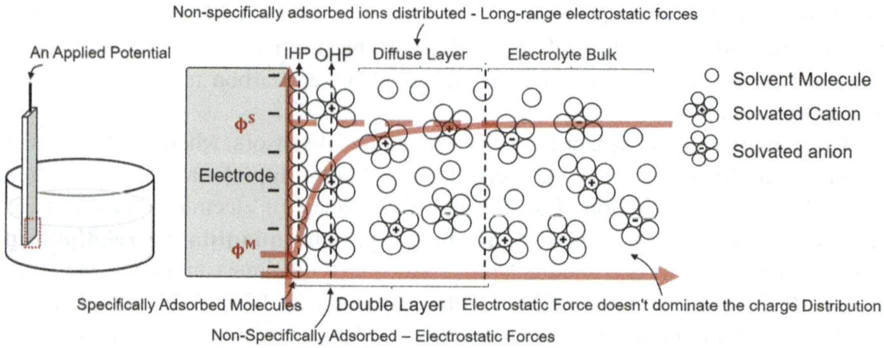

Fig. 3 The electrode-electrolyte interface for a single electrode

When a potential is applied to the electrode relative to the electrolyte, the solvent molecules and sometimes other ions (not shown in the figure for simplicity) are specifically adsorbed to the surface of the electrode, forming a layer known as the inner Helmholtz plane (IHP). The ions with opposite polarity with respect to the potential at the electrode (cations in this case) are adsorbed nonspecifically due to the electrostatic forces, whose locus of the center of the ions generates another layer known as the outer Helmholtz plane (OHP). Long-range electrostatic forces cause the cations and anions to be distributed at longer distances from the electrode surface, which is known as the diffuse layer, and the electrolyte bulk is the area where the long-range electrostatic force no longer dominates the ion distribution.

The IHP, OHP, and diffuse layer make up the double layer (DB) at the electrolyte-electrolyte interface [13].

As a result of the charge distribution at the electrode-electrolyte interface, a potential difference arises between the electrode and the electrolyte bulk. This difference is defined as the electrode potential [12], as shown in Fig. 3. The electrode potential E is described by (1), where ϕ^M is the potential at the electrode and ϕ^S is the potential of the electrolyte bulk:

$$E = \phi^M - \phi^S \tag{1}$$

In conclusion, the electrode-electrolyte contact produces a double layer, and a potential is created across this interface. The analysis presented in this chapter will center mostly on these two aspects.

3.2 *The Electrochemical Sensor System*

When measuring the electrode potential of a single-electrode system with a voltage meter, the probe of the voltage meter becomes the other electrode, as shown in

Fig. 4, creating a potential difference between the surface of the probe and the electrolyte bulk. Therefore, the measured potential, E_{mea}, is the difference between the electrode potential, E, and the probe potential E_{probe}:

$$E_{mea} = E - E_{probe}. \tag{2}$$

Equation (2) shows that the value of the electrode potential E, which is of interest, cannot be estimated owing to the uncertainty of the potential E_{probe}, whose value is determined by Nernst equation [12]. Besides, the applied potential change ΔE_{app} is not equal to that at the electrode ΔE, as expressed in Eq. (3):

$$\Delta E_{app} = \Delta E - \Delta E_{probe} \tag{3}$$

Therefore, with a single-electrode sensor system, it is not feasible either to directly detect the absolute potential of a single electrode or to apply a potential change to the electrode of interest.

In Fig. 5, the probe of the voltage meter or potentiometer is replaced with a reference electrode (RE) whose electrode potential is constant and dependent on the electrode material. In a conventional Ag/AgCL RE, for instance, the electrode potential is $E_{RE,AgCL} = 0.199\,\text{V}$ [12].

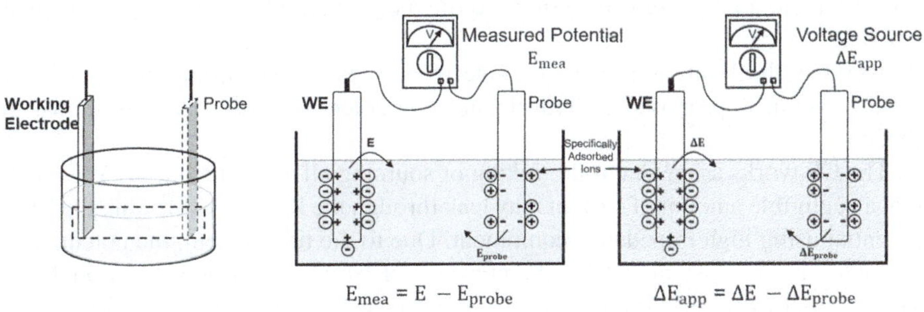

Fig. 4 The single-electrode system: sensing or applying the electrode potential

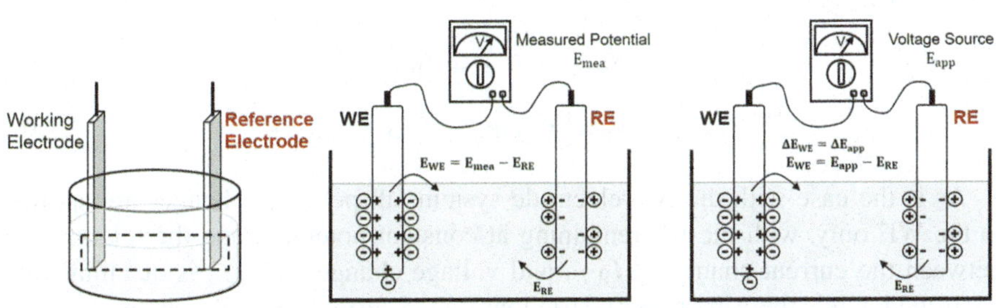

Fig. 5 The two-electrode sensor system: sensing or applying the electrode potential

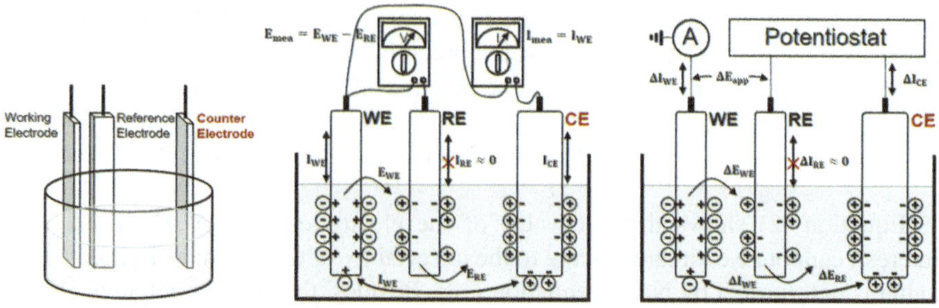

Fig. 6 The three-electrode sensor system: sensing or applying the electrode potential

This kind of electrode serves as a reference for the electrode under investigation, the working electrode (WE), so that the control of the potential at WE, as defined by Eq. (4), is possible. However, if the current through the RE is very high, the potential at RE might change; thus, the potential at the WE is no longer controlled:

$$\Delta E_{WE} = \Delta E_{app} \tag{4}$$

In conclusion, the RE functions as a reference, such that any applied voltage changes will only affect WE, if RE conducts a tiny current. This limits the use of this sensor type to applications that don't require a high current for the electrochemical reaction.

To break the limitation of the two-electrode sensor system, the three-electrode sensor system is proposed by introducing a counter electrode (CE), as shown in Fig. 6.

The CE works as an electrode sinking or sourcing all current from the WE, such that a negligible amount of current may leak through the RE by establishing the RE's potential using high impedance equipment. Due to the tiny current, the potential of RE will remain constant even in the presence of large current flowing through the WE. In order to provide sufficient charges (electrons), the area of CE is often greater than the one of WE and RE [9]. The operation of the three-electrode sensor system can be represented by Eqs. (5) and (6):

$$I_{mea} = I_{CE} = I_{WE} \tag{5}$$

$$V_{mea} = V_{WE} - V_{RE} \tag{6}$$

As is the case with the two-electrode system, all potential changes are applied to the WE only, with the RE remaining at constant potential, and the relationship between the current change, ΔI_{WE}, and voltage change, ΔV_{WE}, is quantitatively controlled, while there are fewer restrictions on the current strength. Three-electrode sensor systems are typically employed in electrochemical research and applications due to these advantages. A three-electrode sensor system requires a piece of specialized device called a potentiostat, which, on the one hand, is used to maintain the potential at RE while blocking the inflow of current and, on the other hand, transports the current through CE. The usage of the potentiostat increases the complexity of the sensor interface integrated circuits.

4 Typical Electrochemical Techniques

Potentiometry measurement (PM), voltammetry measurement (VM), coulometry measurement (CM), and electrochemical impedance spectroscopy (EIS) are four of the most common electrochemical measurement methods in the development of electrochemical sensor interface design. This section gives an overview of those topics.

4.1 Potentiometric Analysis

During the potentiometry measurement, a reference electrode and an indicator electrode are submerged in the electrolyte under test. The indicator electrode is utilized to detect the change of the concentration of a redox couple. It is usually an inert noble metal, such Pt, and the potential difference between the reference electrode and the indicator electrode is measured. This potential difference offers information on the concentration of a certain analyte of interest [14] and can be calculated using the Nernst equation (7) [12]:

$$E_{cell} = E^0_{cell} - \left(\frac{RT}{nF}\right) \cdot \ln Q \tag{7}$$

where E_{cell} is the recorded cell potential, E^0_{cell} is the standard cell potential, R is the universal gas constant, T is the temperature in Kelvin, n is the amount of transferred electrons, F is the Faraday constant, and Q is the reaction quotient reflecting the instantaneous ratio of oxidation/reduction concentrations between the two electrodes.

4.2 Coulometric Analysis

In coulometry, a constant voltage is delivered to the WE with respect to the RE in an electrochemical cell, or a constant current is delivered between the WE and RE for two-electrode systems, therefore entirely changing the oxidation state of the analyte. The total current passed is then recorded to calculate the number of electrons passed, which, according to Faraday's law, reveals the concentration of a certain analyte [15]:

$$Q = nFN_A \tag{8}$$

where N_A is the moles of the analyte and Q represents the total charges.

4.3 *Voltammetric Analysis*

The objective of a voltammetry measurement is to examine the resulting current versus the applied time-varying/constant voltage, which is then used to record the Faradaic current and comprehend the kinetics or processes of redox reactions of a particular kind of analyte [16]. Cyclic voltammetry (CV), whose applied signal is a ramp signal with a predetermined scan rate (SR), initial potential (IP), and switching potential (SP), as shown in Fig. 7, is a widely used voltammetry method.

The SR range varies greatly from 1 mV/sec all the way up to 50 V/Sec, depending on the size of the electrode, material of the electrode, and the analyte under study. The SR is usually selected as a trade-off to provide a fast enough measurement while maintaining the lowest amount of acceptable charging (capacitive) currents. A typical value of SR is 100 mV/sec.

The switching potential is normally dependent on the analyte of interest, such that the voltage is swept back after the oxidation (or reduction) of the molecule is completed. The initial potential could be selected to eliminate interfering species or to reduce the voltage window (lower measurement time). Usually, this is based on doing wider ranges and narrowing these in as necessary.

The cell voltage, namely, $V_{WE} - V_{RE}$, typically starts at 0 V, but it can be set to positive or negative voltages and swept back, ranging from −3.0 V to 2.5 V for carbon electrodes and −1.7 V to +1.7 V for metal electrodes.

The SR can be converted to a frequency, as the ramp signal is a periodic signal, as expressed in Eq. (9):

$$f_{CV} = \frac{\text{Scan Rate}}{2 \times |\text{Switching Potential} - \text{Initial Potential}|} \tag{9}$$

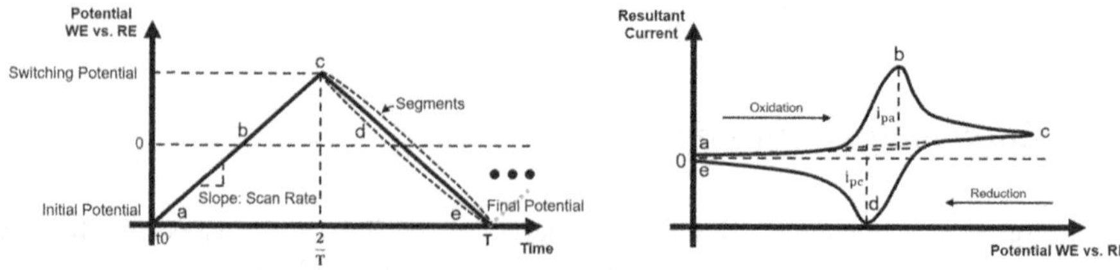

Fig. 7 Cyclic voltammetry

It can be observed from Eq. (9) that the frequency of CV is quite low; hence, this is a low-frequency application in the IC design sense. Amperometry measurements, whose applied potential is a constant voltage at the WE in terms of RE, are a subset of voltammetry. This form of measurement focuses on the resulting current versus time and is a commonly required readout circuit topology for IC designers.

4.4 Electrochemical Impedance Spectroscopy (EIS)

EIS, which measures the impedance of the electrochemical cell, is one of the most commonly used electrochemical methods. EIS is superior to other electrochemical measuring methods in investigating signals across a broad frequency range, from less than 1 mHz to higher than 1 MHz, by applying small-signal analysis to the electrochemical cell under test [17].

In EIS, the sensor electrode-electrolyte interface is linearized as a pseudo-linear time-invariant (LTI) system, as shown in Fig. 8, by applying small signal (less than 5 mV) to the cell.

When a sinusoid signal, $x_n(t)$, with an initial magnitude $|A|$ and an initial phase, $\phi_{n,i}$, is applied to the RC circuit, the response current $i_n(t)$ will have a scaled magnitude $|A||H(j\omega_n)|$ and shifted phase, $\phi_{n,i}$, where $n = 0, 1, 2, 3 \cdots$ indicates different frequencies. Then the impedance of the circuit, $Z_n(j\omega_n)$, can be calculated by Eq. (10):

$$Z(j\omega_n) = \frac{V_n(j\omega_n)}{I_n(j\omega_n)} = |Z_n(j\omega_n)|\angle(\phi_{n,v} - \phi_{n,i}) \tag{10}$$

where, $V_n(j\omega_n)$ and $I_n(j\omega_n)$ are the phasor representation of the sinusoid signal $x_n(t)$ and, $i_n(t)$, respectively.

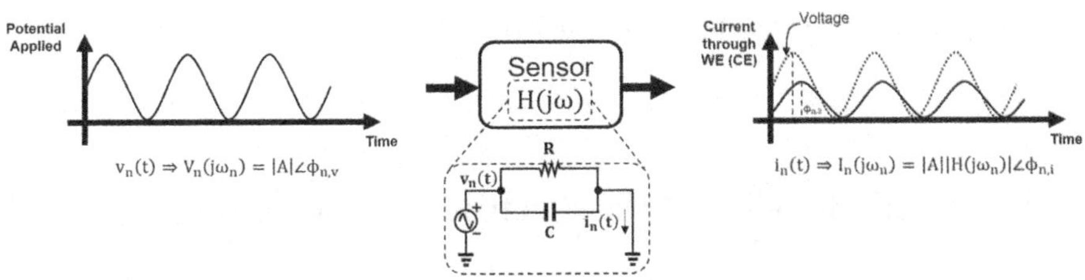

Fig. 8 Electrochemical impedance spectroscopy

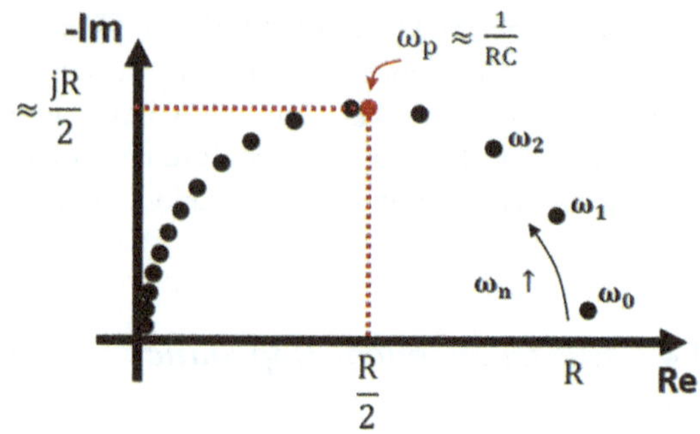

Fig. 9 The Nyquist diagram for a parallel RC circuit

For an RC parallel network, its impedance, $Z_{RC}(j\omega_n)$, is given by Eq. (11):

$$Z_{RC}(j\omega_n) = \frac{V_n(j\omega_n)}{I_n(j\omega_n)} = \left|\frac{R}{1 + j\omega RC}\right| \angle tan^{-1}\left(\frac{Im\left[\frac{R}{1+j\omega RC}\right]}{Re\left[\frac{R}{1+j\omega RC}\right]}\right) \tag{11}$$

If the $Z_{RC}(j\omega_n)$ is plotted in a Nyquist plot, as shown in Fig. 9, the R and C can be readily obtained from the chart:

$$R = Re\left[\frac{R}{1 + j\omega_n RC}\right]|_{\omega_n \approx 0} \tag{12}$$

$$C \approx \frac{1}{\omega_p R} \tag{13}$$

As a consequence of the higher operating frequency and very low resultant current (nA level for nanosensors), designing interfaces for EIS is a significant challenge. Parasitic capacitors and resistors, in particular, may have a large impact on the end outcomes, and developing new topologies for EIS with better accuracy, sensitivity, and higher resolution but smaller-scale IC area has been an active area of research.

5 Electrical Modeling of the Electrochemical Sensor

For the interface design, the modeling of the electrode-electrolyte interface and the two-electrode and the three-electrode electrochemical sensor systems is of the utmost importance. In order to develop circuit specifications and novel topologies, an understanding of the electrochemical equivalent circuit model is required. This section will begin by modeling the electrode-electrolyte interface before moving on to the two- and three-electrode systems.

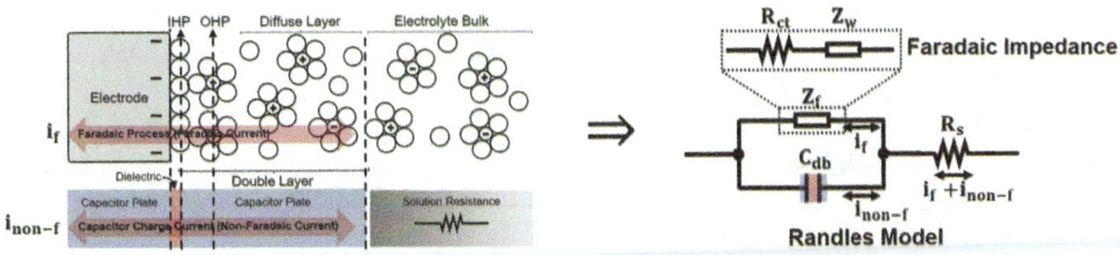

Fig. 10 The modeling of the electrode-electrolyte interface

5.1 *Modeling of the Electrode-Electrolyte Interface*

The interface between the electrode surface and the electrolyte bulk consists of a double layer, which comprises the IHP, OHP, and diffuse layer, as discussed in Sect. 3.1. As shown in Fig. 10, the structure of the double layer, when combined with the electrode, generates a capacitor known as the double-layer capacitor, C_{db}, with the dielectric IHP and the two plates of the capacitor being the electrode and the double layer, respectively.

When a certain voltage is developed across the electrode and double layer, two different kinds of current will flow across the interface: a faradaic current and a non-faradaic current. Non-faradaic current, such as the current used to charge a capacitor, does not adhere to Faraday's law, whereas the Faradaic current, which is approximated by chemical processes, does. Both types of current occur concurrently and hence should be represented by separate current paths. One branch consists of a single double-layer capacitor, and the other is represented with a faradaic impedance, Z_f. Considering the resistance due to the solution, R_s, a typical model for the electrode-electrolyte interface, Randles model was proposed in [18], as shown in right-hand side of Fig. 10.

The Faradaic impedance can be regarded as a combination of a charge-transfer resistor, R_{ct}, in series with a Warburg impedance that models the impedance caused by the mass transport of the electroactive species [19–21].

5.2 *Modeling of the Two-Electrode System*

Randles model is useful to model the structure of a single electrode-electrolyte interface, while a two-electrode sensor system has two separated electrode-electrolyte interfaces. As the oxidation reactions and reduction reactions take place at the RE and WE simultaneously, the electrode-electrolyte interfaces for the two electrodes should have the same structure, as shown in Fig. 11.

In a two-electrode sensor system, the RE should be placed as close as possible to WE, so that the IR drop across R_s is minimized. In addition, as aforementioned in Sect. 3.2, the change of the applied potential is only seen at the WE, and therefore the structure of RE can be simplified. In nanosensor-based electrochemical integrated

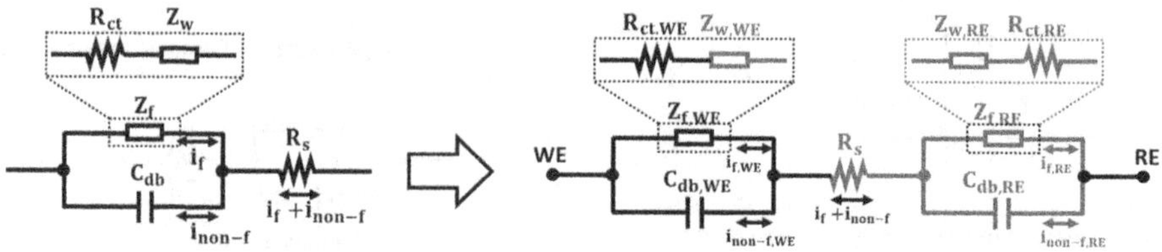

Fig. 11 The modeling of the two-electrode sensor system

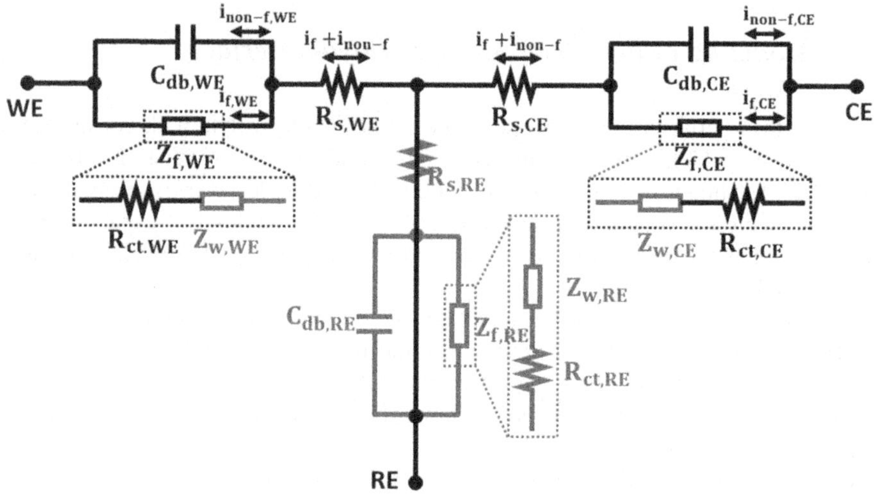

Fig. 12 The modeling of the three-electrode sensor system

systems, the mass transport increases, and thus the current is no longer limited by the diffusion of ions toward the WE [22]. As a result, the mass transfer-dominated Z_W becomes negligible, and the model can be simplified to a simple RC Randles circuit.

5.3 Modeling of the Three-Electrode Sensor System

The model of the three-electrode sensor system is an extension from one of the two-electrode systems, as shown in Fig. 12. Since there is negligible or nearly no current flowing though the RE, the structure of the electrode-electrolyte interface at the RE is simplified. The CE introduces another electrode-electrolyte interface. As described in Sect. 3.2, the WE and CE have different surface areas and therefore the double-layer capacitor, charge-transfer resistor, and solution resistor.

The final model is shown in Fig. 13. This model is commonly used when modeling electrochemical sensors in the context of readout interface design. Some

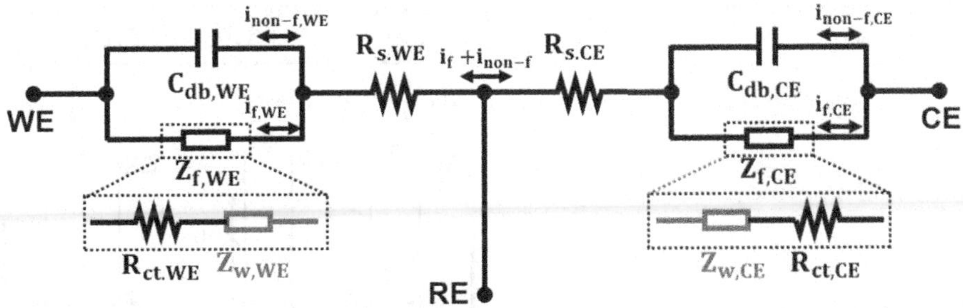

Fig. 13 The modeling of the three-electrode sensor system

practical assumptions and considerations can be made: (1) the solution resistance R_s is much smaller than the charge-transfer resistor R_{ct}. This helps to simplify the step when performing EIS: R_s can be ignored, and the frequency requirement can be relaxed. (2) In the case of nanosensors, the cell current will be very small, due to the high charge-transfer resistance, which may cause high small-signal output resistance of the transistors. (3) The double-layer capacitor at the CE is much larger than that of the WE, and the solution resistance, $R_{s,WE}$, is smaller than $R_{s,CE}$. This may require consideration when performing stability analysis.

6 Typical Electrochemical Sensor System Interface Architectures

The electrochemical sensor interface consists of two design targets, the potentiostat and the current readout circuit, as shown in Fig. 14. The potentiostat acts to stabilize a known potential at RE by sinking or sourcing the current at the CE while preventing any current flow into the RE. In other words, the current only flows between CE and WE, so that the RE electrode potential will not shift. The other part, the current readout circuit, converts the current into a signal (e.g., voltage) which can be digitized and further processed in the digital domain.

This section will discuss two typical sensor interfaces. One is an operational amplifier (Op-Amp)-based potentiostat forming a trans-impedance amplifier (TIA) as the current readout circuit, which converts the cell's current to voltage. The other is the current mirror-based potentiostat with current mirrors as current readout devices. The current-to-frequency converter (CFC) is applied to digitize the sensor current.

Fig. 14 The potentiostat and current readout

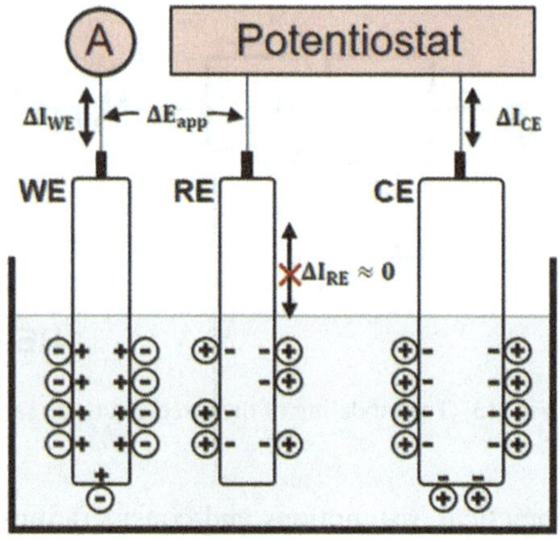

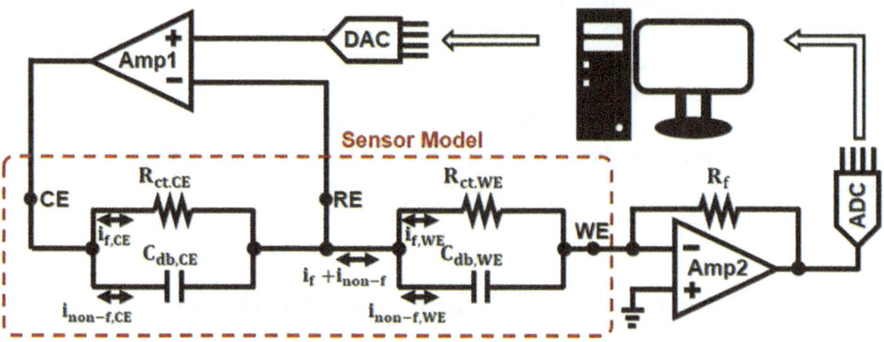

Fig. 15 Op-Amp-based potentiostat and TIA

6.1 *Example 1: Op-Amp-Based Potentiostat + TIA*

In this topology, an operational amplifier is utilized to construct a negative feedback loop with the CE impedance in the path, as shown in Fig. 15. Since $Amp1$ and $Amp2$ are capable of dealing with bidirectional current, this type of electrochemical sensor interface can handle both oxidation and reduction chemical reactions. The potential at RE is biased by the negative feedback loop with nearly no current flowing through it, due to the high DC impedance looking into the amplifier input. The application of this circuit is widespread, including voltammetry and EIS.

For nanosensor applications, the charge-transfer resistor, $R_{s,WE}$, ranges from approximately 1 $M\Omega$ to 200 $M\Omega$, and the double-layer capacitor, $C_{db,WE}$, can be hundreds of pF to the tens of nF [23, 24]. Therefore, the corner frequency ranges from a few Hz to a few kilo Hz, according to Eq. (14).

$$f_c = \frac{1}{2\pi R_{ct,WE} C_{db,WE}} \tag{14}$$

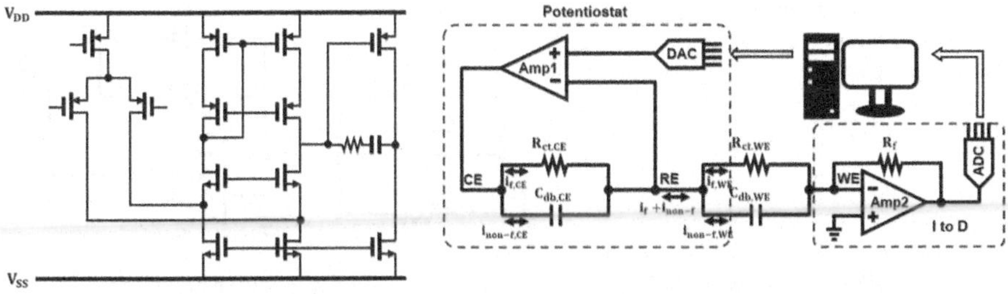

Fig. 16 The PMOS-based two-stage folded-cascode amplifier

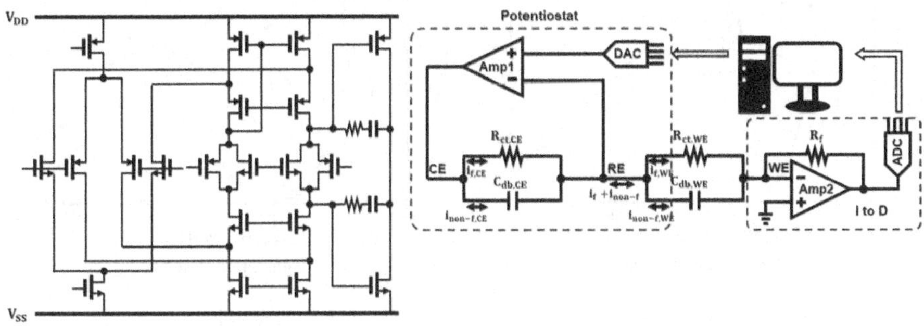

Fig. 17 The rail-to-rail folded-cascode with floating class AB controlled output stage

$Amp1$ and $Amp2$ are often implemented as two-stage folded-cascode designs [25], as shown in Fig. 16, for example, or a rail-to-rail folded cascode with floating class AB controlled output stage for larger input and output range as the one shown in Fig. 17 [26]. Since power dissipation and gain bandwidth product (GBW) tend to be the target specifications for these amplifiers, current sources should be in strong inversion and input devices in weak inversion.

When various cell voltages are applied, the values of the double-layer capacitor and charge-transfer resistor fluctuate, which significantly affects the TIA's stability. $Amp2$ is considered as a first-order system, as shown in (15), and therefore introduces a pole at $\frac{1}{2\pi R_{op} C_{op}}$, where R_{op} and C_{op} are the output resistance and load capacitance of $Amp2$, respectively, and A_{op} is the DC gain of $Amp2$:

$$A(j\omega) = \frac{A_{op}}{1 + j\omega R_{op} C_{op}} \tag{15}$$

The feedback factor $\frac{1}{K_{no,C_f}} = 1 + j\omega R_f C_{WE}$ causes another pole at the frequency $\frac{1}{2\pi R_f C_{WE}}$, and the instability occurs at the intercept of $A(j\omega)$ and $\frac{1}{K_{no,C_f}}$. To solve this issue, as indicated in Fig. 18, the TIA should be overcompensated by inserting a feedback capacitor C_f.

Fig. 18 The schematic for TIA with/without the feedback capacitor

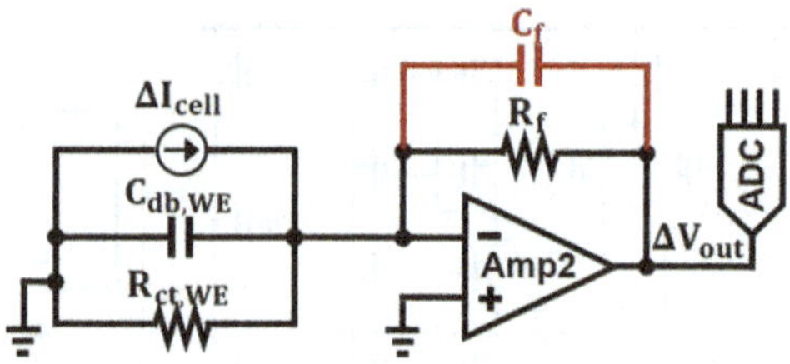

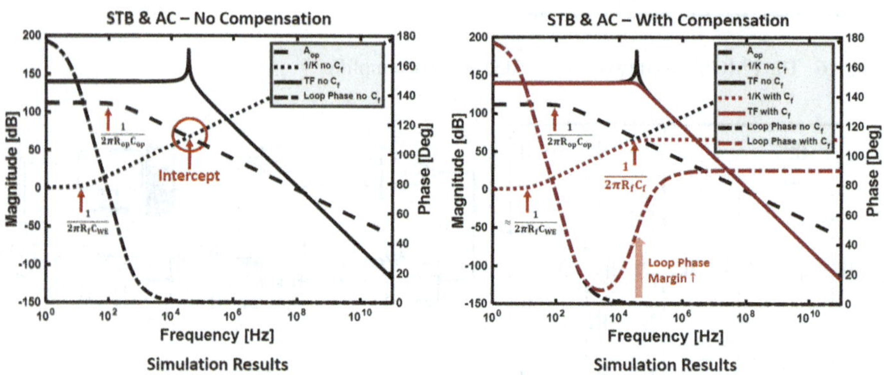

Fig. 19 The stability analysis for TIA with/without the feedback capacitor

By introducing a feedback capacitor C_f, a zero is placed at the frequency $\frac{1}{2\pi R_f C_f}$, whose value should be chosen at or before the intercept of $A(j\omega)$ and $\frac{1}{K}$, which compensates the phase margin to achieve higher stability, as shown in Fig. 19. Here, a practical assumption can be made: $C_f << C_{WE}$.

$$\frac{1}{K_{no,C_f}} = 1 + j\omega R_f C_{WE} \Rightarrow \frac{1}{K_{with,C_f}} \approx \frac{1 + j\omega R_f C_{WE}}{1 + j\omega R_f C_f}|_{C_f << C_{WE}} \tag{16}$$

The drawback of this topology is that the amplifier needs to be strong enough to sink/source the current from/to the sensor. This current ranges from the nA level to the hundreds of μA level, depending on the sensor type.

6.2 *Example 2: Current Mirror-Based Potentiostat + Current-to-Frequency Converter*

The current mirror-based potentiostat was first proposed in [27] for three-electrode electrochemical sensor systems, as shown in Fig. 20, and the one for a two-electrode system was proposed in 1987 [28].

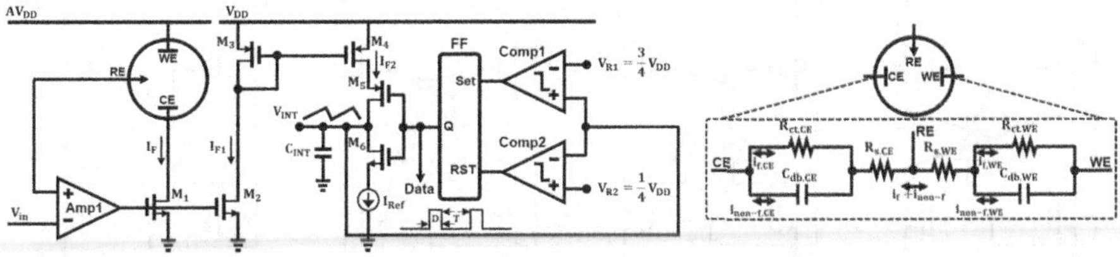

Fig. 20 The current mirror-based potentiostat and current-to-frequency converter

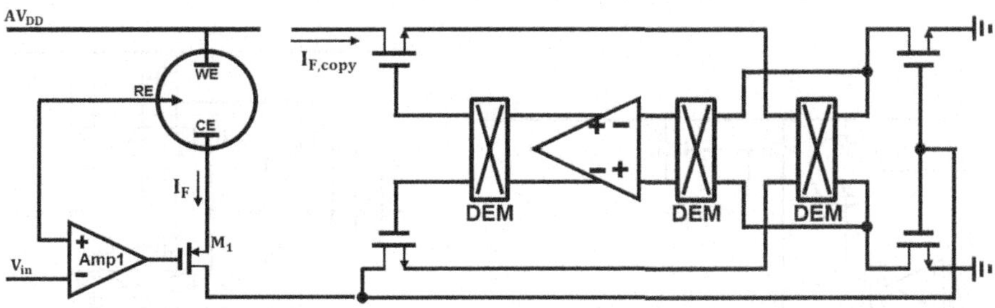

Fig. 21 The current mirror-based potentiostat with DEM and GB cascode current mirror

In this architecture, a bypass transistor M_1 is introduced to pass the resultant electrochemical sensor current and control the potential at the RE to match V_{in}, by establishing a negative feedback loop with an operational amplifier $Amp1$. The current readout is achieved by a current-to-frequency converter, transforming the magnitude of the current to time delay pulses, as expressed in Eq. (17) by comparing the preset potential and the current-charged potential, V_{INT}, at the capacitor C_{INT}. The benefits of this topology are that this potentiostat relaxes the difficulties in designing amplifiers in terms of heavy loads and strong current, compared to the TIA-based solution. Therefore, this topology may be more suitable for low-power applications, such as implantable or mobile devices:

$$T = \frac{V_{DD}C_{INT}}{2I_F}, \quad D = \frac{V_{DD}C_{INT}}{2I_{Ref}} \tag{17}$$

In this architecture, mismatch between current mirror branches is the primary design consideration. Mismatch can be caused by process variation, including the aspect ratio mismatches $\frac{W_1}{L_1} \neq \frac{W_2}{L_2}$ and threshold voltage mismatches $V_{th1} \neq V_{th2}$, as shown in Fig. 20, between M_1 and M_2. The second type of mismatch is caused by the different overdrive voltages, $V_{od1} \neq V_{od2}$, between M_1 and M_2, due to the channel length modulation of the transistor if the transistor is biased in the strong inversion region. Aspect ratio mismatch reduction is discussed in [26].

In this work, the dynamic element matching (DEM) technique is employed to the current mirror current branches, as shown in Fig. 21. The DEM operates in four phases, as shown in Fig. 22, which up-converts the mismatch between the two

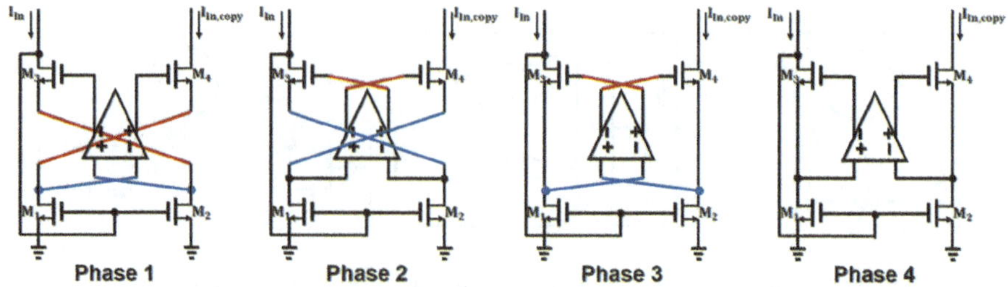

Fig. 22 The dynamic element matching configuration [26]

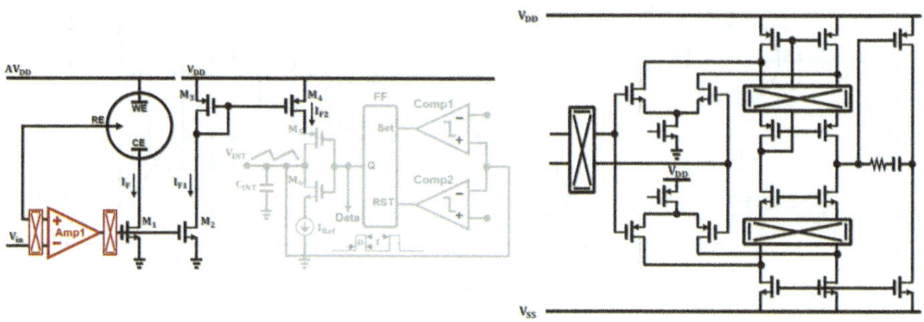

Fig. 23 The current mirror-based potentiostat with a chopping amplifier

branches, yielding $\frac{W_1}{L_1} \approx \frac{W_2}{L_2}$. In addition, a gain-boosted (GB) cascode current mirror is adopted to reduce the difference between V_{od1} and V_{od2}.

Regarding the mismatches of the threshold voltages, a trade-off is considered between the parasitic capacitor of the transistor, in particular the total parasitic capacitor at the drain C_{DD} [29], and the threshold variance, as expressed in (18) [30], where $\sigma_{\Delta V_{th}}$ is the threshold standard deviation. The mismatch is inversely proportional to area, but increased area comes at the cost of increased parasitic capacitance:

$$\sigma_{\Delta V_{th}} \propto \frac{1}{\sqrt{WL}}, \quad C_{DD} \propto WL \tag{18}$$

Since nA resolution is required, noise must be a key consideration. Since electrochemical sensors operate at low frequencies, flicker noise (FN) is of particular importance. The amplifier contributes to the majority of the FN, and a chopper stabilization (CS) approach is used to reduce the FN, as shown in Fig. 23 [31]. In this example, a two-stage rail-to-rail amplifier with Miller compensation is used, along with three choppers designed to attenuate the FN from the input and output stages.

For performing stability analysis, the circuit can be linearized to its small-signal equivalent model, as seen in Fig. 24. The loop gain is approximated by Eq. (19) [27].

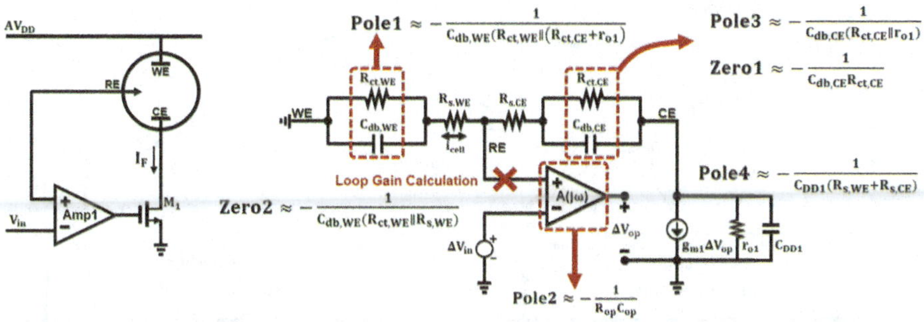

Fig. 24 The equivalent small-signal model for stability analysis

The loop is broken at the non-inverting input. The amplifier used here is modeled as a first-order low-pass filter with a DC gain A_{op}, as expressed in (15).

Some practical assumptions are made for doing this analysis as previously mentioned: (1) The solution resistance R_s is much smaller than the charge-transfer resistance R_{ct}. (2) The parasitic capacitance C_{DD1} of M_1 is much smaller than the double-layer capacitor. (3) For nanosensor systems, the cell current is very small yielding a large output resistance r_{o1} of M_1 and $r_{o1} >> R_{ct}$.

$$LG \approx -\frac{g_{m1}r_{o1}A_{op}(1+sR_{ct,CE}C_{db,CE})(R_{ct,WE}+R_{s,WE}+sR_{ct,WE}R_{s,WE}C_{db,WE})}{(1+sR_{op}C_{op})\{1+sC_{db,WE}[R_{ct,WE} \parallel (R_{ct,CE}+r_{o1})]\}[1+sC_{db,CE}(R_{ct,CE} \parallel r_{o1})][1+sC_{DD1}(R_{s,WE}+R_{s,CE})]} \tag{19}$$

According to the loop gain transfer function, the system has two zeros and four poles as described in Eqs. (20) and (21).

It can be seen that $Zero1 \approx Pole3$ when $r_{o1} >> R_{ct,CE}$ and $Pole4$ is far away from other poles due to the small value of the parasitic capacitor of the transistor C_{DD1}. $Pole1$ is dominated by the sensor electrode interface at the WE, which will be the dominant pole. The second pole, $Pole2$, is the nondominant pole caused by the amplifier. Their relationship is shown in (22):

$$Zero1 \approx -\frac{1}{R_{ct,CE}C_{db,CE}}; \quad Zero2 \approx -\frac{1}{(R_{ct,WE} \parallel R_{s,WE})C_{db,WE}} \tag{20}$$

$$\begin{aligned} Pole1 &\approx -\frac{1}{C_{db,WE}[R_{ct,WE} \parallel (R_{ct,CE}+r_{o1})]}, \quad Pole2 \approx -\frac{1}{R_{op}C_{op}}, \\ Pole3 &\approx -\frac{1}{C_{db,CE}(R_{ct,CE} \parallel r_{o1})}, \quad Pole4 \approx -\frac{1}{C_{DD1}(R_{s,WE}+R_{s,CE})} \end{aligned} \tag{21}$$

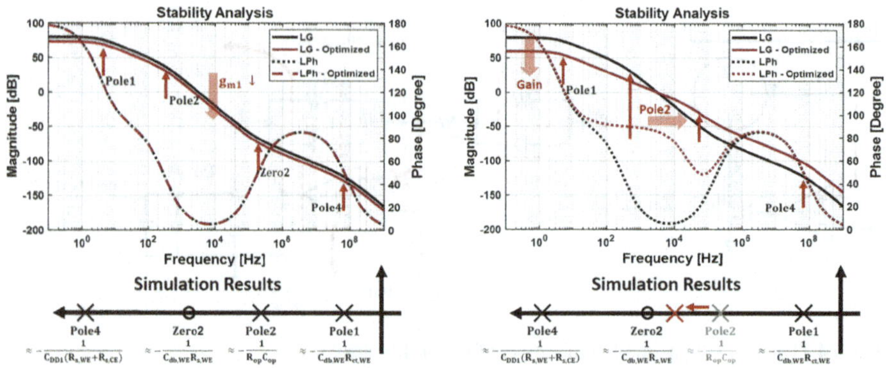

Fig. 25 The stability analysis with modeled amplifier for the circuit

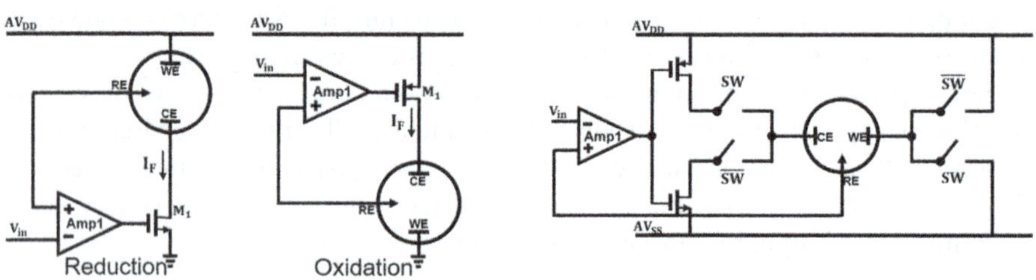

Fig. 26 The current mirror potentiostat for reduction and oxidation reactions and differential configuration

$$Zero1 \approx Pole3; \quad Pole1 < Pole2 << Pole4 \tag{22}$$

To ensure stability, the second pole, referred to as the amplifier pole, must be located away from the dominant pole introduced by the sensor, as shown in Fig. 25. The output resistance and parasitic capacitance of the amplifier must be selected with care, trading off against gain and phase margin, forcing the second pole to a higher frequency. At the same time, the reduction of the size of M_1 contributes to ensuring stability by pushing the zero-crossing to lower frequency, but this will sacrifice the gain and GBW.

In electrochemical measurements, the electrochemical sensor has two configurations, reduction and oxidation, as shown in the right-hand side of Fig. 26. This results in two types of the current: the reduction current and the oxidation current.

The current mirror-based potentiostat cannot handle the reduction and oxidation processes simultaneously. Tsai et al. [32] presented a differential architecture to address this issue, as shown in Fig. 26.

The differential current mirror-based potentiostat is a combination of the current mirror potentiostat for reduction and oxidation reactions. The measurement mode of the sensor, for reduction current or for oxidation current, can be switched by controlling the switches SW and $\overline{SW}$ alternatively, and they can be implemented with simple pass transistors or transmission gates. This potentiostat is limited to

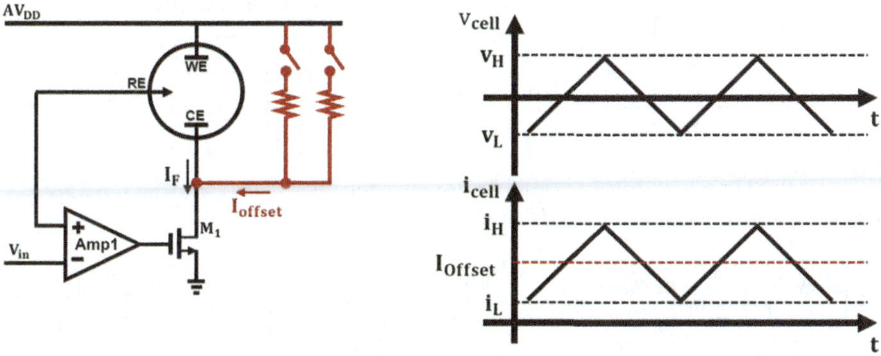

Fig. 27 The current mirror potentiostat with an offset current introduced by using bypass resistors

measuring the oxidation current and reduction current separately, and therefore is not suitable for CV measurements, etc., whose reduction and oxidation current transits from one to the other.

To solve this problem, an offset current can be added to the path of M_1 [31], to shift the DC level upward to deal with oxidation and reduction currents. This may introduce extra thermal noise, as shown in Fig. 27.

7 Future Trends

The model introduced in Sect. 5.3 is a static model that can be used used for amperometric or coulometric analysis, as well as EIS when the small-signal assumption is adopted. These measurement approaches presume that the model parameters are constant and consider the model to be an LTI system. In practice, the double-layer capacitor, charge-transfer resistor, and solution resistor are strongly connected not only to the sensor type but also to the concentration of the target species and applied voltage strength [12]. This will reduce the effectiveness of circuit analysis, such as stability analysis. Consequently, a more accurate model of the electrochemical sensor is required.

Similar to the analysis and modeling of transistors, the behavior of the electrochemical sensor electrode-electrolyte interface can be modeled in terms of large-signal and small-signal behaviors. In the large-signal model, as is pertinent in CV, the values of the C_{db} and R_{ct} vary significantly. They are linearized around their operating point for small-signal analysis.

As shown in Fig. 28, a possible approach for a more realistic model is to represent the sensor using machine learning (ML)-based techniques, such as a neural network (NN), decision tree (DT), etc.

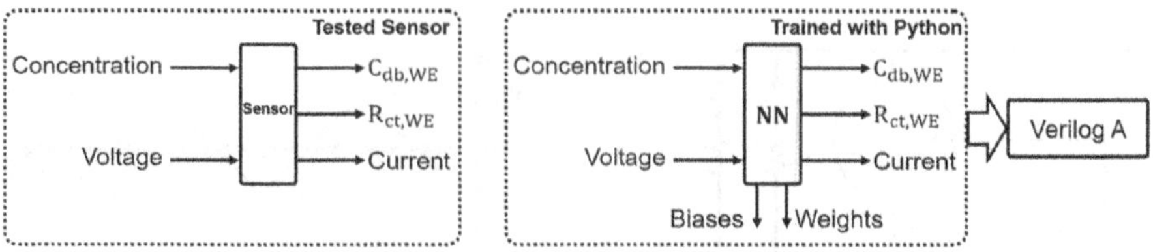

Fig. 28 The diagram for the electrochemical sensor model with ML

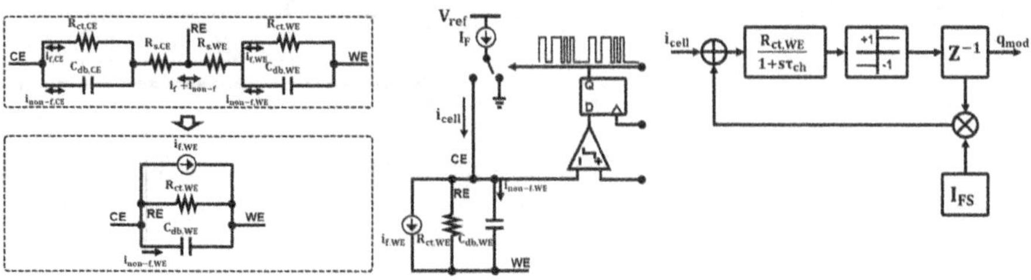

Fig. 29 The potentiostatic delta-sigma ADC

The model can be implemented with Verilog-A so that the simulator, like Cadence Virtuoso, can perform the simulation.

Another challenge in electrochemical sensor interface IC design is the power consumption and area of the circuit. As mentioned in Sect. 1, the electrochemical sensor is widely adopted in the areas of implantable or mobile devices. Therefore, ultra-low power and minimal size are of importance.

Some applications like glucose detection require the electrochemical sensor interface to be configured for amperometry [33], which shows DC signals for processing. The sensor can be highly integrated into the sensor interface to achieve amperometry, by utilizing the linearized equivalent model, as proposed in [34], which used a potentiostat based on delta-sigma modulation (DSM). The schematic of this topology is shown in Fig. 29.

In this design, the equivalent electrical model of the electrochemical sensor is further simplified to a current source with its internal impedance R_{ct}, which reflects the activities at the electrode-electrolyte interface at the WE, and a capacitor in parallel, modeling the non-faradaic current. The simplified model is placed in a delta-sigma feedback loop, where the potential at the RE is biased by a comparator. A single-bit digital-to-analog (DAC) and a delay element are employed to achieve first-order delta-sigma modulation. The DAC reference current and the time constant of the lossy integrator are given by:

$$\tau_{ch} = R_{ct,WE}C_{db,WE}, \qquad I_{FS} = \frac{V_{ref}}{R_{ct,WE}} \tag{23}$$

The speed of this design may limit its application in electrochemical systems, but the design is highly compact and low power, suitable for both low-power and implantable applications.

8 Conclusions

This chapter has built a model of the electrochemical sensor from the single-electrode system to the double-electrode system and finally the three-electrode system. Two readout circuit designs are discussed to illustrate the specific challenges involved in electrochemical readout circuit design. Applications of these sensors are also given. A deep understanding of the sensor model will contribute to proposing new topologies, for example, the highly integrated delta-sigma potentiostats.

In addition, the demand of a more practical electrochemical sensor model reflecting the large-signal behavior is mentioned, which shows a promising chance to make contributions to developing more advance sensor interfaces.

Acknowledgments This publication has emanated in part from research supported by a research grant from Science Foundation Ireland and the Department of Agriculture, Food and the Marine on behalf of the Government of Ireland under the Grant 16/RC/3835 (VistaMilk) and supported from research conducted with the financial support of Science Foundation Ireland (SFI) and is co-funded under the European Regional Development Fund under Grant Number 13/RC/2077 (Connect).

References

1. A. Murphy, I. Seymour, J. Rohan, A. O'Riordan, I. O'Connell, Portable data acquisition system for Nano and Ultra-micro scale electrochemical sensors. IEEE Sensors J. **21**(3), 3210–3215 (2021). https://doi.org/10.1109/JSEN.2020.3021941
2. A. Murphy, K. Hanley, N. Creedon, A. O'Riordan, I. O'Connell, Advanced data acquisition for emerging nano-electrochemical sensors, in *2018 IEEE International Instrumentation and Measurement Technology Conference (I2MTC)*, (2018), pp. 1–6. https://doi.org/10.1109/I2MTC.2018.8409766
3. K. Hanley, A. Murphy, N. Creedon, A.O. Riordan, D.O. Hare, I.O. Connell, Current readout circuit for point-of-care infectious disease diagnostics in animal health, in *2018 IEEE Nordic Circuits and Systems Conference (NORCAS): NORCHIP and International Symposium of System-on-Chip (SoC)*, (2018), pp. 1–5. https://doi.org/10.1109/NORCHIP.2018.8573507
4. A.M. Niamh, C. Ryona Sayers, S.B. Alan O'riordan, Novel single gold nanowire-based electrochemical Immunosensor for rapid detection of bovine viral diarrhoea antibodies in serum. J Biosens Bioelectron **6**(3) (2015). https://doi.org/10.4172/2155-6210.1000174
5. K. Dawson et al., Fully integrated on-chip nano-electrochemical devices for electroanalytical applications. Electrochim. Acta **115**, 239–246 (2014). https://doi.org/10.1016/j.electacta.2013.10.144
6. B. Patella et al., Electrochemical detection of dopamine with negligible interference from ascorbic and uric acid by means of reduced graphene oxide and metals-NPs based electrodes. Anal. Chim. Acta **1187**, 339124 (2021). https://doi.org/10.1016/j.aca.2021.339124

7. Global electrochemical sensor market | 2022–27 | industry share, size, growth – mordor intelligence, 2022/9/12. https://www.mordorintelligence.com/industry-reports/global-electrochemical-sensors-market-industry
8. Global electrochemical sensor market – 360 market updates, 2022/9/12. https://www.360marketupdates.com/global-electrochemical-sensor-market-12883892
9. Z. Wang, A. Murphy, A. O'Riordan, I. O'Connell, Equivalent impedance models for electrochemical nanosensor-based integrated system design. Sensors **21**(9) (2021)
10. A. Hulanicki, S. Glab, F. Ingman, Chemical sensors: definitions and classification. Pure Appl. Chem. **63**(9), 1247–1250 (1991)
11. Pooja, M. Rana, P. Chowdhury, Modern applications of quantum dots: environmentally hazardous metal ion sensing and medical imaging, in *Handbook of Nanomaterials for Sensing Applications* (Elsevier, Amsterdam, 2021), pp. 465–503
12. A.J. Bard, L.R. Faulkner, *Electrochemical Methods: Fundamentals and Applications/Allen J. Bard, Larry R. Faulkner*, 2nd ed. (John Wiley, New York; Chichester, 2000)
13. M. Dunwell, Y. Yan, B. Xu, Understanding the influence of the electrochemical double-layer on heterogeneous electrochemical reactions. Curr. Opin. Chem. Eng. **20**, 151–158 (2018)
14. M.S. Ozdemir, M. Marczak, H. Bohets, K. Bonroy, D. Roymans, L. Stuyver, K. Vanhoutte, M. Pawlak, E. Bakker, A label-free potentiometric sensor principle for the detection of antibody-antigen interactions. Anal Chem **85**(9), 4770–4776 (2013)
15. G. Kreysa, K.-i. Ota, R.F. Savinell (Eds.), *Encyclopedia of Applied Electrochemistry* (Springer New York, New York, 2014)
16. F. Scholz, Voltammetric techniques of analysis: the essentials. ChemTexts **1**(4) 2015
17. H.S. Magar, R.Y.A. Hassan, A. Mulchandani, Electrochemical impedance spectroscopy (EIS): principles, construction, and biosensing applications. Sensors **21**(19) (2021)
18. J.E.B. Randles, Kinetics of rapid electrode reactions. Discuss. Faraday Soc. **1**, 11 (1947)
19. E. Warburg, Ueber das verhalten sogenannter unpolarisirbarer elektroden gegen wechselstrom. Annalen der Physik und Chemie. **303**(3), 493–499 (1899)
20. S.R. Taylor, E. Gileadi, Physical interpretation of the warburg impedance. CORROSION **51**(9), 664–671 (1995)
21. J. Huang, Diffusion impedance of electroactive materials, electrolytic solutions and porous electrodes: warburg impedance and beyond. Electrochim. Acta **281**, 170–188 (2018)
22. M.J. Madou, R. Cubicciotti, Scaling issues in chemical and biological sensors. Proc. IEEE **91**(6), 830–838 (2003)
23. P.M. Levine, P. Gong, R. Levicky, K.L. Shepard, Active cmos sensor array for electrochemical biomolecular detection. IEEE J. Solid State Circuits **43**(8), 1859–1871 (2008)
24. N. Creedon, R. Sayers, B. O'sullivan, E. kennedy, P. Lovera, A. O'Riordan, Label-Free Impedimetric Nanoband Sensor for Detection of Both Bovine Viral Diarrhoea Virus (BVDV) and Antibody (BVDAb) in Serum. ChemRxiv: Analytical Chemistry (2018)
25. K.A.A. Mamun, N. McFarlane, A CMOS potentiostatic glucose monitoring system for VACNF amperometric biosensors, in *2015 IEEE International Symposium on Circuits and Systems (ISCAS)* (IEEE, Piscataway, 2015), pp. 477–480
26. K. Nam, G. Choi, H. Kim, M. Yoo, H. Ko, A potentiostat readout circuit with a low-noise and mismatch-tolerant current mirror using chopper stabilization and dynamic element matching for electrochemical sensors. Appl. Sci. **11**(18), 8287 (2021)
27. M.M. Ahmadi, G.A. Jullien, Current-mirror-based potentiostats for three-electrode amperometric electrochemical sensors. IEEE Trans. Circuits Syst. Regul. Pap. **56**(7), 1339–1348 (2009)
28. R. Turner, D.J. Harrison, H.P. Baltes, A CMOS potentiostat for amperometric chemical sensors. IEEE J. Solid State Circuits **22**(3), 473–478 (1987)
29. B. Razavi, *Design of Analog CMOS Integrated Circuits*, 2nd edn. (McGraw-Hill Education, New York, 2017)
30. M. Pelgrom, A. Duinmaijer, A. Welbers, Matching properties of MOS transistors. IEEE J. Solid State Circuits **24**(5), 1433–1439 (1989)

31. Y.-C. Chen, S.-Y. Lu, Y.-T. Liao, A microwatt dual-mode electrochemical sensing current readout with current-reducer ramp waveform generation. IEEE Trans. Biomed. Circuits Syst. **13**(6), 1163–1174 (2019)
32. J.-H. Tsai, Y.-C. Chen, Y.-T. Liao, A power-efficient bidirectional potentiostat-based readout IC for wide-range electrochemical sensing, in *2018 IEEE International Symposium on Circuits and Systems (ISCAS)* (IEEE, Piscataway, 2018), pp. 1–5
33. M.M. Ahmadi, G.A. Jullien, A wireless-implantable microsystem for continuous blood glucose monitoring. IEEE Trans. Biomed. Circuits Syst. **3**(3), 169–180 (2009)
34. S. Sutula, J. Pallares Cuxart, J. Gonzalo-Ruiz, F.X. Munoz-Pascual, L. Teres, F. Serra-Graells, A 25-μw all-MOS potentiostatic delta-sigma ADC for smart electrochemical sensors. IEEE Trans. Circuits Syst. Regul. Pap. **61**(3), 671–679 (2014)

Next-Generation Molecular Detection with a CMOS Capacitive Sensor

Tim Cummins and Brian O'Farrell

Abstract CMOS biosensors hold great promise for the high volume/low-cost scalability of molecular diagnostics. However, despite many publications in this area over the past decades, mainstream consumer adoption examples are limited. This chapter explores some of these biosensor opportunities and inherent challenges. It begins with an introduction to molecular detection basics and reviews current laboratory and point-of-care detection methods. Some historical biosensor approaches and recent CMOS biosensor examples are reviewed. We present initial results of a fully synthetic SARS-CoV-2 PNA-probe assay with a standard-CMOS capacitive bead detector which aims to overcome some of the enzyme and process complexities of the previous methods.

1 Introduction

Decentralization of diagnostics and healthcare from hospitals into community care settings, and ultimately into the home, has been proposed to reduce the burden on overloaded central hospitals and laboratories. The COVID-19 pandemic and the implications for routine health maintenance during periods of nonpharmaceutical interventions, such as "lockdowns," has driven a new awareness of the value of digital health and telemedicine. This is now recognized by the World Health Organization in their 2022 *Consolidated Telemedicine Implementation guide* [1], which states: "the delivery of health-related services and information using information and communications technologies (ICTs) is a critical driver for expanding access to services and promoting continuity of care."

Modern wireless electronics and biosensors are helping to accelerate this trend, with over-the-counter home biosensors now readily available for monitoring of blood pressure, blood-glucose levels, heart rate, ECG, blood oxygen, and other

T. Cummins (✉) · B. O'Farrell
Altratech Ltd, Cork, Ireland
e-mail: tim.cummins@altratech.com

P. Harpe et al. (eds.), *Biomedical Electronics, Noise Shaping ADCs, and Frequency References*, https://doi.org/10.1007/978-3-031-28912-5_6

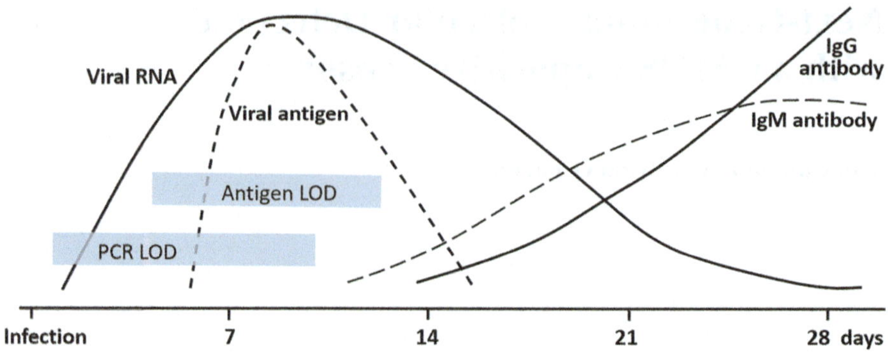

Fig. 1 Pathophysiology and timeline of viremia, antigenemia, and immune response during acute SARS-CoV-2 infection

vital signs. Regularly, these devices may not be as accurate as hospital-based diagnostics, but they nevertheless fulfil an important role in widescale screening and citizen empowerment in taking greater ownership over their own individual health journey. Use of home diagnostics, specifically, was brought to worldwide public consciousness during the recent COVID-19 pandemic, caused by the SARS-CoV-2 individuum of the species *severe acute respiratory syndrome–related coronavirus*. Lateral-flow SARS-CoV-2 rapid antigen self-test kits were found in almost every household. This was despite some initial concerns of clinicians that these were not as accurate as hospital and laboratory-based SARS-CoV-2 PCR molecular tests, or Point-of-Care LAMP SARS-CoV-2 molecular tests. Analytical sensitivity (or Limit-of-Detection, LOD) is shown for each of these in Fig. 1, which also shows the typical timelines of viral RNA, viral antigen, and antibody development in the human body following an infection. PCR is the "gold standard" test method due to its lower LOD and earlier detection of infection by detecting the actual RNA viral genome. A positive antigen test detection may not occur until a few days later, due to its higher LOD and the body's antigen response also occurring typically a few days after the viral RNA peak. Against this, subjects may test positive for SARS-CoV-2 using PCR tests long after their infectiousness is problematic, complicating clinical decisions.

The ideal for an infectious disease diagnostic would be to replicate, in a home setting, the sensitivity and specificity of laboratory DNA and RNA PCR molecular tests, but with the robustness, simplicity, and the low cost of the lateral-flow antigen test. This is illustrated conceptually in Fig. 2.

Additionally, the home test would ideally reciprocate the range of tests available in the central lab, for example, conducting both serological (antigen/antibody) tests and molecular tests (DNA/RNA).

CMOS biosensors offer great promise toward achieving this goal, with the ability to combine two key functions of biosensors: transducing and signal processing.

On the other hand, Hassibi [2] notes that CMOS biosensors require complex and upfront capital-intensive manufacturing/assembly processes and convergence

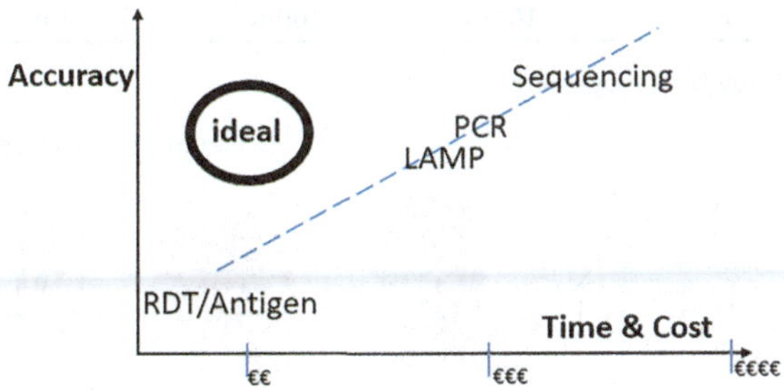

Fig. 2 Simplified representation of test accuracy versus time and cost

of multiple disciplines beyond engineering. He cautions that CMOS biochips are "an overpromised field with lots of unproven technologies and failed projects, and limited successful commercial products."

In this chapter, we delve into some of the possible reasons for this and review some recent biosensor and CMOS biochip examples. We conclude the chapter with initial results for a new approach, employing a CMOS fringe-field bead capacitance sensor, fabricated on a standard CMOS process, where the bead is a proxy for a DNA or RNA molecule captured in the upstream assay by peptide nucleic acid (PNA) probes.

2 Molecular Detection Methods

Optical microscopy of micrometer cells and bacteria has been in existence since the 1600s. However, the nanometer dimensions of DNA, RNA, and antibody/antigen proteins mandate the use of electron microscopy or X-ray crystallography in research laboratories to identify and observe them directly (Fig. 3). In this section, we review the medical diagnostic methods which have emerged in recent decades of detecting and identifying these proteins and molecules, directly and indirectly.

2.1 Antigen and Antibody Lateral Flow Devices

Lateral flow device (LFD) immunoassays are common, user-friendly, inexpensive, and readily available testing devices that are used for detection of analytes, directly or indirectly, to aid in diagnosis of medical conditions. Direct detection of human chorionic gonadotropin (HCG) hormone in the home pregnancy test is perhaps the most well-known example. Although over 50 years old, LFDs nonetheless proved a most valuable technology during the recent COVID-19 pandemic, indirectly detecting SARS-CoV-2 positivity by detecting the body's antigenic response to the virus. They are also called rapid-diagnostic tests (RDT), giving a result usually within 15 to 30 minutes. A derivation of classical chromatography, they work by binding specific antibodies conjugated to nanoparticles with a specific target in

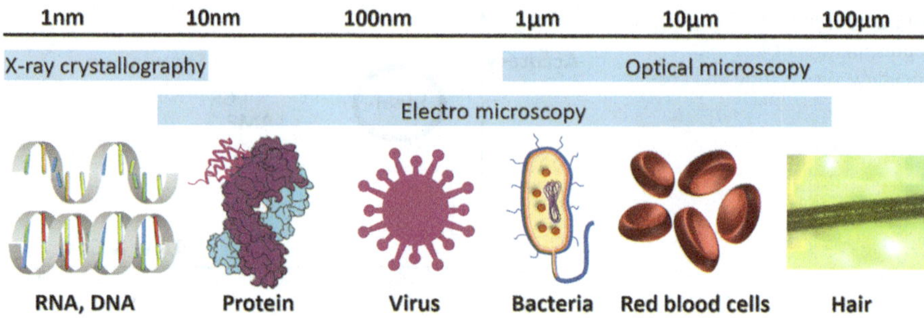

Fig. 3 Relative molecular dimensions of various biological structures

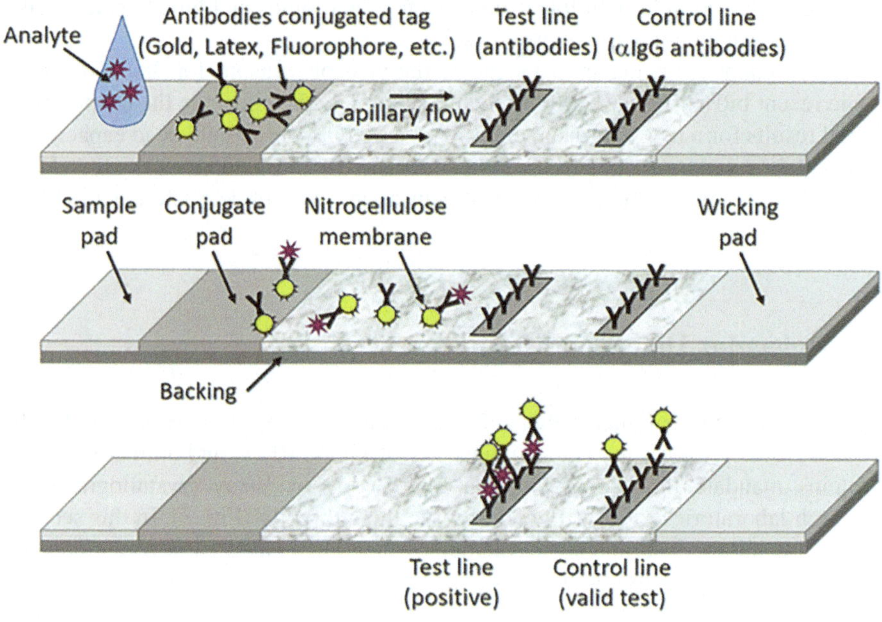

Fig. 4 Lateral flow assay architecture

the sample. In the case of SARS-CoV-2, this is an antigen target. This antibody-antigen complex moves via capillary flow to a test area. Where an antigen is bound, the complexes become captured on another antibody, which forms a sandwich across the antigen, holding the complex in place and indicating a positive test. A control line downstream captures nanoparticles with no antigen attached through interactions with the initial antibody. The control and positive lines are identified by the presence of a colored line due to bead-labelling (Fig. 4). The reader is referred to reference [3], which gives a detailed description of LFD types and methods of operation.

2.2 DNA Sequencing

Sequencing is the process of determining the nucleic-acid sequence – the order of nucleotides – of the five standard nucleobases, the fundamental units of the genetic code of all living matter: adenine (A), cytosine (C), guanine (G), thymine (G), and uracil (U). The bases A, G, C, and T are found in DNA, while A, G, C, and U are found in RNA (Fig. 5). Note the binding or hybridizing patterns: A-T and C-G, that is, adenine always binds to thymine, and cytosine always binds to guanine. This is important for probe and primer design in molecular tests.

The reader is referred to reference [4] for an overview of sequencing methods and technologies.

Knowledge of DNA sequences has become indispensable for basic biological research, medical diagnosis, and virology. However, sequencing is available only in advanced university and hospital research facilities, or in very few central virus-reference laboratories. Despite much optimism over the years, it remains a stubbornly expensive method, ranging from hundreds (e.g., for short-reads or single gene) to thousands of dollars for full-genome sequences, for example, in cancer analysis [5]. The cost variation is largely explained by complicated molecular biology upstream, which can be used across samples to maximize the output of a sequencing run. This works by applying sequence tags, which function as a code, which can later flag to bioinformaticians which sequences came from which sample. This lowers the cost per sample but at the considerable cost of extra processing, which again requires a sophisticated central laboratory.

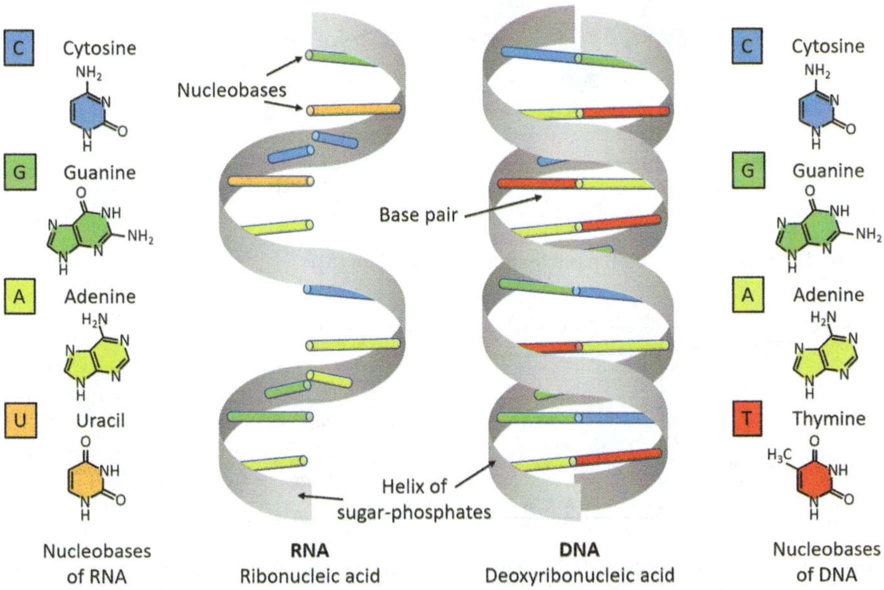

Fig. 5 Nucleobase structures of double-stranded DNA and single-stranded RNA

Sequencing became very important in late 2019, at the start of the COVID-19 pandemic, in identifying the causative novel coronavirus (nCOV). This showed it as a novel strain of the *severe acute respiratory syndrome* virus. Sequencing enabled the rapid availability of the genome of the causative agent to researchers and diagnostic vendors. Due to its resulting severe acute respiratory symptoms, it became known as SARS-CoV-2. A viral particle of approx. 100 nm diameter, its first complete sequence, known as the "Wuhan reference genome," was uploaded to the public GenBank database (accession number NC_045512) in December 2019 [6]. This was a new individuum of *severe acute respiratory syndrome–related coronavirus* which had earlier caused the SARS epidemic. This information allowed researchers and biopharma companies to quickly assess available diagnostic and vaccine technologies designed for related coronaviruses. An RT-PCR test was provided to the WHO by January 14, 2020 (Corman et al. [7]).

The SARS-CoV-2 structure is shown in Fig. 6, and an expanded portion of its sequence is shown in Fig. 7.

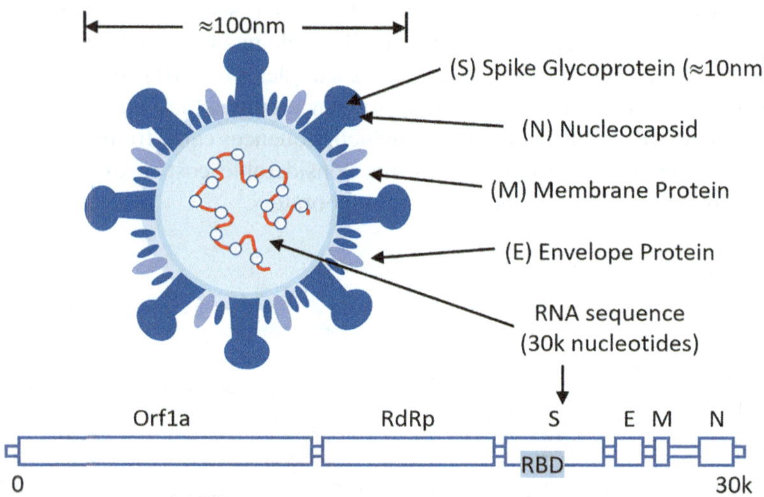

Fig. 6 The SARS-CoV-2 viral particle structure (Based on [8])

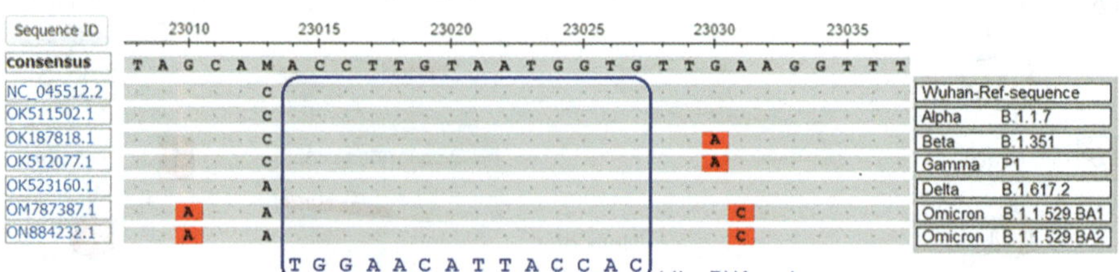

Fig. 7 Portion of the SARS-CoV-2 spike gene receptor binding domain (RBD), showing some mutations, and a PNA probe targeting a region conserved across many variants

2.3 PCR Testing

The polymerase-chain-reaction (PCR) method gets its name from the amplification of a target DNA or RNA sequence into billions of copies. This is enabled by the heat-resistant enzyme *Taq* DNA polymerase, which was first discovered in a thermophilic bacterium *(Thermus aquaticus)* living successfully in boiling geyser waters in Yellowstone National Park [9]. Theoretically, a single copy of the target DNA or RNA could be amplified and detected. This form of low copy number PCR is traditionally common in forensics but was problematic as that level of sensitivity could be caused by very incidental or even historic contact with evidence in a case. It can also cause problems where detection of DNA or RNA is reflective of past rather than current infection. In practice, however, the LOD of commercial PCR tests ranges from a few dozen to several hundred copies/mL, and even thousands of cp/mL for some highly mutated viruses. This lesser sensitivity commonly reflects inefficiency in the many steps during sample processing and/or inefficiencies during the many cycles of PCR, explained further below.

PCR can be employed for detection once the DNA or RNA sequence of a target virus or pathogen is already known. This allows design of combinations of short sequence-specific probes which target the DNA or RNA sequence and allow subsequent enzymatic processing events to occur allowing a detectable reaction.

A target DNA sequence can be detected by first breaking apart the double-stranded DNA in the sample ("denaturing") to single strands, usually at high temperature, for example, 95 °C. This makes available the single-stranded sequences for probing by chemically synthesized complementary probes or primers which target specific regions of the target sequence.

PCR probes are typically >25 nucleotides or base pairs (nt or bp) in length. However, for illustrative purposes, an 8-bp example (c) is given below. If the complementary probe sequence (c) and target sequence (b) match, they should bind with 100% specificity. If even only one nucleobase has mutated, as in (a), then they should not match and not bind to each other. This single-based mutation (SBM) is also known as a single-nucleotide-polymorphism (SNP). Good "SBM specificity" is a key measure of a good molecular test:

(a) A**G**TTCCGG – a SNP or SBM mutation which should not match.
(b) AATTCCGG – The 'target' sequence to match/bind and be detected.
(c) TTAAGGCC – Complementary probe/primer sequence to detect the target.

Figure 8 shows the first two cycles of a PCR reaction. The target DNA is denatured (at ~95 °C typically), that is, it is split into two separate strands. In the annealing step (45 °C to 60 °C), the two primer probes "match," that is, bind or hybridize to the target region. Then the temperature is raised (from ≈72 °C to 75 °C), and the polymerase enzymes then promote double-stranded extension of the sequence. This results in two double-stranded copies of the original target sequence portion. The denature-anneal-extend temperature cycle then repeats, creating four copies, as shown in Fig. 8.

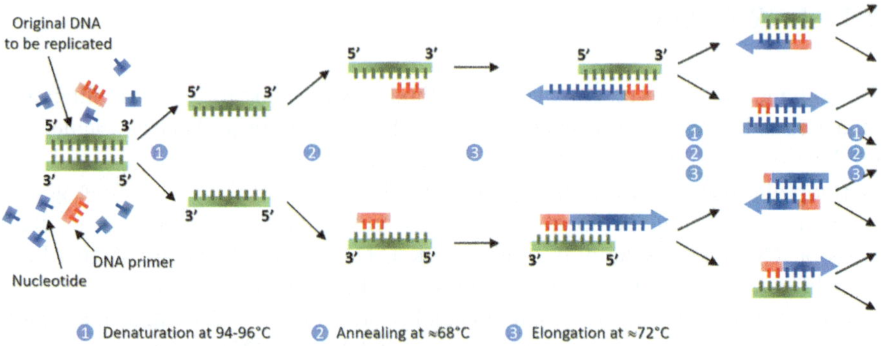

Fig. 8 PCR (polymerase chain reaction) cycles

Each further temperature cycle therefore doubles the number of copies. Thus, 30 temperature cycles will create 2^{30} or 1 billion copies from the original short target region. If the probes are labelled with a fluorescent dye marker, then the billion fluorophores create a faint glow, which is detected by an optical scanner or plate reader. A single PCR reaction will typically have three or four nonoverlapping fluorophore dye colors, enabling detection of two DNA targets and a positive and negative control. Higher multiplex testing of many targets is possible by splitting the sample into different wells or channels in more complex laboratory instruments. However, this adds complexity and makes the technology even less suitable for decentralization to point-of-care and home settings.

These PCR laboratory instruments typically have a four-figure to six-figure cost, and the PCR cost per test remains high, often \$100+, or perhaps lower in certain high-throughput situations [10]. The PCR equipment costs reflect the complexity of heating/cooling with Peltier cells, chambers, valves, actuators, and bulky laser optical detection systems comprising photomultiplier tubes, filters, lenses, and array illumination for 96-well or 384-well multiplex plates. The per-test costs reflect the significant overheads of trained specialist staff required to run the tests, as well as extensive laboratory infrastructure of air conditioning, isolation fume hoods, centrifuges, and other sample preparation equipment and reagents, together with refrigerators and freezers for storing these. The pre-PCR sample preparation stages of filtration/centrifuging, separation, lysing, DNA extraction (e.g., with magnetic beads), and purification are quite complex. Many of the errors in PCR testing occur during the pre-PCR sample preparation phase. Agencies like FindDx have negotiated lower (four-figure) equipment and per-test costs for HIV and TB viral tests in low- and medium-income countries, in conjunction with donor body subsidies [11]. However, trained staff are still required for sample preparation, extraction, and pipetting. This inhibits true widescale community deployment.

An interesting development during the COVID-19 pandemic was the Emergency Use Authorization (EUA) for Visby Medical's SARS-CoV-2 portable test [12].

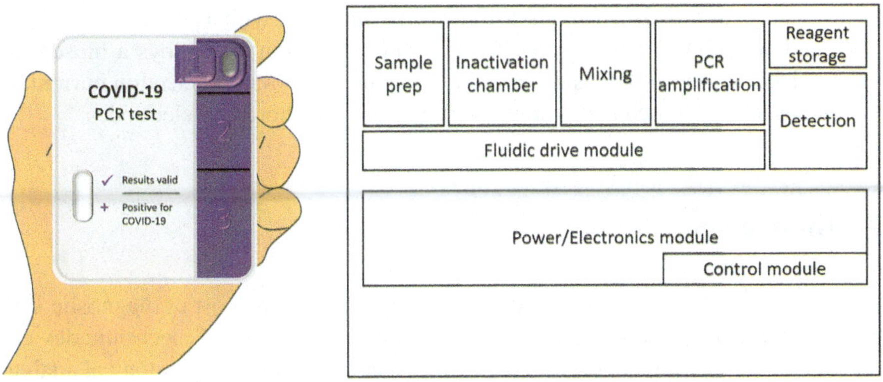

Fig. 9 Visby Medical SARS-CoV-2 "PCR in the palm of your hand". (Based on https://www.visbymedical.com/covid-19-test/ and [13])

It miniaturizes and integrates the entire sample-handling, thermocycling, and detection process, into "PCR in the palm of your hand", as shown in Fig. 9:

Its lower limit of detection is 1112 copies/mL [14]. It is currently priced at $155 for a single-use disposable test [15].

2.4 *LAMP/Isothermal Testing*

Loop-mediated isothermal amplification (LAMP) has emerged as an alternative to PCR in recent years. It also uses polymerase enzymes for replication and amplification of the target DNA. However, there is no temperature cycling, which simplifies the equipment required. Instead, it requires six primer probes for specificity (typically >50 bp) and employs a single ("isothermal") temperature (60 °C typically) to promote the enzymatic amplification reaction. Reference [16] provides a more detailed explanation and review of LAMP diagnostics.

Because of their simpler equipment and operation, LAMP tests are becoming prevalent in point-of-care settings, for example, the Abbott IDnow system. However, its LOD (3900 to 20,000 cp/mL) [17] is not as sensitive as PCR. Critically, the colorimetric detection principle used with some LAMP assays is very sensitive to sample pH which was found to be a barrier to LAMP during the pandemic. The large numbers of probes involved in LAMP may also cause more problems compared to PCR, when variants and sub-variants mutate creating novel mismatches to probes designed for the original wild-type genome. Multiplexing for variant level detection or for ruling in or out other viruses with similar symptoms is also more complicated due to problems with the LAMP probes interfering with each other.

Cue Health received an EUA for their isothermal amplification single-plex SARS-CoV-2 home test during the COVID-19 pandemic [18]. It has a palm-sized heater/reader unit ($159), a single-use cartridge and swab with a per-test cost of $52 [19], and an LOD of 1300 copies/mL [20]. For detection it employs a biosensor, where electrochemical labels are used to generate a nano-amp detection current as amplicons are generated and bind to capture probes anchored on electrodes.

3 Biosensors

Biosensors offer good promise toward reducing the size and cost of diagnostic systems. Biosensors combine a physicochemical detector (often an electronic device) with a biological component, which allows a specific chemical or biological analyte to be detected. Frequently such biosensors feature a self-assembled monolayer (SAM) where the biological component has been attached to the surface of the sensor. Lei at al [21] list many published research examples of biosensor transducers (gold electrodes, ISFET, photodetector, cantilever, nanowire, hall-sensor, SAW, spiral coil, silicon nanowire) and the various sensing parameters employed in these (magnetism, fluorescence, mass, nuclear-spin, charge, capacitance, impedance, and current-voltage cyclic-voltammetry). In this section, we focus on electrical biosensors.

3.1 Electrochemical Glucose Sensor

The best-known and most commercially successful biosensor is the blood-glucose electrochemical biosensor (Fig. 10). It consists of a reader with control electronics, and single-use disposable test strips, for manufacturability in high volume and at low cost. Its working principle, perfected over many decades, is to immobilize glucose oxidase enzyme on a working electrode. The enzyme catalyzes the conversion of glucose in the blood sample to gluconic acid and hydrogen peroxide (H_2O_2). Glucose is quantified by the electrochemical measurement of the H_2O_2, either by current-flow amperometry or cyclic voltammetry (CV, sweeping of the working-electrode voltage versus the reference-electrode voltage with DACs).

Known as a potentiostat, the working principle and typical CV circuit is shown Fig. 11. A silver-chloride reference electrode (RE) holds the assay liquid at a fixed potential. The voltage on the working electrode (WE) is then varied with respect to RE, resulting in an analyte-indicative current flow in the counter electrode (CE).

Less invasive versions of the glucose sensor have recently been approved in the market, for example, the Abbott Libre arm-mounted sensor (Fig. 12). This has a subcutaneous electrochemical sensor on the tip of a 5 mm plastic filament, a $0.1 mm^2$ carbon working-electrode, a carbon counter electrode, and an Ag/AgCl reference electrode. The PCB has a thermistor to measure body temperature, a battery, and a single-chip microcontroller. This has analog-to-digital (A-to-D) converters and I-V amplifiers to measure glycemia amperometrically, nonvolatile memory for

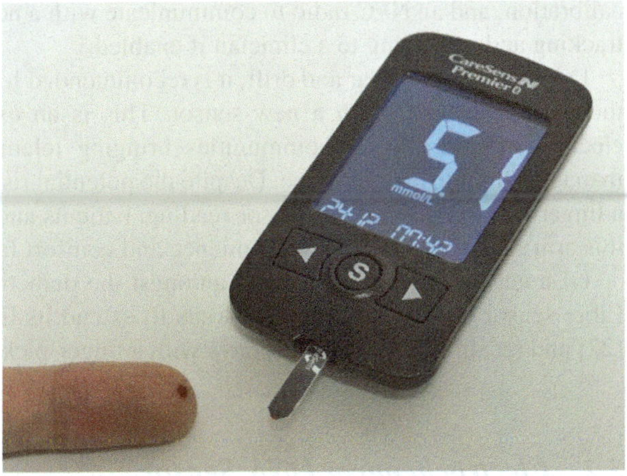

Fig. 10 Blood glucose biosensor reader and test strip

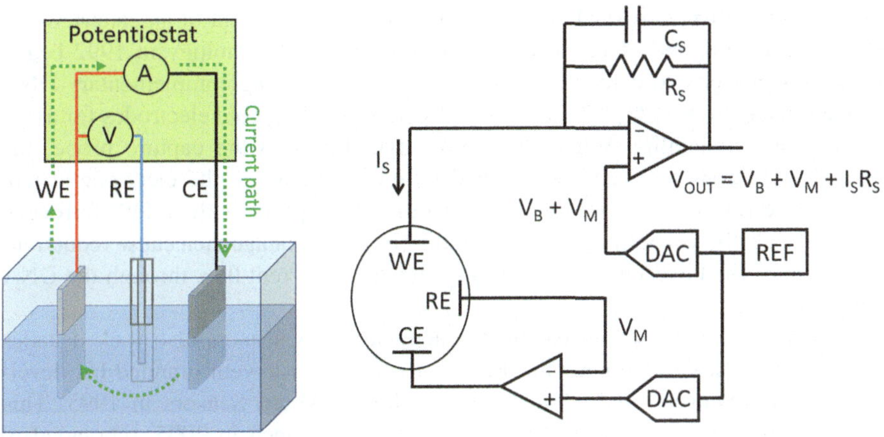

Fig. 11 Electrochemical sensor working principle (left) and potentiostat control circuitry (right, based on www.analog.com)

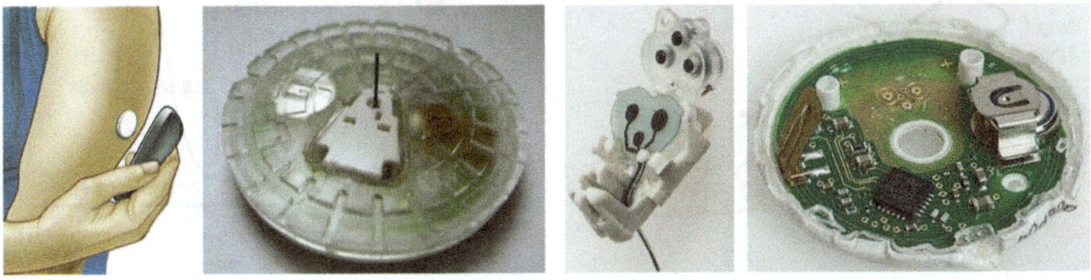

Fig. 12 Libre arm-mounted sensor, sensor tip, and internal PCB

calibration, and an NFC radio to communicate with a nearby smartphone for results tracking and uploading to a clinician if enabled.

Due to electrode wear and drift, it is recommended for only 2 weeks of operation, then to be replaced with a new sensor. This is an excellent example of sensor, electronics, and clinical communities bringing telemedicine and remote health management into everyday use. Despite the potential risk of not being as accurate as a finger-pick blood-glucose sensor reading, patients and users have readily adopted this arm sensor for its sheer convenience and comfort factors.

Of note, there has been a drive amongst the diabetic community to "hack" the Libre sensor employing online tutorials to extend its lifetime to 28 days or longer [22] and recalibrate it every few days with a finger-pick blood glucose reading.

3.2 *Electrochemical DNA Sensor*

DNA is a negatively charged conductive molecule, reminiscent of an electric wire. Barton, Kayyem, and others at California Institute of Technology in 1992 began using this property to detect a target DNA by immobilizing complementary DNA probes on electrodes [23]. These are usually noncorroding gold electrodes for assay stability and reliability. When the DNA target binds to the capture probe, the resulting H^+ ions released are detected by electron flow in the electrode. There are many variations, for example, the labelling of the probe with an Fe^+ ferrocene molecule. Cyclic voltammetry sweeps of the electrode voltage then cause reduction-oxidation of the ferrocene, with resulting change of current flow through the DNA (Fig. 13).

Just like the glucose sensor, electrochemical DNA sensors also took decades to reach large-scale commercialization. For example, Kayyem branded his developments as eSensor® and spun it out as Clinical Micro Sensors in 1995. This was acquired by Motorola in 1999, divested to Osmetech in 2005, rebranded to GenMark-Diagnostics for IPO in 2010, and acquired by Roche in 2021.

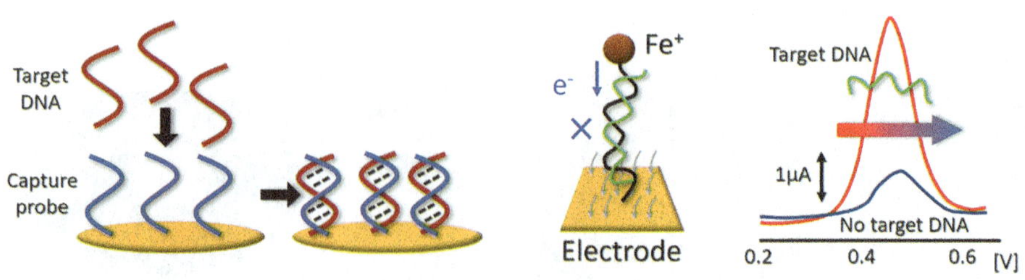

Fig. 13 Electrochemical DNA detection sensor principle

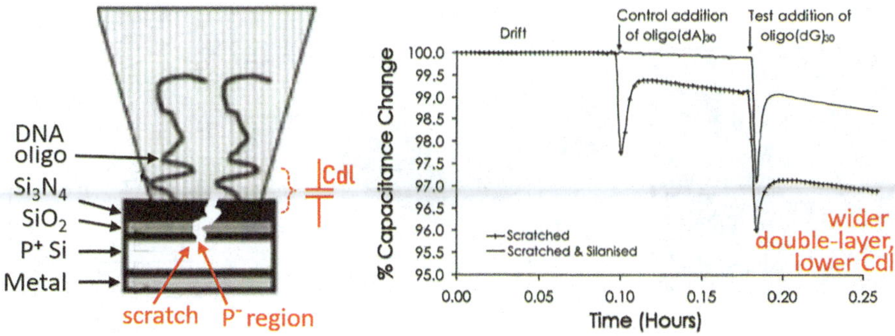

Fig. 14 Capacitive DNA sensing principle (From [23], with notes added)

3.3 Capacitive DNA Biosensors

In 1998, Berggren et al. demonstrated capacitive DNA detection with attomolar sensitivity of immobilized DNA probes on gold rods with a self-assembled monolayer (SAM) of oligonucleotides [24]. The detection principle relies on DNA being a negatively charged molecule. This repels anions in the surrounding electrolyte. When immobilized on a surface, a further layer of ions agglomerates at the surface (Helmholtz/Stern layer). This results in a double-layer (Debye layer) charge-depletion zone (similar to a P-N junction depletion region). Typically, a few nanometers wide, it has a capacitance C_{dl} ranging from pF to nF per cm^2.

Nano-rods, nano-pillars, and carbon-nanotubes for DNA detection have also been patented and published, leveraging the high surface-to-volume ratio of these structures for sensitivity. However, these structures are difficult to manufacture in high volume. Few, if any, of these have been commercialized.

Berney et al. in 2000 demonstrated capacitive DNA detection on a planar P^+ doped silicon substrate with an oxynitride passivation surface [25]. Figure 14 shows its C_{dl} reducing as DNA is added/immobilized, due to the double-layer being widened as the negatively charged DNA repels anions in the surrounding liquid.

Figure 14 also shows higher sensitivity being achieved by scratching the sensor surface. Berney attributes this to a lightly doped P^- region being formed by the scratch. Just like a lightly doped P-N junction, this results in further widening of the charge-depletion region and even more reduction in C_{dl}. The limit of detection achieved is 100 pmol.

3.4 Magnetic Bead GMR Biosensor

Paramagnetic beads are used in many molecular biology assays, for example, in DNA and RNA extraction, purification, and labelling. They are magnetic only in the

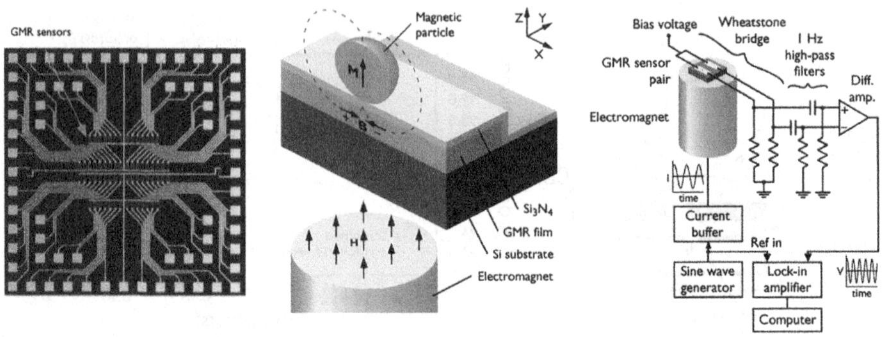

Fig. 15 Magnetic bead GMR biosensor (From [24])

presence of a magnetic field and lose their magnetism when the field is removed. This prevents agglomeration and clumping effects, which could interfere with the assay. Their diameters range from a few tens of nanometers to a few microns. Different surface coatings, chemistries, or capture probes give each type of bead its own binding properties. The bead, with a captured DNA or RNA target attached, can be removed from the lysed sample by an external magnetic field and moved through further wash and purification steps.

The concept of using the large paramagnetic bead itself as a "label" for detecting its captured DNA has been proposed, for example, Baselt et al. in 1998 with a giant magneto resistor (GMR) biosensor [26]. This has two identical GMR sensors, operating at 2.5 mA each, configured in a Wheatstone bridge detection circuit, as shown in Fig. 15. The operating principle is to detect a small differential variation of GMR sensor resistance in response to a magnetic field change due to the presence of a bead on one sensor. A downside is the requirement for a large external magnet to create the magnetizing field. This negates some of the advantage of miniaturizing the sensor itself.

4 CMOS Biosensors

The NPN base-emitter ΔV_{be} temperature sensor (Widlar 1965, [27]) and the ISFET pH sensor (Bergveld 1970, [28]) are among the earliest examples of semiconductor sensors. In both cases, it took years or decades of co-integration of signal-conditioning, calibration, and digital readout circuits for these to reach high-volume commercial success. In this section, we look at the ISFET, and some other CMOS biosensor examples, and discuss their commercialization journeys.

4.1 The ISFET

The basic structure of the ISFET (Ion-Sensitive Field-Effect Transistor) is shown in Fig. 16 (L). It is similar to a standard MOSFET, except it has liquid over the gate area. This liquid is held at a fixed voltage by an immersed reference electrode, typically of silver-chloride construction. In this example, the reference electrode is tied to the source voltage ($V_{GS} = 0$ V). The basic principle is that pH changes in the liquid alter the charge in the gate region. The effect appears as a modulation of the threshold voltage, of approximately 50 mV/pH. This causes changes in drain-source current, as shown in Fig. 16(R):

- I_{ds} = max for low pH of 2, when H^+ ions in the liquid are a maximum
- I_{ds} = min for high pH (10), when H^+ ions in the liquid are a minimum

The reader is referred Bergveld's 2003 paper "30 years of ISFETOLOGY" [29], which is an interesting and even sometimes wistful look back at the ISFET's commercialization history – or lack thereof. Despite hundreds of published ISFET papers over the decades and over 100 patents granted to various inventors, it achieved only small-scale commercialization in a few niche areas. It did not see adoption in the biomedical applications initially envisioned, due to a variety of biocompatibility, packaging, and repeatability issues, including reference electrode reliability. It also had difficulty competing against "cheap and cheerful" paper-strip and glass pH sensors. He notes the integration of an ISFET in a CMOS process in 1999 [30], and discusses the issues and obstacles which universities, innovators, and small-companies face in further development, such as finding grant-support for an "old" technology, or whether it needed "big-players" and large "market-pull" applications to justify further investments. He concludes his 2003 paper with thoughts on what may or may not happen in "possibilities for the next 30 years."

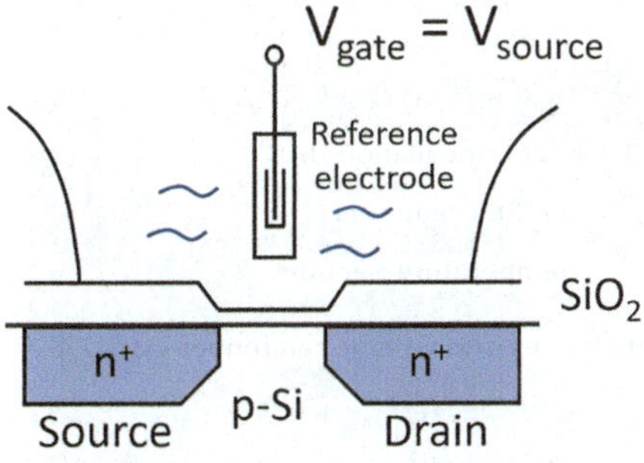

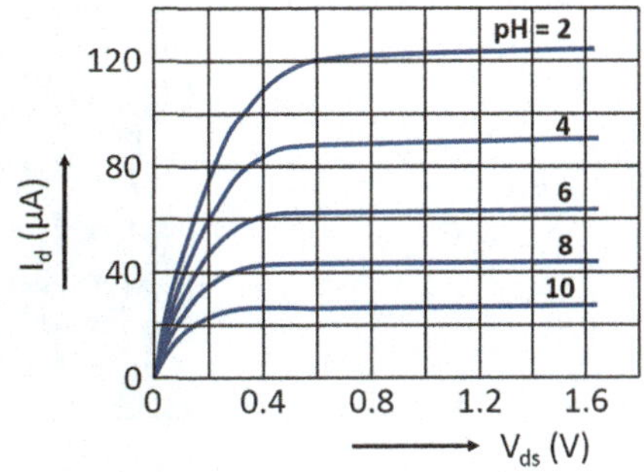

Fig. 16 (L) ISFET structure, and (R) I_d-vs-pH (with $V_{gs} = 0$ V) (Based on [29])

4.2 ISFET DNA Detection and Ion Semiconductor Sequencing

A few years later, DNA detection and sequencing emerged as big "market-pull" applications, with ISFET detection of the hydrogen ions generated during DNA polymerization, that is, the hybridizing of a single-stranded DNA with its complementary probe sequence. In 2004, for example, Toumazou patented readout circuits for CMOS ISFETs [31], culminating in a 19-SNP DNA detection chip presented at ISSCC 2010 [32] (Fig. 17).

In 2003, Hassibi et al. proposed DNA polymerization and H^+ ion-generation as a method of DNA sequencing [33]. Known as "ion-semiconductor-sequencing" and "sequencing by synthesis," the unknown target DNA sequence is determined by the detection of the hydrogen ions that are released when a complementary strand is extended based on the sequence of a template strand.

In 2007, Ion Torrent Inc., based on the work of Toumazou, Hassibi, and others, took this concept further by integrating over one million ISFETs on a CMOS chip (ION 314) [34]. Each ISFET has its own microwell with its own unique template strand, from a large library of strands covering large parts of the DNA genome to be identified. Other chips followed, with many more ISFETs to enable sequencing of longer genomes, for example, ION 316 (6.2 M ISFETs) and ION 318 (11.1 M ISFETs) for 10^8 and 10^9 bases sequenced, respectively.

The multi-ISFET and microwell principle is illustrated in Fig. 18 (Huang [35]):

Figure 19 shows Huang's implementation of a high-speed 12-bit pipelined A-to-D converter for detection of the multiple ISFET H^+-induced currents, via row and column decoders, sample-hold, and pre-amplifiers.

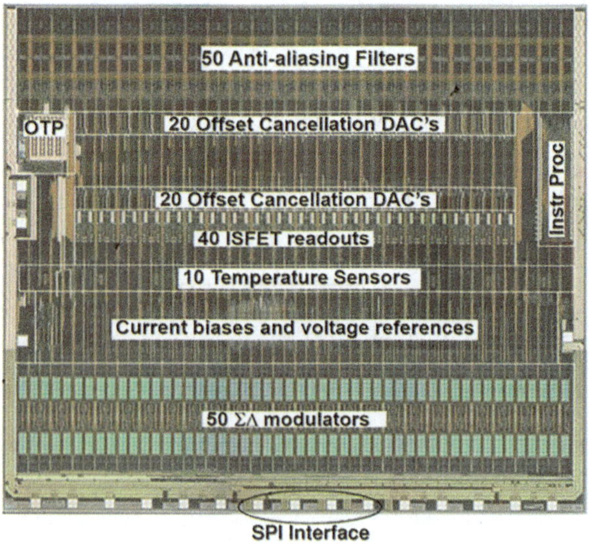

Fig. 17 0.35 μm CMOS chip, 5.5 × 4.7 mm, with 40 ISFETs, for DNA 19-SNP detection (From Ref. [32])

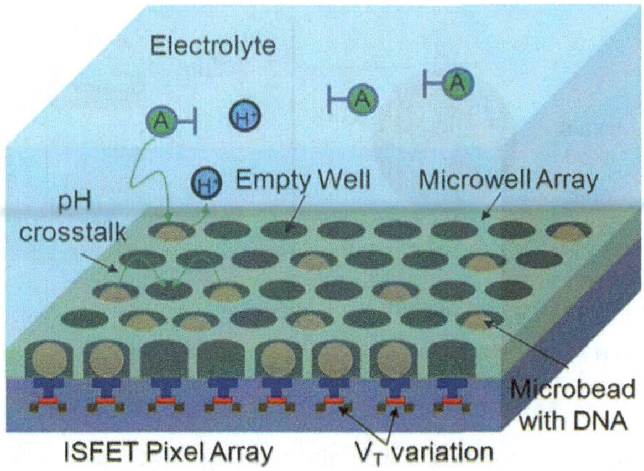

Fig. 18 CMOS multi-ISFET/microwell/microbead chip for DNA sequencing (From [35])

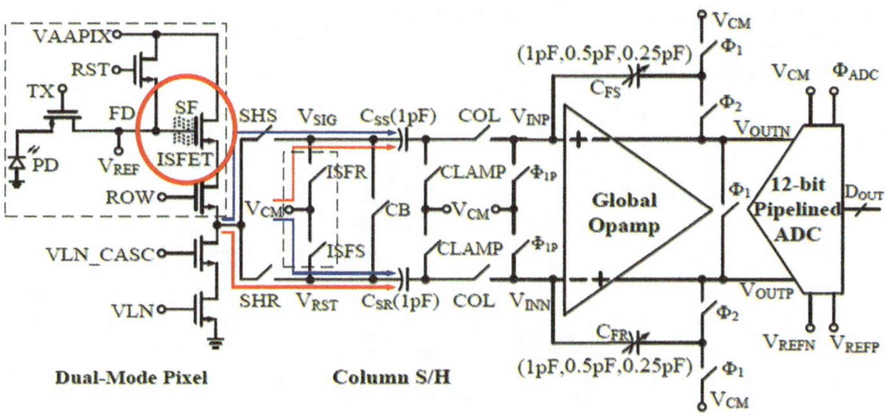

Fig. 19 Multi-ISFET H+ detection, amplification, and conversion circuits (From [35])

4.3 CMOS Hall Sensor Biochip

Boser and Florescu addressed bead detection on a CMOS biochip by integrating a microcoil to generate the magnetizing field and an N-well Hall sensor for bead detection in a fully integrated solution that required no external magnets [36]. Figure 20 (L) shows the microcoil Hall sensor arrangement, and (R) the CMOS process modifications required: aluminum wet etching, RIE dry plasma trench etching to optimize bead location on the sensors, and CR/Au deposition for gold-coating to allow protein adsorption for assay specificity. Figure 21 shows an enzyme-linked immunosorbent assay (ELISA) sandwich assay conducted atop the biosensor. The gold-surface is first coated with surface antibodies and polyclonal goat IgG specific to the F_{ab} region of Human IgG. Human IfG antigen is then introduced and incubated. A successful binding event results in the magnetic bead being captured atop the Hall sensor and detected, its large mass providing a form

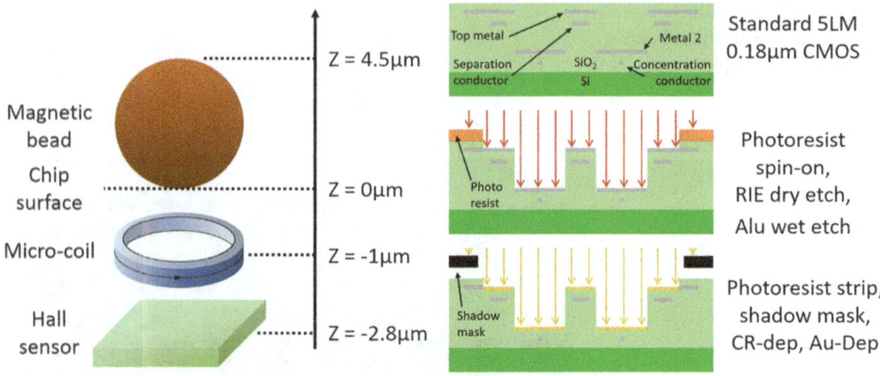

Fig. 20 (L): Microcoil and Hall sensor; (R): CMOS process modifications (Based on [36])

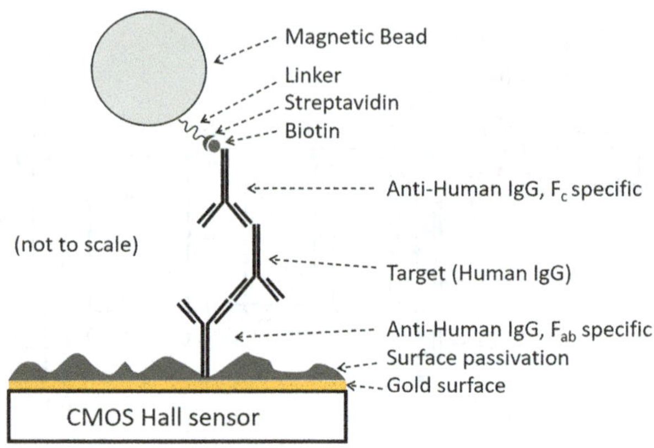

Fig. 21 ELISA-type assay conducted on the surface of the Hall sensor (Based on [36]])

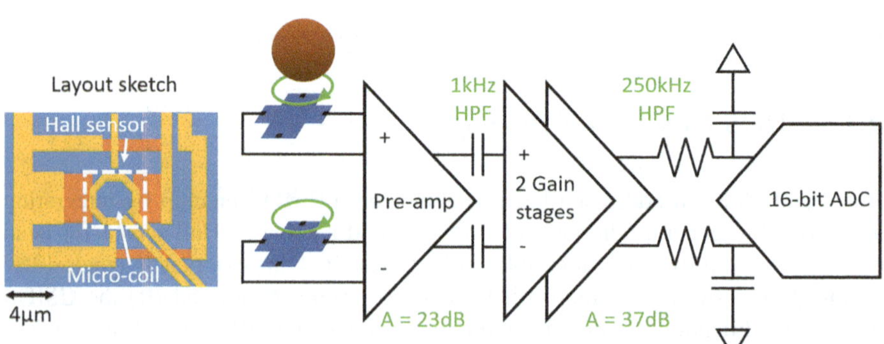

Fig. 22 Coil/Hall sensor layout and bead detection circuits (Based on [36])

of dendritic amplification of the tiny antibody-antigen proteins. Figure 22 shows its differential sensor operation method, the coil/hall-sensor layout, and bead-detection circuits.

This integrated CMOS biochip solution did not reach clinical trials and commercialization. Spun-out as Silicon Bio-Devices Inc., it received various grants and published results of some blood-analysis assays. It was rebranded as XipDx in 2018 but was reported as "out of business" in 2022 [37]. This could be due to simple statistics: most high-risk startups fail. This authors' view is that nonstandard CMOS process modifications (often in a research laboratory) can present obstacles to investment, to foundry selection, and to production scalability. Also, the choice of ELISA/immunoassay demonstrators pits it against the "cheap-and-cheerful" ELISA lateral-flow test strips, making clinical and market differentiation of this CMOS implementation more difficult.

4.4 CMOS Capacitive DNA Biosensor

At ISSCC 2012, Lee et al. [38] demonstrated capacitive DNA detection integrated on a CMOS process, with a limit of detection of 100 pmol, similar to Berney's discrete capacitive sensor of Fig. 14. Lee modified the CMOS process to integrate gold electrodes. The device locates two electrodes in a single current source as shown in Fig. 23. Oligomer DNA probes, complementary to the target DNA, are immobilized on the gold electrode surfaces.

The detection principle is the same as in Fig. 14, in which C_{dl} reduces when a target DNA binds or hybridizes to the capture probe on one electrode. This CMOS capacitive sensor is titled "Label-free DNA detection," meaning that no labels (fluorescent) are attached to the oligomer probes, unlike as in PCR. This simplifies the oligomer assay, while transferring the sensitivity challenge to the CMOS

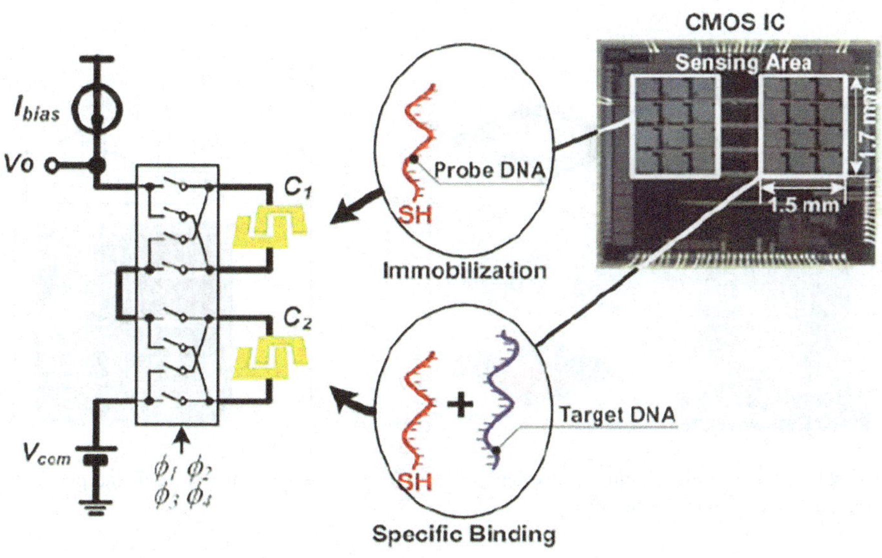

Fig. 23 CMOS capacitive DNA biosensor (From [38])

domain. In the differential architecture, the device detects the relative difference in the capacitance ($\Delta C, |C_1\text{-}C_2|$) between a bare probe-functionalized working electrode and a hybridized working electrode. This difference is accumulated by a switched-capacitor-type parasitic-insensitive discrete-time integrator, which increases the signal-to-noise ratio (SNR) of the device significantly.

It is unclear if this CMOS chip has been commercialized. Its authors are currently focused on non-CMOS nanoplasmonic PCR microfluidic methods.

4.5 CMOS Bioluminescence Assay Sensor

At ISSCC 2017, Hassibi et al. (Insilixa) demonstrated bioluminescent DNA detection integrated on a CMOS process, as shown in Fig. 24 [39]. A limit of detection of target DNA in a sample is not given, since this architecture is detecting the billions of PCR amplicons from an upstream PCR reaction.

The CMOS process is modified by integrating an array of photodiodes, and long-pass multi-dielectric (TiO_2 and SiO_2) optical interference filters on the chip surface Multiple capture probes are designed to target different portions and mutations of the target DNA. The probes are then spotted onto individual sensors of the array. Amplicons from an upstream PCR reaction are applied to the sensor. The sensors with a positive optical signal then indicate which mutations are captured and detected. A high-dynamic-range photosensor detects small signals in the presence of a large background, with a unipolar $\Delta\Sigma$ photo-sensor circuit and capacitive transimpedance amplifier (CTIA) in each pixel integrate the photocurrent.

At VLSI 2021, Hassibi expanded the array to 1024 pixels [40]. This enabled simultaneous detection of many genetic mutations in the sample, important for

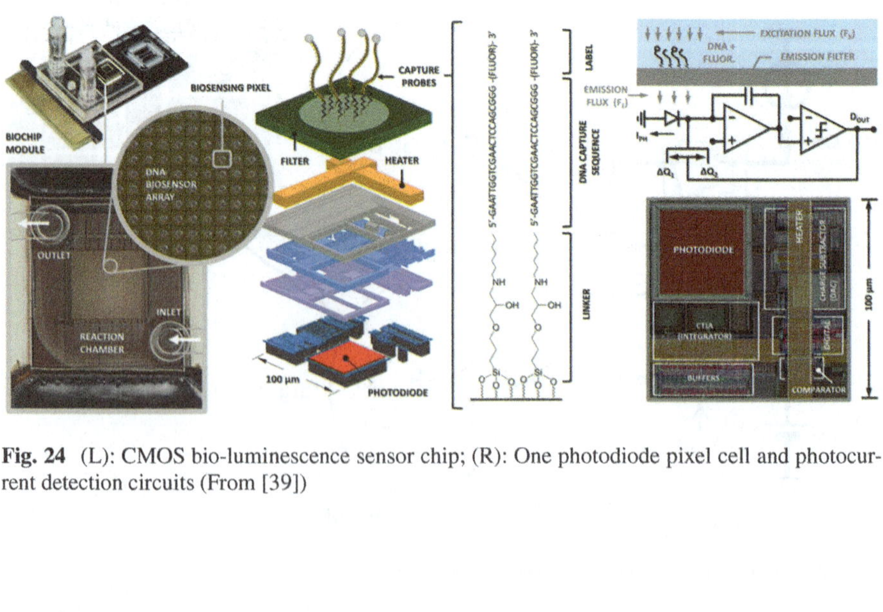

Fig. 24 (L): CMOS bio-luminescence sensor chip; (R): One photodiode pixel cell and photocurrent detection circuits (From [39])

medical diagnostic applications in variant tracking and antibiotic resistance. The technology was acquired by a large diagnostics company in 2021 [41].

5 The PNA-BeadCAP® Molecular Detection Assay

We developed the PNA-BeadCAP® nonenzymatic assay to simplify molecular detection of DNA and RNA targets. It employs synthetic peptide nucleic acid (PNA) probes, a novel bead-based assay, and capacitive detection of the beads with a standard-CMOS chip. The reader is referred to patents [42, 43, 44] for a description of the assay architecture and operating principles. Briefly, the sequential, highly specific in-solution hybridization events in the assay result in the release of a synthetic proxy bead over the biosensor. Subsequent capture of the proxy on the correct complementary biosensor leads to the transduction of the proxy presence into a capacitance signal, representing the amount of RNA in the sample.

5.1 PNA Probes

The PNA probes of the assay are synthesized by solid-phase FMOC peptide synthesis [45] and contain the standard A, T, C, and G monomer bases as in PCR DNA oligomer probes. However, whereas DNA probes have a negative charge due to their (deoxy)ribose-phosphate backbone, PNAs have a neutral backbone, due to substitution of N-(2-aminoethyl) -glycine units linked by amide bonds (Fig. 25):

This gives PNA probes many distinct advantages in a molecular assay:

PNA probe features	**Advantages**
Neutral backbone	Hybridize in any pH, in any crude sample ➔ eliminates sample-prep; the probes can capture RNA directly in whole blood or saliva
Very high specificity	High binding strength. Excellent single-base-mismatch (SBM or SNP) detection
Short (10–15 nt typically)	Can target more viruses and narrower regions (than PCR/LAMP probes of >25 nt typically). See PNA-probe illustrative example in Fig. 7
Higher melt temperature (tm) (10 °C–15 °C)	Stronger affinity than DNA/RNA sequences of same length. Better for Single Base Mismatch (SBM) & SNP detection, e.g., COVID-19 variants
Synthetic	Stable. Synthetic. No refrigeration. Minimal degradation/inhibition issues. Long shelf life
Bioorthogonal-PNA's	Synthetic. An abiotic replacement for the Biotin-Streptavidin linkers of other assays

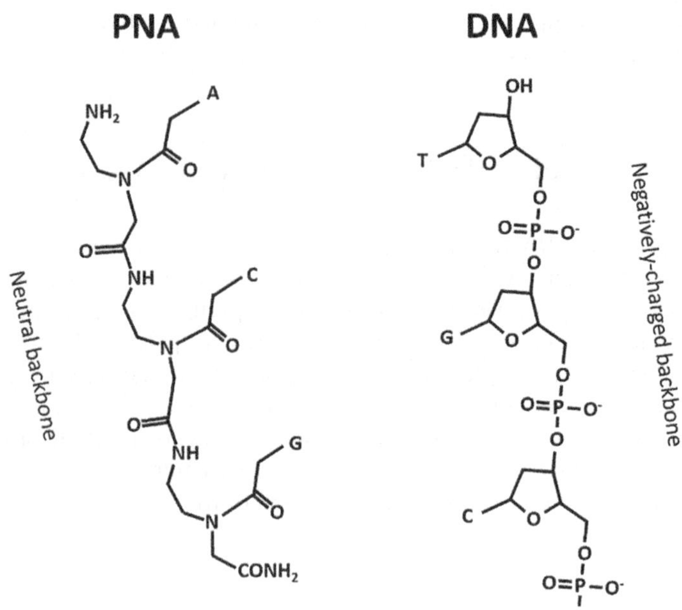

Fig. 25 PNA (neutral backbone) vs DNA (negatively charged backbone) structural differences (Based on [46])

5.2 *BeadCAP® Capacitance Detection Method and Circuits*

The beads from the upstream assay are detected and counted by capacitive sensing, in the fringe-field of interdigitated electrodes (IDE) fabricated in the CMOS top-metal layer, which are protected beneath the silicon nitride passivation layer (Fig. 26), that is, there is no contact between the assay biology and electronics. This removes corrosion risks and a potent source of interference. While most electric field lines are contained in the silicon, a small fringe portion of field lines protrude above the silicon nitride (Si_3N_4) passivation. These are the transducer portion, providing up to 100 mV electric field across a bead. This causes a dipole response in the bead, which correspondingly increases the charge on the electrodes, thereby increasing the capacitance as seen by the capacitive-to-digital converter.

The overall end-to-end assay performance (in vitro) is shown in Fig. 27 for three different concentrations of RNA spiked in biological samples. The top SEM photo shows beads correctly captured on the target sensor, spotted with a complementary PNA probe. The bottom photo shows no beads are captured for an off-target RNA. The limit of detection of this in vitro manual implementation of the assay is approximately 10 fmol. This is four orders of magnitude superior to the 100 pmol observed with the prior-art capacitive DNA sensors of Figs. 14 and 23.

Further work is underway to reduce the LOD to attomolar levels by assay miniaturization into a hand-held cartridge, automation, and other optimization steps

Fig. 26 Beads on chip surface and bead capacitance fringe-field sensing method

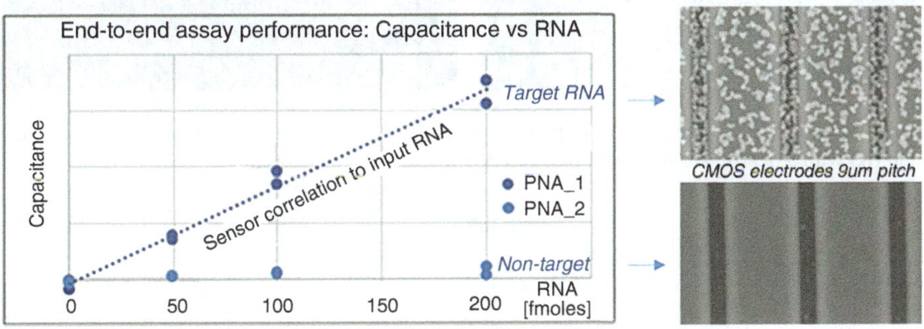

Fig. 27 Digital capacitance readout of RNA amount in lab biological samples

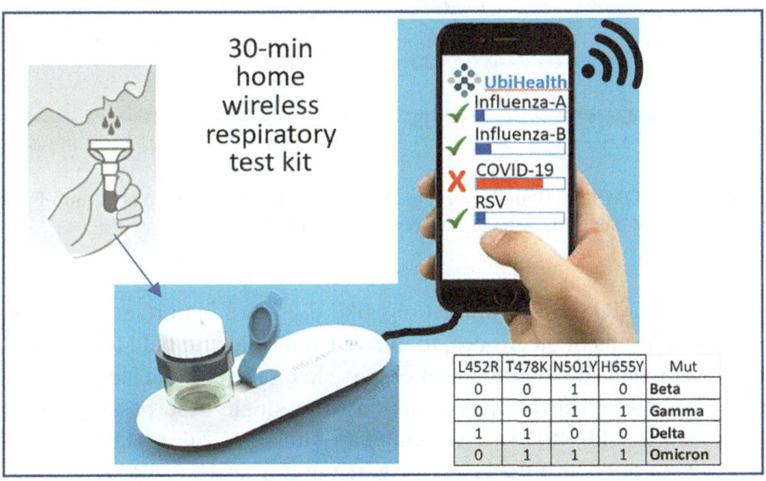

Fig. 28 Respiratory product profile of the miniaturized assay

[47]. The present sensor chip has 30 sensors, which enables multiplex detection of many sequences and variants in the target sample.

The target product profile for the fully integrated assay is shown in Fig. 28.

Resolving atto-Farad bead capacitances is key to the operation of this assay. Figure 29 shows simulations of the electric field around two electrodes, (L) in free space, and (R) with a ground-plane just beneath, the latter being representative of

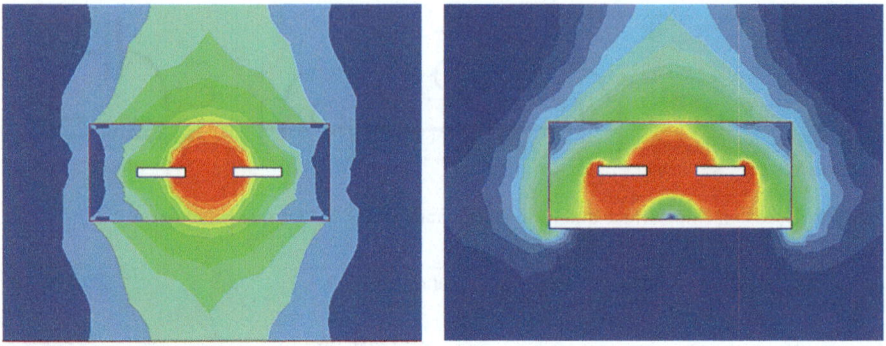

Fig. 29 Maxwell 2D electrostatic simulation of the field around two electrodes, (L) in free space and (R) with a nearby ground plane (red = 1 V/μm)

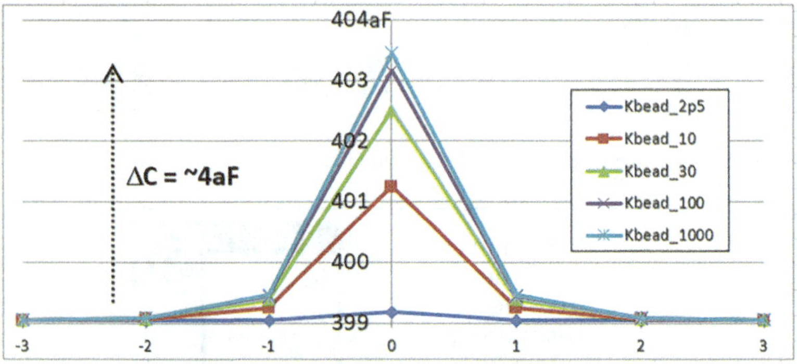

Fig. 30 Maxwell COMSOL Boundary Element Model (BEM) simulations of atto-Farad bead capacitance. At 1.6 V electrode voltage, 4aF equates to ~40 electrons

electrodes on a CMOS chip. Simulations like these are important to optimize the shape and dimensions of electrodes, to maximize the electric field to interrogate the magnetic or nonmagnetic nearby bead or particle on the chip surface.

Figure 30 shows electrostatic simulations of a single bead passing through the electrode electric field [48]. The resulting capacitance is 2aF to 4aF, depending on the dielectric constant K of the bead, for K ranging from 10 to 1000:

Figure 31 shows the architecture of the Candy second order ΣΔ converter for detecting the bead capacitance, and Fig. 32 shows the detailed modulator circuit implementation [49]. In this architecture, the IDE bead-sensing capacitor C_s (circled) is the input variable. The same reference voltage is applied to both the reference capacitor, C_{ref}, and the sensing capacitor C_s. This significantly reduces the effect of any reference noise. It also eliminates the requirement to calibrate or use any curvature correction within the on-chip bandgap voltage reference. Note the first integrator utilizes a fully floating input structure, with the input common mode to the amplifier being set by the feedback DAC. As the applied voltages to the

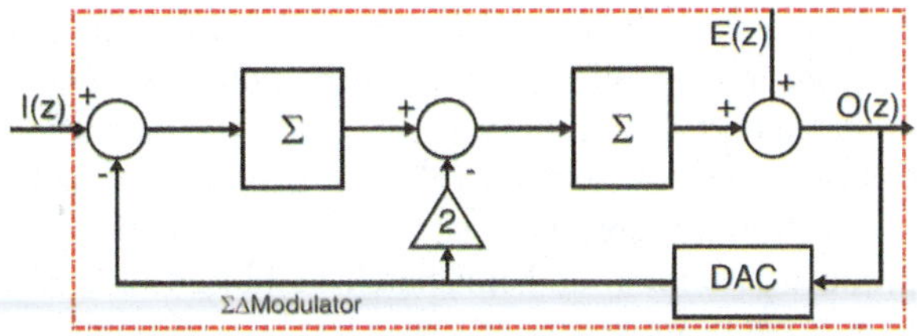

Fig. 31 Architecture of the sigma delta ΣΔ converter

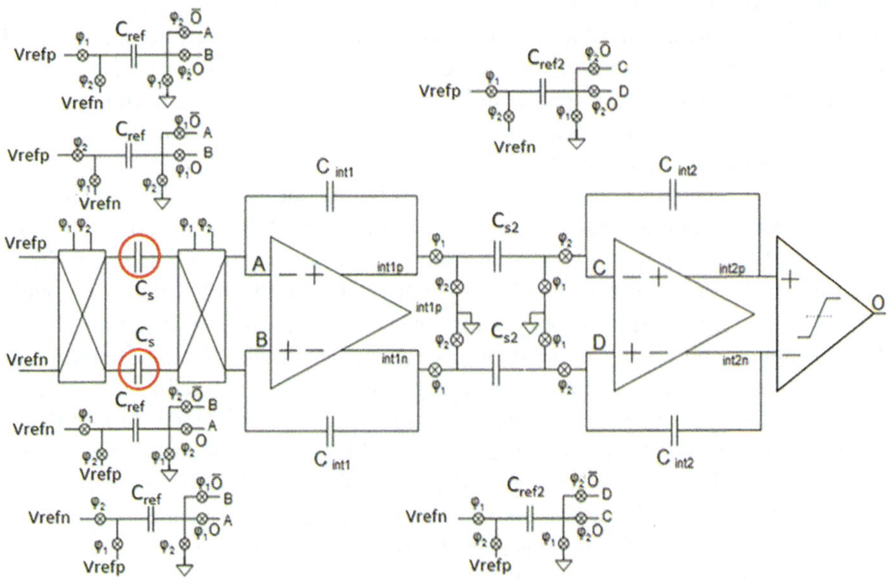

Fig. 32 ΣΔ modulator schematic

reference and sensor capacitors are the same, it is necessary to consider the thermal noise charge when examining the repeatability of the system.

The sampled thermal noise, Q_n, is correlated and is given by:

$$Q_n = CV_n = \sqrt{\mathrm{kTC}}$$

and the sampled thermal noise of the 1^{st}integrator is:

$$Q_n = \sqrt{\mathrm{kT}\left(16.C_S + 8.C_{ref}\right)},$$

where T is temperature, k is Boltzmann's constant and C is the unit capacitance.

6 Discussion and Conclusions

In this chapter, we reviewed several different molecular detection methods and presented some biosensor and CMOS biochip detection examples. As Hassibi noted [1], it is a mixed picture of some commercial successes and many noncommercialized solutions. The ISFET and the wearable glucose sensor have reached commercial success and widescale adoption, although this took decades. Reasons for noncommercialized examples are many and varied: difficulty in finding ongoing grant support or investments, especially for capital-intensive CMOS developments; difficulty in manufacturing (of coils, nano-rods, nanotubes), miniaturization negated by the need for large external components (GMR biosensor and magnets), incorrect product definition, and mismatch of biosensor technology to market application (e.g., CMOS immunoassay with unclear advantage versus incumbent lateral-flow immunoassay strip tests). Regulatory delays and inertia in the medical and clinical communities must also be factored in. These communities tend to be careful and conservative in adopting new technologies. CMOS process modifications (etching, nonstandard layers, or post-processing) may also be a factor, limiting foundry choice, affecting investment decisions (risk of process obsolescence), and future high-volume/low-cost scalability.

We have presented a nonenzymatic assay "PNA-BeadCAP," which aims to avoid many of these complexities. It employs synthetic PNA probes and a standard unmodified CMOS foundry process for the capacitive-sensor bead detector. There are no CMOS process modifications, which reduces cost and facilitates a wide choice of foundries. This is important for high-volume production scalability. Being fully synthetic and having no enzymes or fluorescent labels, the assay eliminates the need for cold-chain shipping and refrigeration. This will facilitate storage or operation at temperatures up to 40 °C and theoretically unlimited shelf-life, enabling volume building and stockpiling of tests for any future potential emergencies. PNA probes give higher specificity than PCR probes. The CMOS detector chip has 30 sensors, which will enable multiplexing and genotyping/variant-identification in this portable test. The in vitro assay has an LOD of 10 fmol, several orders of magnitude superior to previous capacitive DNA sensors. Work is now underway to reduce this to attomolar levels to approach PCR sensitivity levels, by assay optimization, automation, and miniaturization to a hand-held kit format, intended for home use and self-testing of respiratory viruses from a self-sampled saliva sample.

References

1. S. Swaminathan, P. Bento, et al., *Consolidated Telemedicine Implementation Guide* (World Health Organization, Geneva, 2022) WHO https://www.who.int/publications/i/item/9789240059184. Retrieved 9 Jan 2023
2. A. Hassibi, *CMOS Biochips: The Good, the Bad, and the Hype* (TWEPP Workshop 14 Sep 2018). https://indico.cern.ch/event/608587/contributions/2704610/attachments/1523865/2381946/Hassibi_TWEPP_2017_final_v2.pdf. Retrieved 5 Dec 2022
3. B. O'Farrell, *Evolution in Lateral Flow-Based Immunoassay Systems* (Springer, Lateral flow immunoassay, 2009), pp. 1–33
4. B.E. Slatko et al., Overview of next-generation sequencing technologies. Curr. Protoc. Mol. Biol. **122**(1), e59 (2018). https://doi.org/10.1002/cpmb.59. PMID: 29851291; PMCID: PMC6020069
5. L.G. Gordon et al., Estimating the costs of genomic sequencing in cancer control. BMC Health Srv. Res. **20**, 492 (2020). https://doi.org/10.1186/s12913-020-05318-y

6. F. Wu et al., A new coronavirus associated with human respiratory disease in China. Nature **579**(7798), 265–269 (2020). https://doi.org/10.1038/s41586-020-2008-3. Epub 2020 Feb 3. Erratum in: Nature. 2020 Apr;580(7803):E7. PMID: 32015508; PMCID: PMC7094943
7. V. Corman et al., Detection of 2019 novel coronavirus (2019-nCoV) by real-time RT-PCR. Eur. Secur. **25**, 2000045 (2020). https://doi.org/10.2807/1560-7917.ES.2020.25.3.2000045
8. A. Hassibi, *Mass-deployable Molecular Diagnostics (MDx), including COVID-19 Testing: An IC Designer's Perspective* (Keynote at Annual TxACE Symposium, October 19, 2020)
9. T. Brock, H. Freeze, Thermus aquaticus gen.n. and sp.n., a nonsporulating extreme thermophile. J. Bacteriol. **98**(1), 289–297 (1969). https://doi.org/10.1128/jb.98.1.289-297.1969. PMID: 5781580; PMCID: PMC249935
10. Skytrax, London, UK, *"The varying costs of a PCR test at airports around the world"*,https://skytraxratings.com/the-varying-costs-of-a-pcr-test-at-airports-around-the-world. Retrieved 5 Dec 2022
11. FIND-Dx, *Negotiated Prices, GeneXpert II R2 $9920*, https://www.finddx.org/pricing/genexpert/. Retrieved 5 Dec 2022
12. FDA Emergency Use Authorization EUA203089, *Visby Medical COVID-19 Point of Care Test*, https://www.fda.gov/media/145914/download. Retrieved 5 Dec 2022
13. B. Andreyev, et al., *Devices and Methods for Molecular Diagnostic Testing*, (US patent application US, 2016/0186240, June 30, 2016)
14. A. Renzoni et al., Analytical evaluation of Visby medical RT-PCR portable device for rapid detection of SARS-CoV-2. *Diagnostics**11***, 813 (2021). https://doi.org/10.3390/diagnostics11050813
15. *Visby Medical 10-pack tests $1550,*https://www.meenta.io/product/visby-rapid-covid-19-pcr-test-power-cord-2/. Accessed 12 Mar 2022
16. D. Thompson and Yu Lei, "Recent progress in RT-LAMP enabled COVID-19 detection", Mini review:, Sensors and Actuators Reports, Volume 2, Issue 1, 2020, 100017., ISSN 2666-0539, https://doi.org/10.1016/j.snr.2020.100017
17. T. O'Leary et al., Sensitivity of ID NOW and RT-PCR for detection of SARS-CoV-2 in an ambulatory population. elife **20**(10), e65726 (2021). https://doi.org/10.7554/eLife.65726. PMID: 33876726; PMCID: PMC8081522
18. FDA Emergency Use Authorization E210180, *Cue COVID-19 Test for Home and Over The Counter (OTC) Use*, https://www.fda.gov/media/146467/download. Retrieved 5 Dec 2022
19. Cue Health price list, https://shop.cuehealth.com/. Retrieved 5 Dec 2022
20. Cue Health, *Cue COVID-19 Test: Instructions for Use*, https://www.fda.gov/media/138826/download. Retrieved 5 Dec 2022
21. KM Lei et al, CMOS biosensors for in vitro diagnosis – Transducing mechanisms and application https://doi.org/10.1039/C6LC01002D (Critical Review) Lab Chip, 2016, **16**, 3664–3681
22. K. Kramp, *How to Hack a Glucose Sensor*, https://towardsdatascience.com/how-to-hack-a-glucose-sensor-ebaaf2238170. Retrieved 5 Dec 2022
23. J. Barton et al., Long-range photoinduced electron transfer through a DNA helix. Science **262**(5136), 1025–1029 (1993). https://doi.org/10.1126/science.7802858
24. C. Berggren et al., A feasibility study of a capacitive biosensor for direct detection of DNA hybridization. J. Electroanalysis **11**(3), 156–160 (1999)
25. H. Berney et al., A DNA diagnostic biosensor: Development, characterisation and performance. Sensors Actuators B Chem. **68**(1–3), 100–108., ISSN 0925-4005, (2000). https://doi.org/10.1016/S0925-4005(00)00468-8
26. D.R. Baselt et al., A biosensor based on magnetoresistance technology. Biosens. Bioelectron. **13**(7–8), 731–739 (1998). https://doi.org/10.1016/s0956-5663(98)00037-2. PMID: 9828367
27. R. Widlar, Some circuit design techniques for linear integrated circuits. IEEE Trans. Circuit Theory **12**(4), 586–590 (1965). https://doi.org/10.1109/TCT.1965.1082512
28. P. Bergveld, Development of an ion-sensitive solid-state device for neurophysiological measurements. I.E.E.E. Trans. Biomed. Eng. **BME-17**(1), 70–71 (1970). https://doi.org/10.1109/TBME.1970.4502688

29. P. Bergveld, Thirty years of ISFETOLOGY: What happened in the past 30 years and what may happen in the next 30 years. Sens. Actuators B Chem. **88**(1), 1–20., ISSN 0925-4005, (2003). https://doi.org/10.1016/S0925-4005(02)00301-5
30. J. Bausells et al., Ion-sensitive field-effect transistors fabricated in a commercial CMOS technology. Sens. Actuators B Chem. **57**(1–3), 56–62., ISSN 0925-4005, (1999). https://doi.org/10.1016/S0925-4005(99)00135-5
31. C. Toumazou et al., *A sensing circuit utilising an ion sensitive field effect transistor*, GB patent 2416210B
32. D.M. Garner et al., A multichannel DNA SoC for rapid point-of-care gene detection, in *2010 IEEE International Solid-State Circuits Conference-(ISSCC)*, (IEEE, 2010, San Francisco), pp. 492–493. https://doi.org/10.1109/ISSCC.2010.5433834
33. A. Hassibi and N. Pourmand, *Transient Electrical Signal Based Methods and Devices for Characterizing Molecular Interaction and/or Motion in a Sample*, US patent 722,3540, May 29 2007
34. J. Rothberg et al., An integrated semiconductor device enabling non-optical genome sequencing. Nature **475**, 348–352 (2011). https://doi.org/10.1038/nature10242
35. X. Huang, et al., A dual-mode large-arrayed CMOS ISFET sensor for accurate and high-throughput pH sensing in biomedical diagnosis, IEEE Trans. Biomed. Eng., Vol. 62, No. 9, Sep. 2015, pp. 2224–2233
36. O. Florescu, *Fully Integrated Complementary Metal Oxide Semiconductor (CMOS) Bio-Assay Platform*. https://www.researchgate.net/publication/267705427_Fully_Integrated_Complementary_Metal_Oxide_Semiconductor_CMOS_Bio-Assay_Platform. Retrieved 5 Dec 2022
37. Pitchbook https://pitchbook.com/profiles/company/123267-16. Retrieved 5 Dec 2022
38. K.-H. Lee et al., CMOS capacitive biosensor with enhanced sensitivity for label-free DNA detection. *IEEE Int. Solid State Circuits Conf.***2012**, 120–122 (*2012*). https://doi.org/10.1109/ISSCC.2012.6176945
39. A. Hassibi et al., A fully integrated CMOS fluorescence biochip for multiplex polymerase chain-reaction (PCR) processes, in *2017 IEEE International Solid-State Circuits Conference (ISSCC)*, (IEEE, San Francisco, 2017), pp. 68–69. https://doi.org/10.1109/ISSCC.2017.7870264
40. A. Manickam et al., Multiplex PCR CMOS biochip for detection of upper respiratory pathogens including SARS-CoV-2. *Symposium on VLSI Circuits***2021**, 1–2 (*2021*). https://doi.org/10.23919/VLSICircuits52068.2021.9492353
41. https://www.linkedin.com/in/arjang. Retrieved 5 Dec 2022
42. B. O'Farrell, T. Cummins, *Sample Preparation Method and Apparatus*, US patent 10,160,966
43. B. O'Farrell and T. Cummins, *Nucleic Acid Analysis Method and Apparatus*, US patent 10,738,348
44. T. Cummins, B. O'Farrell, *Capacitive Sensor and Method of Use*, US patent 10,746,683
45. K. Oshaben, C. O'Connell, D. Appella, et al, *RNA target enrichment or depletion of biological samples,* USPTO patent-pending application 2021/0277387
46. A. Ray and B Norden, "Peptide nucleic acid (PNA): Its medical and biotechnical applications and promise for the future", FASEB J., Vol 14, Is 9, https://doi.org/10.1096/fasebj.14.9.1041
47. M. McCarthy, P. Free, J. O'Driscoll, J. Walshe, C. O'Sullivan, et al, *Diagnostic Device and System*, US patent 11,459,601
48. T. Dalton, D. McGuire, et al, *Single-Bead Capacitive Detector for Microbiological Applications*, USPTO patent-pending application 63/304,312
49. I. O'Connell et al, *$\Sigma\Delta$ A-to-D Converter for Use in Sensing Applications* (IEEJ Int'l Analog VLSI Workshop, 2007)

Part II
Noise-Shaping ADCs

The second section of this book discusses the development of ADCs that achieve high resolution exploiting noise-shaping techniques.

The main topologies of such ADCs are noise-shaping SAR and sigma-delta modulators. Noise-shaping SAR ADCs have been recently proposed, and in a short time, they achieve large popularity. Three chapters are dedicated to them, moving to zoom ADC and sigma-delta ADCs in the last part of this section for reference.

In Chap. 7, Michael Flynn introduces the noise-shaping SAR concept. A noise-shaping SAR ADC combines the advantages of SAR and sigma-delta architectures, and it is more energy-efficient than sigma-delta and better suited to higher resolution than SAR. Starting from the first noise-shaping SAR ADCs (limited to moderate resolution and moderate speed), Mike Flynn describes the NS-SAR evolution up to recent work with high-order noise-shaping structures to extend the SNR to audio performance levels. On the other hand, he discusses how interleaving of noise-shaping SAR effectively could tackle the speed bottleneck.

In Chap. 8, Eric Thompson describes recent advances in the development of high-performance (in terms of power, noise, and linearity) noise-shaped SARs. Improvements are possible, thanks to advanced solutions such as dynamic amplifiers, kT/C noise suppression, and low OSRs. The chapter presents how higher performances require careful analysis of implementation non-idealities such as mismatch and flicker noise among others.

In Chap. 9, Hanyue Li presents the aspects regarding the implementation of a continuous-time NS-SAR ADC, an improvement with respect to the previous sampled data NS-SAR ADCs. The different implementation critical points (like opamp design, DAC linearization, etc.) are studied, and proper solutions are demonstrated with experimental results from the comparison of a DT-NS-SAR with a CT-NS-SAR ADC.

In Chap. 10, Efraïm Eland introduces zoom ADCs, which combine a coarse SAR ADC with a fine delta-sigma modulator ($\Delta\Sigma$M) to efficiently obtain high energy efficiency and high dynamic range. While they are well suited for use in various instrumentation and audio applications, they have drawbacks. An overview of recent advances in zoom ADCs that tackle these challenges while maintaining high energy efficiency is then proposed and validated by experimental results from a silicon prototype.

In Chap. 11, Spyros Kalogiros studies the possibility to further reduce thermal noise, which limits the achievable performance of a noise-shaping ADC. The chapter reviews some of the recent advancements in relation to the thermal noise reduction for the case of switched-capacitor delta-sigma ADCs. In addition, the chapter addresses the challenges associated with breaking the 192 dB Schreier FoM performance barrier.

In Chap. 12, Roberto Modaffari presents the industrial development of a second-order 5 bit hybrid CT/DT delta-sigma modulator. The implementation aspects together with the architectural solutions used to limit the impact of the flicker noise on ADC accuracy, to save current consumption, and to reduce testing time and area occupation are proposed and discussed.

The Evolution of Noise-Shaping Successive Approximation (SAR) ADC

Michael P. Flynn and Seungjong Lee

Abstract The noise-shaping successive approximation (SAR) is only a decade old but has quickly become a dominant ADC architecture. A noise-shaping SAR ADC combines the advantages of SAR and sigma-delta architectures. Noise-shaping SAR is more energy efficient than sigma-delta and better suited to higher resolution than SAR. Noise-shaping SAR ADCs are also very compact. The first noise-shaping SAR ADCs were limited to moderate resolution and moderate speed. Recent work applies high-order noise shaping to extend the SNR to audio performance levels. At the other extreme, the interleaving of noise-shaping SAR effectively tackles the speed bottleneck. Noise-shaping SARs are also valuable in hybrid ADCs. As a backend quantizer, a noise-shaping SAR very efficiently increases the order of a continuous-time sigma-delta.

1 Introduction

Both the performance and efficiency of analog-to-digital conversion circuits have improved by orders of magnitude over the last two decades. CMOS technology scaling has driven much of this dramatic improvement. However, new architectures and circuit techniques have played a critical role in exploiting the advantages of scaling. Once popular architectures such as pipeline ADC are now much less popular. Instead, noise-shaping converters (traditionally sigma-delta ADCs) play an ever-larger role, covering the medium- to high-resolution space with improving efficiency and increasing bandwidth. Successive approximation ADCs made the transition from specialty to dominance. SAR converters are an ideal match for digital CMOS technology and have benefited hugely from technology scaling. Some of the most efficient converters are low-speed SAR ADCs. At the other extreme, interleaved SAR ADCs dominate very-high-speed conversion.

M. P. Flynn (✉) · S. Lee
University of Michigan, Ann Arbor, MI, USA
e-mail: mpflynn@umich.edu; andrea.baschirotto@unimib.it

P. Harpe et al. (eds.), *Biomedical Electronics, Noise Shaping ADCs, and Frequency References*, https://doi.org/10.1007/978-3-031-28912-5_7

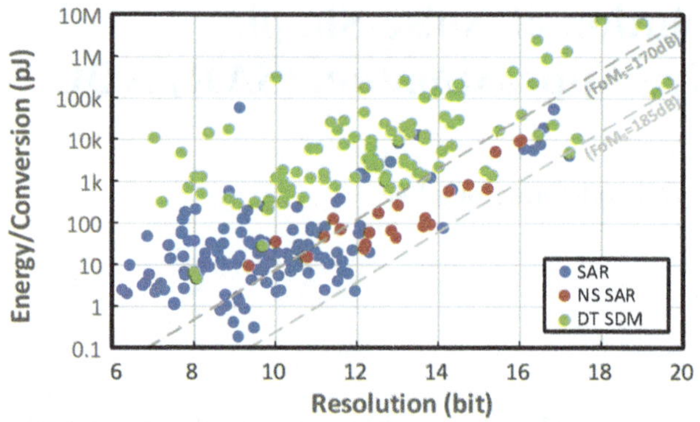

Fig. 1 Energy efficiency versus resolution for SAR, NS SAR, and DT SDM ADCs. (Based on data from Ref. [3])

The noise-shaping SAR ADC was first introduced in 2012 [1, 2] and quickly became a dominant ADC architecture. The noise-shaping SAR architecture combines the advantages of the SAR and sigma-delta architectures. Noise-shaping SAR is more energy efficient than sigma-delta and better suited to higher resolution than SAR. The first noise-shaping SAR ADCs were limited to moderate resolution and moderate speed. Recent work applies high-order noise shaping to extend the SNR to audio performance levels. At the other extreme, the interleaving of noise-shaping SAR effectively tackles the speed bottleneck. Noise-shaping SARs are also valuable in hybrid ADCs – as a backend quantizer, a noise-shaping SAR very efficiently increases the order of a continuous-time sigma-delta.

Figure 1 plots the energy efficiency versus the resolution for SAR ADCs, noise-shaping SAR (NS SAR) ADCs, and discrete-time sigma-delta modulators (DT SDM). SAR ADCs dominate at lower resolutions. At one time, the discrete-time sigma-delta modulator was the architecture of choice for moderate to high resolutions. However, NS SAR ADCs have come to dominate at moderate resolution. Recently, NS SAR ADCs have extended to audio resolutions so that DT SDM is only attractive at very high resolutions. Figure 2 plots the bandwidth versus the resolution for SAR ADCs, NS SAR ADCs, and discrete-time sigma-delta modulators. NS SAR is highly competitive at moderation resolution. Noise-shaping SAR ADCs tend to be far smaller than comparable ADCs. In addition, new techniques such as interleaving further enhance NS SAR's bandwidth.

2 SAR ADC

CMOS SAR ADCs typically comprise a capacitor DAC (CDAC), a comparator, and successive approximation logic (Fig. 3). An early innovation in MOS SAR architecture introduced the use of a single CDAC capacitor array [4] for signal

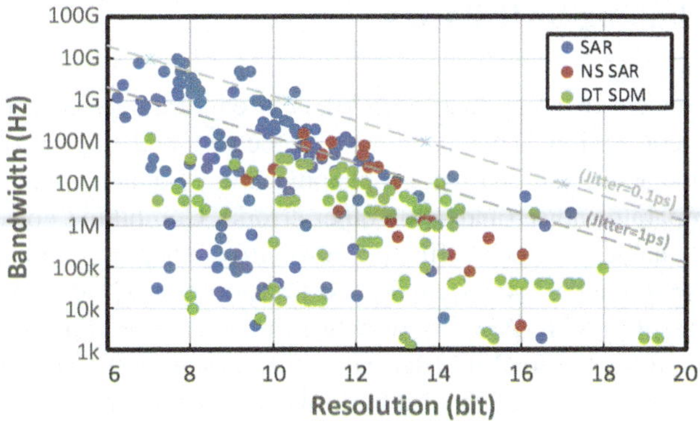

Fig. 2 Bandwidth versus resolution for SAR, NS SAR, and DT SDM ADCs. (Based on data from Ref. [3])

Fig. 3 CMOS SAR ADC

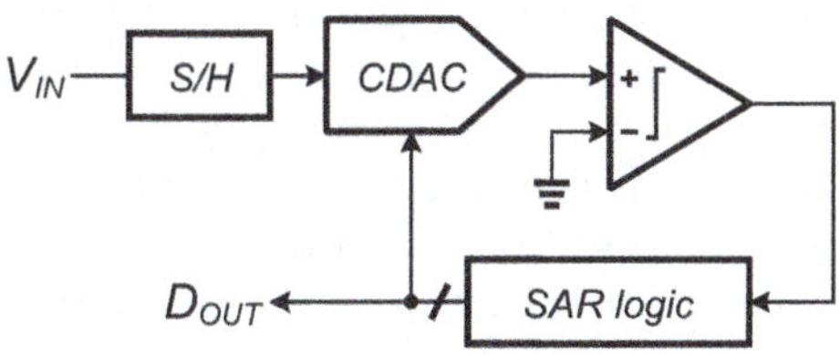

sampling, DAC feedback, and summation. Initially, the CDAC samples the input signal. Then, during the conversion process, trial binary weighted values beginning with half-full scale are added to the sampled value by the CDAC. Typically, a single comparator determines the sign of the bit trials. Finally, the SAR logic drives the successive approximation process and keeps track of the binary results.

A critical advantage of the CMOS SAR architecture is that it greatly benefits from advancements in digital CMOS technology. While the SAR logic is naturally digital, the analog blocks in a SAR ADC also take advantage of digital CMOS technology scaling. Switches and capacitors are the main components of the SAR DAC. CMOS FETs generally make excellent switches. In addition, the fine-line multi-layer metallization in advanced CMOS makes accurate and dense capacitors without resorting to specialized capacitor structures. Finally, many SAR ADC implementations use dynamic comparators, which are very energy efficient and structurally resemble digital latches.

Conventional SAR ADCs are the most energy efficient up to moderate resolutions. This limitation is in part because a SAR does not perform signal amplification. During each approximation step, the CDAC applies a trial voltage, and the comparator must resolve the sign of the resulting error. A higher resolution implies a smaller residue, which places a greater burden on the comparator. A higher resolution demands lower comparator noise and more comparator gain.

3 Noise-Shaping SAR Basics

The critical concept in noise-shaping SAR is to capture and use the final residue from a SAR conversion. This residue is processed and applied in subsequent conversion to shape noise. This noise shaping affects quantization noise but can shape the comparator noise (and some other errors). Combining noise shaping and oversampling enables a higher effective resolution. Shaping comparator noise is critical because it reduces the required noise performance of the comparator.

The residue in a SAR ADC is readily available because it is related to the comparator input for the final bit trial. However, an additional capacitor switching may be required depending on the last bit trial decision. A convenient approach is to sample the final residue onto a residue sampling capacitor. (In some cases, a residue-capturing capacitor is embedded in the CDAC during the entire SAR process.)

Once the residue is captured, it can be processed for noise shaping. There are two distinct approaches: (1) a feedforward approach, often called CIFF after a related approach in conventional sigma-delta ADCs, and (2) an error feedback (EF) approach. Implementing the filter in EF is more straightforward, but the CIFF approach is more robust to component variation. Figure 4 shows both approaches, and a practical implementation may even combine both. V_{RES} is the final residue from the SAR conversion. Filter H1 processes V_{RES} for EF, while H2 processes V_{RES} for CIFF noise shaping. In EF, the output of the H1 filter adds to the ADC input. In contrast, the H2 filter output feeds directly to the quantization node for CIFF.

The natures of the EF and CIFF architectures lead to significant differences in the noise transfer function (NTF) form. Figure 5 shows the signal model, again combining both scenarios. For EF, the NTF is 1-H1(z), so that H1(z) = 1-NTF. We see the advantage of EF if we consider implementing a classical first-order noise shaping with an NTF of $1\text{-}z^{-1}$, which leads to an H1(z) of a simple delay of z^{-1}. More generally, we can implement higher-order NTFs with H1 FIR filters. Although it has a simple implementation, EF is usually not used in conventional sigma-delta ADCs. This limitation is partly because the quantization residue is readily available in an NS SAR. Furthermore, the high quantizer resolution in a SAR ADC means that the residue is relatively small, which helps reduce sensitivity to parameter errors.

The equivalent NTF for CIFF is 1/(1-H2(z)), so H2(z) is (1-NTF)/NTF. Again, considering the straightforward first-order NTF of $1\text{-}z^{-1}$, we arrive at H2(z) of $z^{-1}/(1\text{-}z^{-1})$. So, we see that the CIFF implementation requires an integrator.

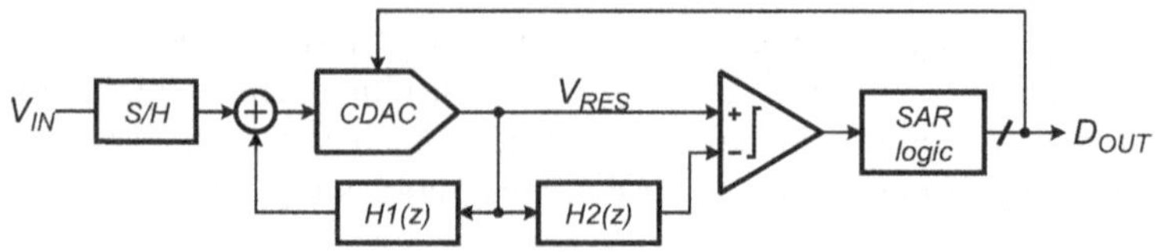

Fig. 4 EF and CIFF noise shaping in an NS SAR [5, 6]

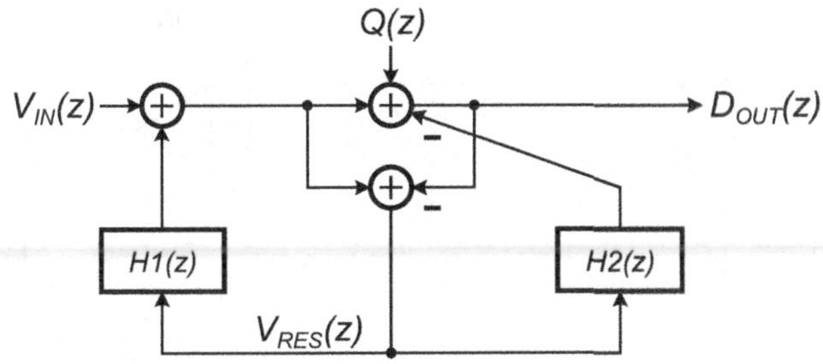

Fig. 5 Signal model for EF and CIFF noise shaping in NS SAR [5, 6]

Although the filter implementation is more complicated than for EF, an advantage is that CIFF is much less sensitive to parameter variation.

4 Implementation

4.1 The First NS SAR: A CIFF Implementation

We begin with a discussion of the CIFF implementation in [1, 2] but start with a much-simplified version to explain some of the underlying concepts. The figure (top) shows the CDAC array and a residue sampling capacitor. The sampling capacitor voltage feeds to the comparator. This configuration is appropriate for CIFF as the comparator is where quantization occurs. Furthermore, a multi-input comparator readily facilitates the combination of multiple inputs and can readily deliver relative gain through the use of different size input transistors. As shown in the signal flow diagram (Fig. 6), this is equivalent to a delay (z^{-1}) in the forward path. Clearly, this configuration lacks the integration we just discussed, and the resulting NTF is $1/(1 + z^{-1})$. This approach provides some noise shaping but hardly enough to be practical.

Introducing integration into the forward path is essential for effective noise shaping. Figure 7 represents the circuit along with the signal flow diagram. An advantage is that the CIFF architecture is tolerant of non-ideal integration, so a low amplifier gain in the integration can still deliver good noise shaping.

Moreover, we can combine a passive discrete-time FIR with the integration for more complex noise shaping. The design in Fig. 8 [1, 2] implements a two-tap FIR filter with two pairs of capacitors, C_{A1} and C_{B1} and C_{A2} and C_{B2}. Sampling is interleaved, so C_{B1} and C_{A1} sample the SAR residue in one cycle, while C_{A2} and C_{B2} sample the residue in the next cycle (Fig. 9). Charge sharing implements the FIR transfer function. C_{A1} and C_{B2} charge share in one cycle, while C_{A2} and C_{B1} charge share in the next cycle. Pairs of A and B capacitors are needed because the

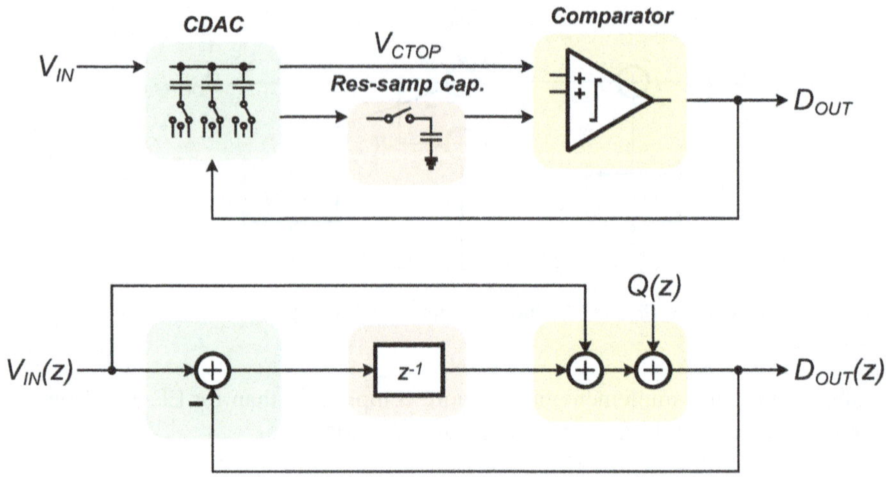

Fig. 6 Much simplified version of CIFF noise shaping in NS SAR [1, 2]

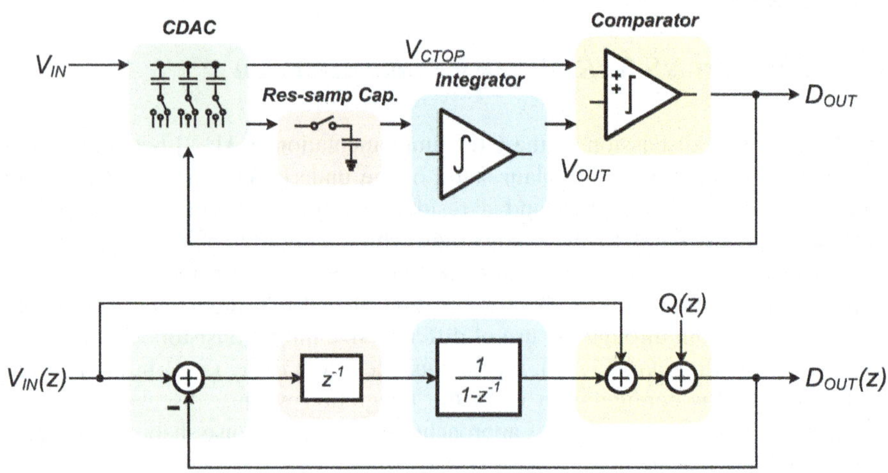

Fig. 7 CIFF noise shaping with integration [1, 2]

sample residue is required for two cycles, and charge sharing destroys the charge sampled on a capacitor.

4.2 EF Implementation

The EF implementation also captures the residue but feeds back to the input after processing with an FIR filter (Fig. 10). The filter can be pretty simple as first-

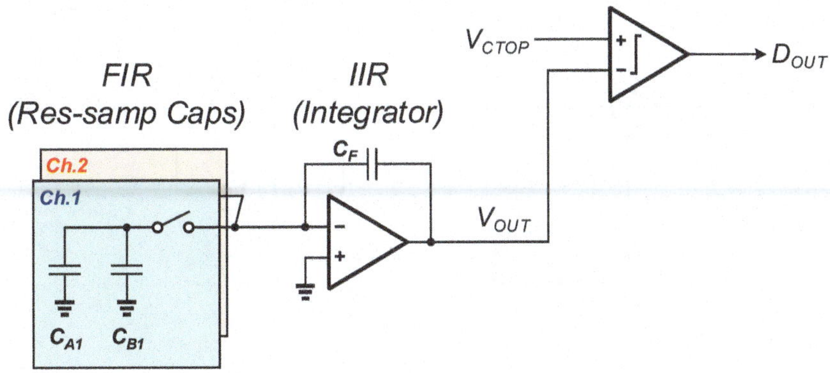

Fig. 8 Combination of FIR and IIR in CIFF NS SAR [1, 2]

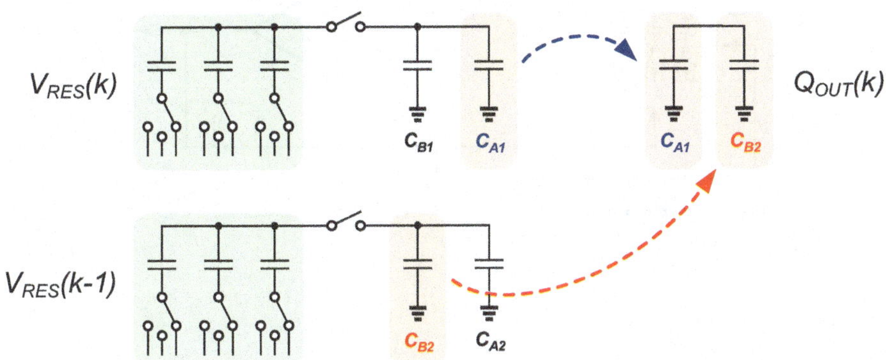

Fig. 9 FIR filtering in NS SAR [1, 2]

order noise shaping with EF only requires a unit delay. A capacitor can sample the residue and one cycle later charge share with the sampled input. A challenge is that the charge sharing leads to attenuation, so it is common to apply some gain. Furthermore, this gain should be well controlled as EF is sensitive to parametric errors.

5 Cascaded Noise Shaping

Higher-order EF filters are typically implemented in the direct form. Although the direct form is easy to implement, cascaded filters are more robust to parameter variation. Recent work [5, 6] shows the advantages of cascaded noise shaping for robustness. The approach, named cascaded noise shaping, or CaNS, cascades sub-loop filters to provide cascaded noise shaping with independent control of the zeros.

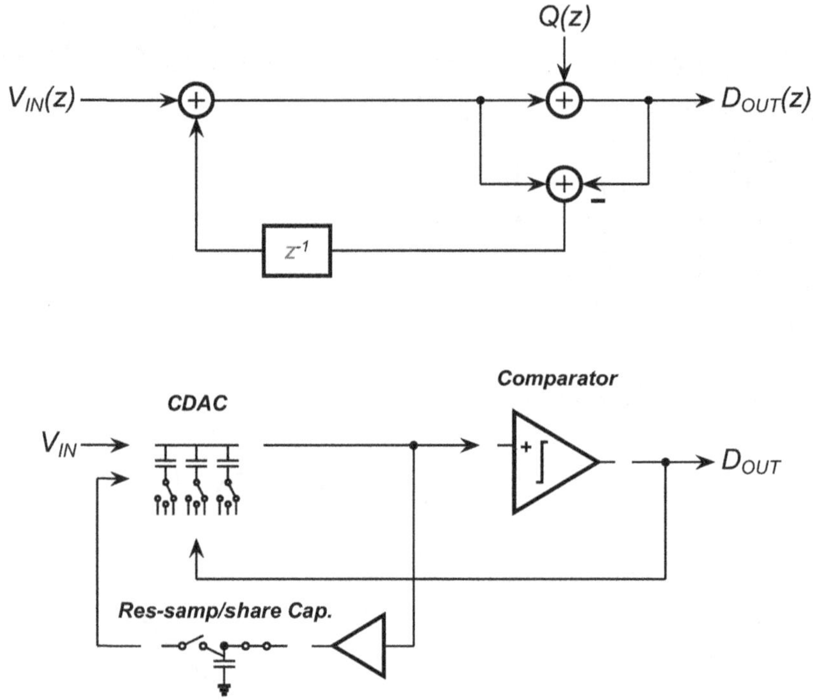

Fig. 10 EF NS SAR

Figure 11 shows a conceptual cascaded noise-shaping system. Both the inner and outer loops employ error feedback. The inner loop captures and processes the quantization error. The outer loop captures and feeds back the shaped error of the inner loop. The overall NTF is:

$$\mathrm{NTF}\,(z) = (1 - \mathrm{FIR}_1\,(z))\,(1 - \mathrm{FIR}_2\,(z))$$

Figure 12 is a simplified depiction of the implementation, which comprises an inner and an outer noise shaping loop. A series summing capacitor (not shown) cascades the noise-shaping loops. Furthermore, feedback in the first loop is by charge sharing the FIR_1 output with the series capacitor. The cascaded structure is resilient to component variation. A further advantage is the outer loop noise shapes the errors of the inner loop.

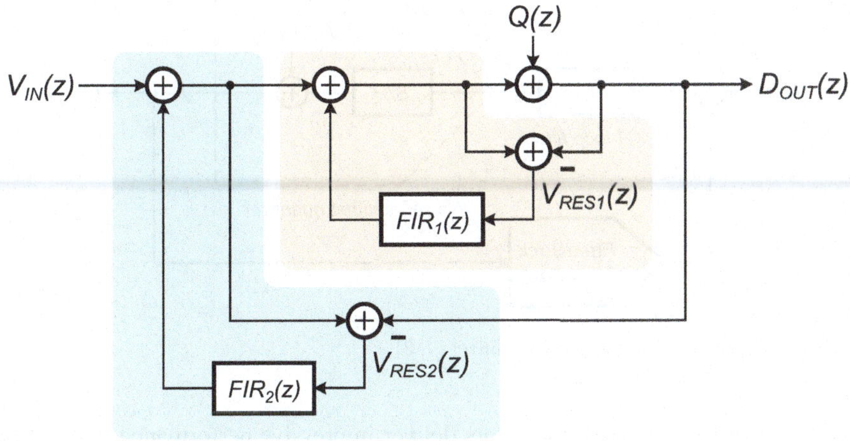

Fig. 11 Cascaded noise shaping [5, 6]

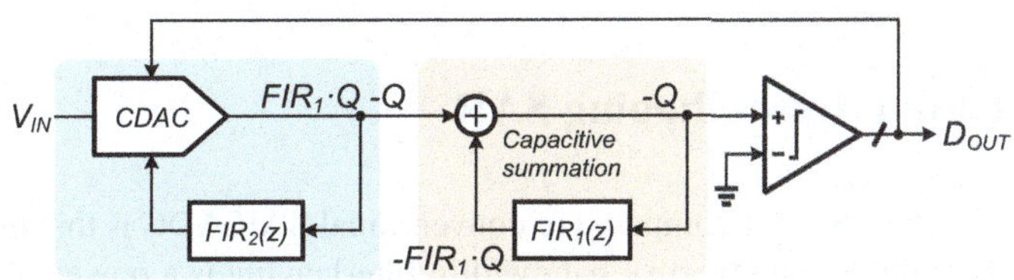

Fig. 12 Simplified depiction of cascaded noise-shaping implementation [5, 6]

6 Hybrid CT Converters with NS SAR Quantizers

Continuous-time sigma-delta modulators (CT SDMs) use continuous-time filters and move sampling to the quantizer input. The continuous-time filtering and the sampling location relax the anti-alias filtering requirements. The continuous-time operation also reduces the bandwidth and power consumption of the filter amplifiers. Finally, the resistive input of a continuous-time converter is relatively easy to drive.

Higher conversion resolution or higher bandwidth generally requires higher modulator order, higher sampling speed, or higher-resolution quantization. Increased modulator order increases filter complexity and power consumption, ultimately leading to stability concerns. Sampling speed is limited by technology, power consumption, and loop delay concerns. While it is advantageous to use the highest possible quantizer, the power consumption and delay of the quantizer ADC place practical limits on the quantizer resolution.

The NS SAR has emerged as a very effective quantizer for continuous-time sigma-delta modulators (Fig. 13). The NS SAR delivers a higher effective resolution than a SAR ADC. Furthermore, the noise shaping in the NS SAR increases the overall converter order without increasing the complexity of the CT filter. As a

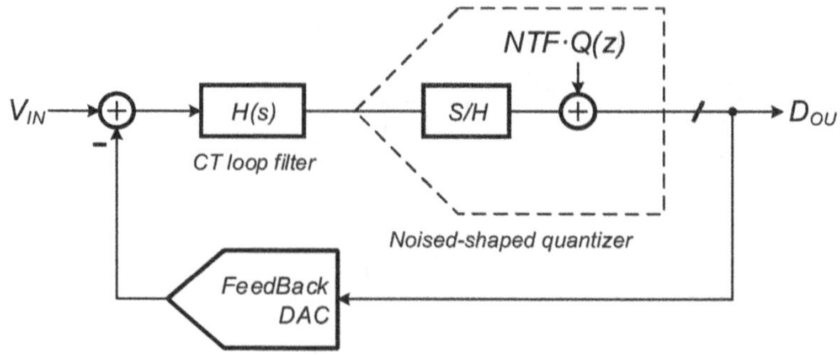

Fig. 13 CT converter with NS SAR quantizer [7, 8]

result, hybrid CT NS SAR converters deliver impressive performance with a first- or second-order CT filter and a first- or second-order NS SAR quantizer.

7 Interleaved Noise-Shaping SAR

A drawback of the NS SAR compared to conventional SAR ADC is that the need for oversampling reduces the effective bandwidth. Interleaving is a powerful technique for improving the bandwidth of Nyquist SAR ADCs. Furthermore, as shown by [9, 10], interleaving is also effective for noise-shaping SAR ADCs. Interleaving can harness the relative delay between channels to implement higher-order noise shaping.

Interleaving of NS SAR ADCs is not immediately straightforward, as direct interleaving of NS SAR ADCs does not produce the desired noise-shaping waveform. This is because a limitation of feedback within each NS SAR unit is that the delay in feedback is equivalent to multiple sample periods of the overall converter. Inter-channel feedback is a promising alternative as the relative delay between each channel is only one sample. However, as shown in Fig. 14, inter-channel feedback presents a causality challenge as the SAR conversion must begin several cycles before the feedback is ready.

Mid-way feedback elegantly avoids the causality problem (Fig. 15). Each sub-ADC SAR conversion process can be viewed as comprising multiple phases. The error feedback signal can be injected before any one of these phases [9, 10]. use a summing preamplifier within each subunit to combine and weigh different feedback signals. The choice of weights and the introduction of redundancy in the SAR algorithm enhance stability. Furthermore, if the number of interleaving channels is less than or equal to the over sample ratio, the channel mismatch artifacts fall out of band. Recent work applies an interleaved NS SAR in a CT SDM [11].

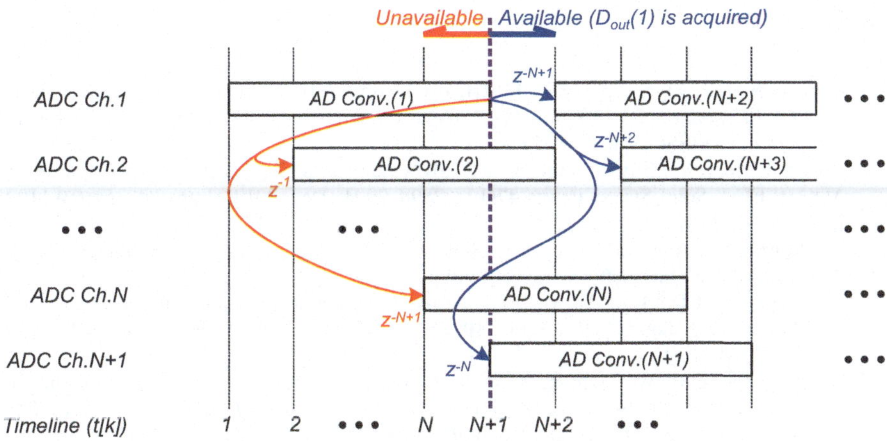

Fig. 14 Causality challenge in interleaved NS SAR [9, 10]

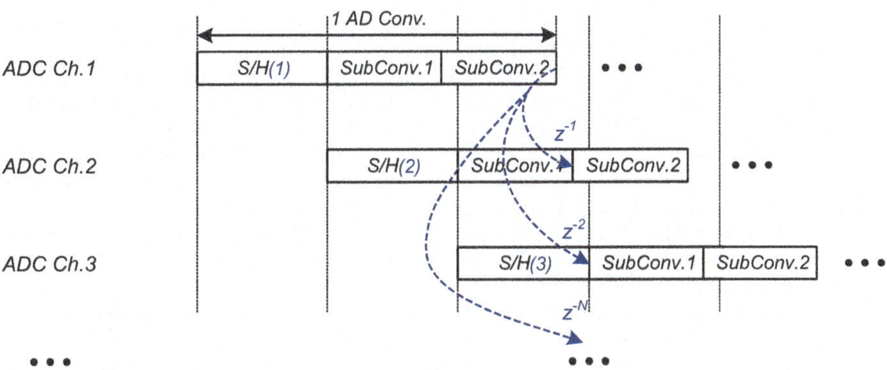

Fig. 15 Mid-way feedback in interleaved NS SAR [9, 10]

8 Conclusions

The NS SAR ADC has become a dominant ADC architecture within a decade of its invention. NS SAR converters provide unequaled efficiency at moderate resolutions and are very compact. Advanced techniques extend conversion resolution and bandwidth so that SAR continues to displace more conventional architectures like DT SDM. NS SAR is also a valuable building block for hybrid ADCs.

References

1. J. Fredenburg, M. Flynn, A 90MS/s 11MHz bandwidth 62dB SNDR noise-shaping SAR ADC, in *2012 IEEE International Solid-State Circuits Conference*, vol. 2012, pp. 468–470. https://doi.org/10.1109/ISSCC.2012.6177094
2. J.A. Fredenburg, M.P. Flynn, A 90-MS/s 11-MHz-bandwidth 62-dB SNDR noise-shaping SAR ADC. IEEE J. Solid State Circuits **47**(12), 2898–2904 (Dec. 2012). https://doi.org/10.1109/JSSC.2012.2217874
3. B. Murmann, ADC Performance Survey 1997–2022, [Online]. Available: http://web.stanford.edu/~murmann/adcsurvey.html
4. J.L. McCreary, P.R. Gray, All-MOS charge redistribution analog-to-digital conversion techniques. I. IEEE J. Solid State Circuits **10**(6), 371–379 (Dec. 1975). https://doi.org/10.1109/JSSC.1975.1050629
5. L. Jie, B. Zheng, H.-W. Chen, R. Wang, M.P. Flynn, 9.4 a 4th-order cascaded-noise-shaping SAR ADC with 88dB SNDR over 100kHz bandwidth, in *2020 IEEE International Solid-State Circuits Conference*, (2020). https://doi.org/10.1109/ISSCC19947.2020.9062905
6. L. Jie, B. Zheng, H.-W. Chen, M.P. Flynn, A cascaded noise-shaping SAR architecture for robust order extension. IEEE J. Solid State Circuits **55**(12), 3236–3247 (Sep. 2020). https://doi.org/10.1109/JSSC.2020.3019487
7. J. Liu, S. Li, W. Guo, N. Sun, A 0.029mm^2 17-FJ/Conv.-step CT $\Delta\Sigma$ ADC with 2nd-order noise-shaping SAR Quantizer, in *2018 IEEE Symposium on VLSI Circuits*, (2018). https://doi.org/10.1109/VLSIC.2018.8502424
8. J. Liu, S. Li, W. Guo, G. Wen, N. Sun, A 0.029-mm^2 17-fJ/conversion-step third-order CT $\Delta\Sigma$ ADC with a single OTA and second-order noise-shaping SAR Quantizer. IEEE J. Solid State Circuits **54**(2), 428–440 (Nov. 2018). https://doi.org/10.1109/JSSC.2018.2879955
9. L. Jie, B. Zheng, M.P. Flynn, 20.3 a 50MHz-bandwidth 70.4dB-SNDR calibration-free time-interleaved 4th-order noise-shaping SAR ADC, in *2019 IEEE International Solid-State Circuits Conference*, (2019). https://doi.org/10.1109/ISSCC.2019.8662313
10. L. Jie, B. Zheng, M.P. Flynn, A calibration-free time-interleaved fourth-order noise-shaping SAR ADC. IEEE J. Solid State Circuits **54**(12), 3386–3395 (Sep. 2019). https://doi.org/10.1109/JSSC.2019.2938626
11. S. Lee, T. Kang, S. Song, K. Kwon, M. Flynn, An 81.6dB SNDR 15.625MHz BW 3rd order CT SDM with a true TI NS Quantizer, in *2022 IEEE Symposium on VLSI Technology and Circuits (VLSI Technology and Circuits)*, (2022), pp. 54–55. https://doi.org/10.1109/VLSITechnologyandCir46769.2022.9830207

Noise-Shaped SAR ADCs: Current Trends and Challenges

Eric Thompson

Abstract Noise-shaped SARs have advanced considerably over the past decade since first published in 2012 (Fredenburg J, Flynn M, A 90MS/S 11MHz bandwidth 62dB SNDR noise-shaping SAR ADC. In: ISSCC, 2012). The trends include dynamic amplifiers, kT/C noise suppression, and low OSRs. These trends help improve power, noise, and linearity. However, a new set of error sources come to the fore. The more prominent error sources include mismatch and flicker noise among others. This chapter will cover the driving reasons behind the new trends and techniques for tackling these more prominent error sources such as mismatch shaping, and calibration will be examined.

1 Introduction

Noise-shaped SARs combine the benefits of delta-sigma and SAR topologies. Delta-sigma topologies employ oversampling and quantization noise shaping to achieve higher accuracy in a band of interest. The SAR topology being a standard Nyquist rate topology can place more stringent constraints on anti-alias filters and driving circuitry. This results in a complex AAF or application in a more noise-constrained, low-interference environment. The advances achieved in SAR speeds, in part due to process scaling, have enabled the addition of more oversampling and the benefits associated with this.

The first noise-shaped SAR ADC was published in 2012 [1], adding a loop filter and summation block to a standard SAR architecture (Fig. 1). The NS SAR takes the following steps: (1) sum the input sample and the loop filter output, (2) quantize the summed value, and (3) update the loop filter.

Since the first publication, there have been advancements in several areas, including loop filter topologies, amplifier design, kT/C suppression, and ease of drive

E. Thompson (✉)
Analog Devices International, Raheen Industrial Estate, Limerick, Ireland
e-mail: Eric.Thompson@analog.com

P. Harpe et al. (eds.), *Biomedical Electronics, Noise Shaping ADCs, and Frequency References*, https://doi.org/10.1007/978-3-031-28912-5_8

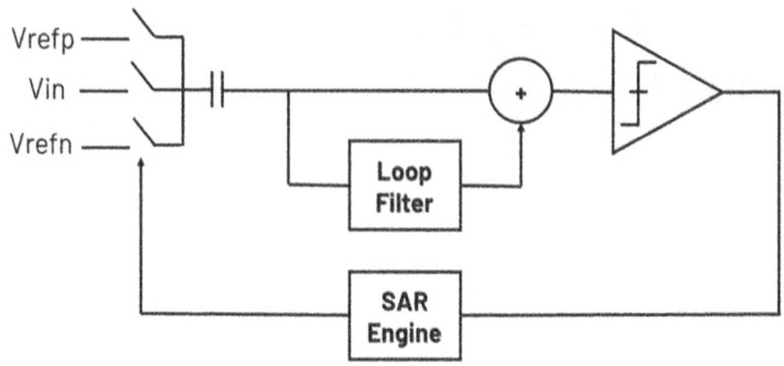

Fig. 1 Generic noise-shaped SAR

techniques. These have improved the performance of the NS SAR but introduce some limitations and trade-offs. Some of the aforementioned techniques push the trade-offs in the same direction making the overall accuracy more dependent on mismatch and flicker noise. This chapter will highlight these trade-offs and solutions to achieve high accuracy.

This chapter is organized as follows: Sect. 2 will give an overview of summation techniques, and Sect. 3 gives an overview of loop filters. These give the necessary background to cover the more advanced techniques. Section 4 will outline some trends, some of which will be covered in more detail in Sects. 5, 6, 7, and 8, finishing up with conclusions in Sect. 9.

2 Summation

There are two main summation approaches which will be covered, charge summation (parallel and series) and current summation.

2.1 Charge-Based Summation

The parallel charge summation is shown below in Fig. 2 [2]. The input is sampled during $\o_S$ onto CDAC. The output of the loop filter and the input sample charge are summed by charge sharing during $\o_{LF}$. The conversion is then performed, and the loop filter updated during $\o_{res}$.

The residue voltage at the end of the conversion is amplified and stored on capacitor Cfb. The charge on Cfb is shared with the input capacitor CDAC. The amplifier must overcome the charge sharing loss, so the amplifier gain is set by equation for A in Fig. 2. This value of amplifier gain, A, is the one required to achieve first-order shaping of the quantization noise.

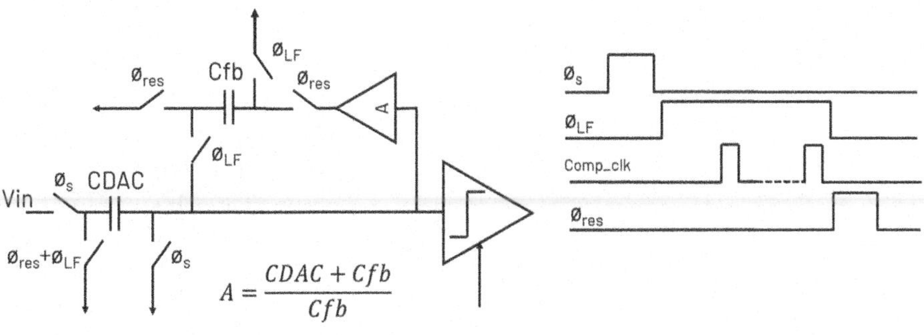

Fig. 2 Parallel charge summation

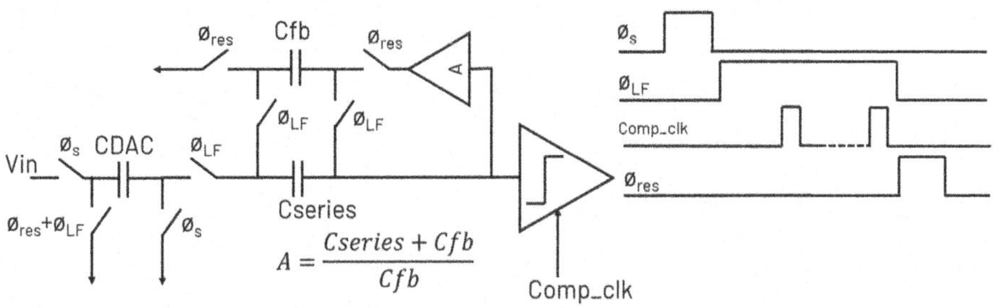

Fig. 3 Series charge summation

The series charge summation works in a very similar manner except the summation is done by placing a capacitor in series with the input capacitor. This is illustrated in Fig. 3 [3]. This approach decouples the amplifier gain requirement from the input capacitor size.

CDAC holds the signal voltage, and the capacitors Cseries and Cfb hold the filter voltage. Placing these in series sums the voltages. If Cseries were reset during ϕ_S, the filter voltage would only depend on the input to the amplifier and act much like an FIR filter. Without a reset, it has memory of its previous values and behaves much more like an IIR filter.

2.2 *Current-Based Summation*

This summation uses differential pairs to sum current, sometimes drawn as a multi-input comparator. It is shown in Fig. 4 [1]. As the input to the second summation node is driven by the loop filter output amplifier, it has the advantage of not suffering from charging losses, at the expense of requiring more active components.

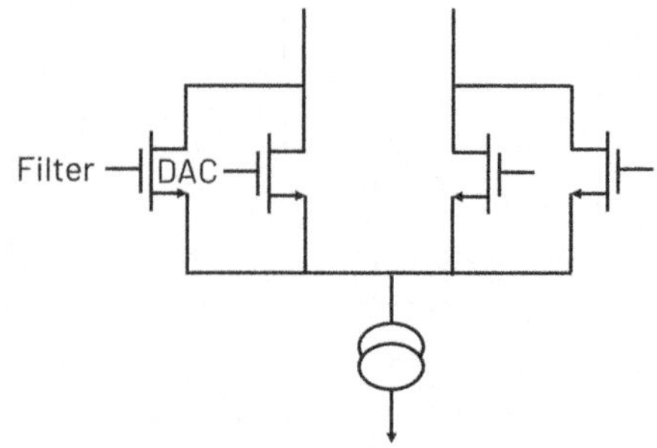

Fig. 4 Current summation

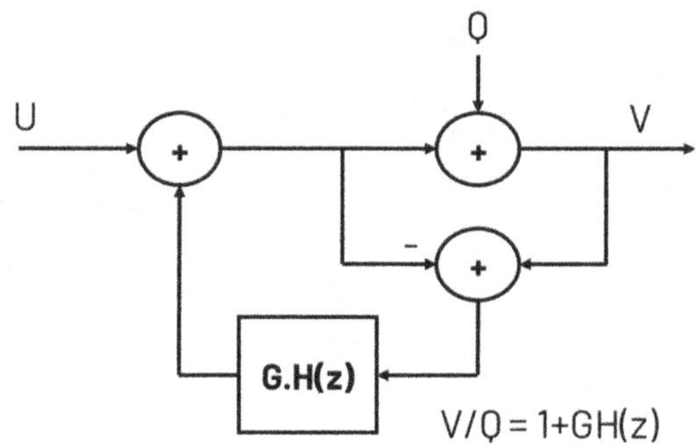

Fig. 5 Error feedback loop topology

3 Loop Filter Topologies

There are two main loop filter topologies used, error feedback (EF) and cascade of integrators feedforward (CIFF).

3.1 *Error Feedback (EF)*

The EF topology is shown below in Fig. 5 [4].

If G.H(z) equals $-Z^{-1}$, then the noise transfer function is a first-order differentiator. An example of this can be seen in Fig. 2. The accuracy of G sets the accuracy of the in-band noise suppression. If open-loop amplifiers and passive summation are used, the drift of these circuits will affect the converter resolution and may require calibration.

3.2 *Cascade of Integrators Feedforward (CIFF)*

The CIFF topology is shown in Fig. 6 [1, 9]. This topology can be used in conjunction with the current summation method in Fig. 4. A loop filter transfer

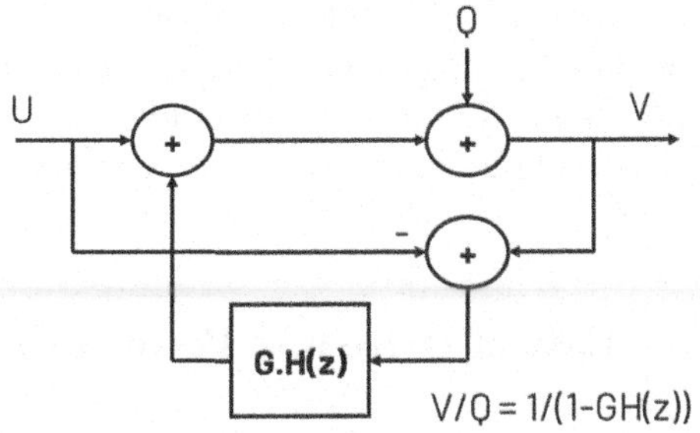

Fig. 6 Cascade of integrators feedforward loop topology

function of an integrator will result in the first-order shaping of the quantization noise.

The CIFF topology is much less prone to mismatch and variations in the summing accuracy.

It should be noted that within the loop filters, open- or closed-loop amplifiers can be used. Open-loop amplifiers have the benefit of not adding kT/C noise by taking an additional sample, at the expense of wider variation in gain over process voltage and temperature, combined with higher non-linearity.

In a delta-sigma, the loop filter is fed with the difference between the current sample and the previously quantized loop filter output, and as the OSR gets lower, more signal is processed by the amplifier. This effect was eliminated by Silva et al. [5], by ensuring the integrators only carry quantization noise and no signal content. In the NS SAR as the loop filter is updated at the end of the conversion based on the current sample and ensures the loop filter also only carries quantization noise, and no signal content.

4 Trends in NS SAR

NS SAR have typically used low oversampling ratios (OSR) since the first publication in 2012. The reason for this is that SAR power has strong digital component, $P = CV^2F$, due to the DAC, latch, and element shaping contributions. The slower the ADC runs, the lower the OSR and power, resulting in an improved FoM. In fact, when reviewing Murmann's ADC survey [6], only two papers with OSRs >25 were observed [7, 8] and had a maximum output sample rate of 2KSPS. To combat the lower OSR, higher-order loops are required.

Smaller geometries lead to higher speeds for the same OSR, lower power, and better FoM. This naturally tends to aid high-speed requirements. However, when looking for increased precision, decreasing OSR does not necessarily aid this endeavor. In fact, only one oversampled SAR ADC has achieved >100 dB SNDR [6], albeit at an output data rate of 2KSPS in the ADC survey.

For comparison with other delta-sigma offerings, the zoom ADC architecture has shown SNRs above 100 dB in a 20 KHz audio bandwidth but utilizes an OSR of 128 [9].

More recent trends have been to include (a) kT/C noise suppression, first proposed by Kapusta [12], to reduce the input capacitor size and reduce power and (b) use of dynamic amplifiers to reduce power and increase speed. These will be discussed further in subsequent sections.

5 Ease of Drive and Power Considerations

It is well known that amplifier gain and bandwidth requirements are set by the accuracy required, the time allotted, and the load to be driven. If more time can be allowed, the bandwidth requirements, power, and noise decrease for a given accuracy. Techniques to allocate more time, such as employing multiple DAC and sampling networks, have been demonstrated for SAR ADCs in literature [11] and will be covered in Sect. 5.1. Pre-charge techniques can also be used to reduce the charge required to settle to the next input level and will be covered in Sect. 5.2.

5.1 Multi-sampling Approach

The multi-DAC approach shown in Fig. 7 gives rise to two areas for power savings. The first is that amplifier driving the converter gets longer to settle the load. The second is that a whole clock cycle can be allocated to the SAR conversion and loop filter update. This also reduces the power requirement in the converter. The drawback of this approach is that there are now more elements introducing mismatch in the system and this can degrade the overall linearity of the converter.

This technique from Kapusta can be extended to NS SAR architectures also as shown in Fig. 8.

The element mismatch shaping in each DAC runs at M times slower, where M is the number of DAC or sampling networks interleaved. This degrades the SNR,

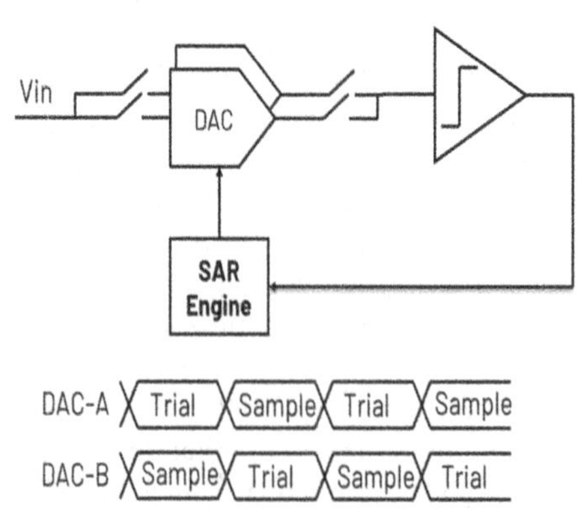

Fig. 7 Multi-sampling approach

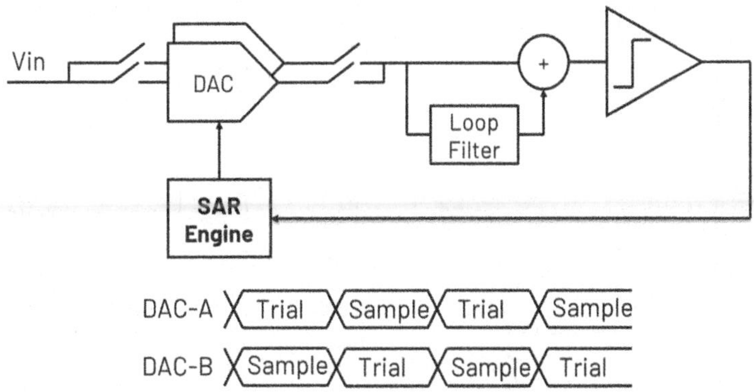

Fig. 8 NS SAR multi-sampling

and having M > 2 does not improve settling constraints on the previous stage and degrades the SNR unnecessarily.

5.2 Pre-charge Techniques

Pre-charge techniques give most benefit in an oversampled system and in fact can be detrimental to settling with fast-moving inputs. When a conversion has completed, it is possible to place a charge or voltage on the input sampling capacitor equivalent to previously quantized value.

Figure 9 illustrates a pre-charge sequence: firstly, the input is sampled (a); secondly, the charge is delivered to the comparator (b); and thirdly, the bit trials are performed (c). At the end of the bit trials, the residue is present at the comparator node. This is the charge or voltage that feeds the loops filter in the NS SAR. At the end of the bit trials, the charge removed by the SAR operation can be replaced by reversing the bit trials (d). The sequence then reverts to input sampling again as required (a).

Multi-sampling combined with pre-charge techniques reduces the requirements on the driving stage and as the driving stage is often the dominant power consumer in the signal chain can be a requirement of many design specifications.

6 kT/C Suppression

Kapusta et al. published techniques that demonstrated kT/C noise suppression [10]. These techniques have more recently been adopted into NS SAR converters [12]. The result is that the input sample kT/C noise can be held on the preamplifier of the comparator and cancelled during the SAR conversion.

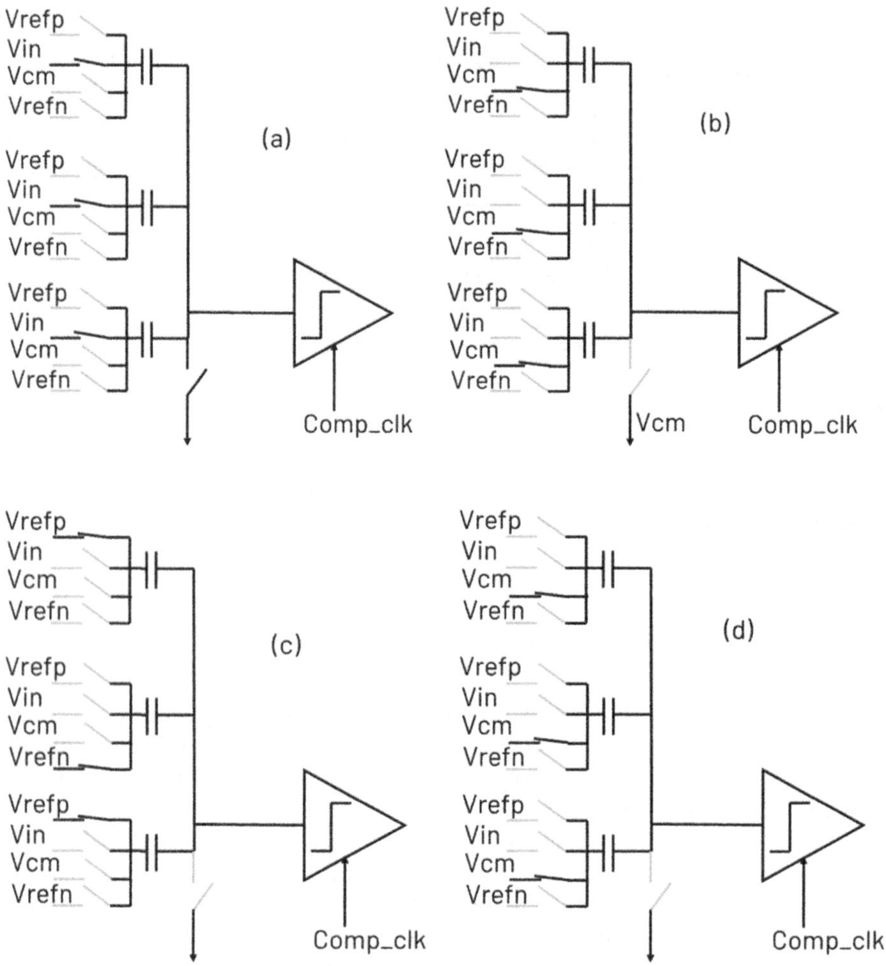

Fig. 9 Pre-charge sequence, (**a**) input sampling, (**b**) charge delivery, (**c**) bit trials, and (**d**) charge return

During ϕ_1, the input is sampled; when ϕ_1 goes low, the kT/C noise is amplified and held on Cl by the gm cell and then ϕ_2 goes low. The gain of the gm stage allows Cl to be reduced below the value of Ca. Care must be taken that the input signal does not vary significantly between the falling edges of ϕ_1 and ϕ_2, as the gain of the gm stage may encounter headroom issues (Fig. 10).

This technique allows for smaller input capacitors to be used and eases the load constraints placed on the prior stage. For example, if the kT/C noise power drops by 4× (corresponding to a 2× drop in noise voltage), the input capacitor can be reduced by 4. This does increase mismatch within the capacitor array and can degrade SNR unless addressed by other techniques. A 4× reduction in capacitor size will translate

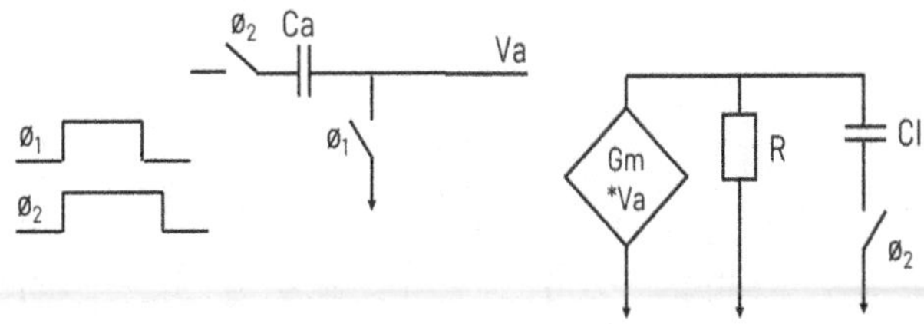

Fig. 10 KT/C noise suppression circuit

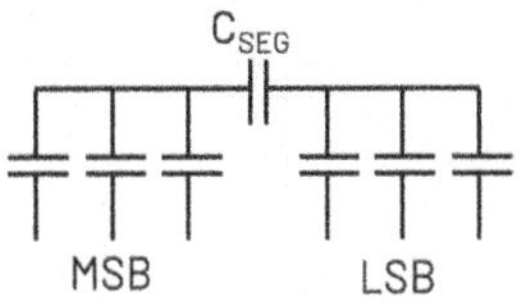

Fig. 11 Segmented CDAC

to a 2× increase in mismatch (based on area and Pelgrom's rule) or 6 dB in the SNR due to mismatch.

7 DAC Element Mismatch

It has been shown that trade-offs to ease the drive or load constraints on the prior stage can degrade the SNR. Using multiple DAC sampling networks reduce the mismatch error shaping rate, and kT/C noise suppression techniques result in smaller capacitors, increasing the magnitude of the mismatch in each element of the DAC.

Binary weighted DACs with a large number of bits lead to smaller capacitors which are more prone to systematic mismatch errors due to parasitic effects, e.g., 5 fF unit cap with 1% mismatch requirement would need to ensure parasitic errors are kept below 0.05 fF. To overcome the issues surrounding smaller capacitors, segmentation of the CDAC is used as illustrated for a 6-bit 3:3 segmentation in Fig. 11.

There are two main techniques used to counteract the effects of mismatch, mismatch error shaping and calibration.

7.1 Mismatch Error Shaping

Delta-sigma converters have used mismatch error shaping schemes since multibit quantizers have been used. These techniques equally find application in NS SARs. As mismatch in the MSBs affects the result more, they are subjected to mismatch error shaping in identical ways to delta-sigma's.

More recently, the LSBs also require shaping, and shaping across a segmentation capacitor needs to account for two items:

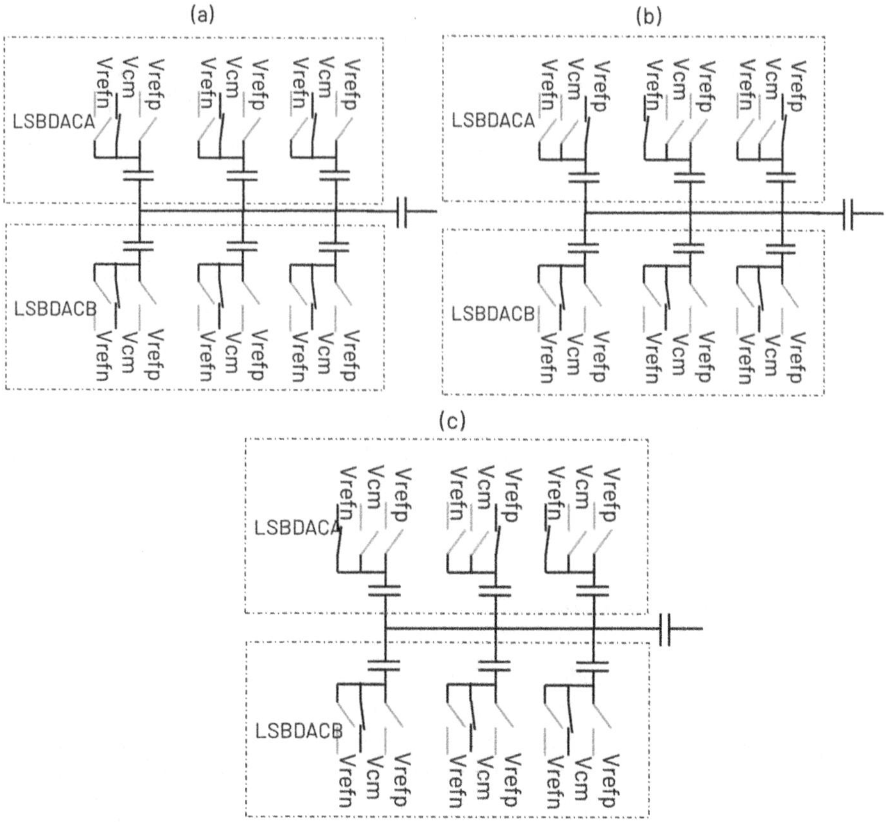

Fig. 12 LSB DAC sequence: (**a**) input sampling, (**b**) bit trials, and (**c**) reverse prior bit trials on LSBDACA and use LSBDACB for bit trials

(a) The error in the coupling capacitor C_{SEG} (Fig. 11): this weights all the LSBs.
(b) The LSB element mismatch.

The technique proposed by Hurrell et al. [13] first-order shapes both errors by high-pass filtering or differentiating the LSB data. Prior techniques used in delta-sigmas [14] place only shaped quantization noise on the LSBs.

The sequence is illustrated in Fig. 12. There are two LSB DACs used. These LSB DACs ping-pong between bit trials and cancelling the previous error from the prior conversion.

LSB DACA is used on the first conversion for bit trials (Fig. 12a, b). On the next bit trial, LSB DACA applies the opposite of the prior result to remove the error applied during the first conversion (Fig. 12c), giving a first-order differentiated error. This is similar to pre-charging the LSBDAC to the opposite of the prior result. During this second conversion, LSB DACB is used for bit trials in order to not lose any ADC range.

The diagram above is for illustration purposes only and assumes that $Vcm = (Vrefp - Vrefn)/2$. The same effect can be achieved in several ways, but this way was chosen for ease of illustration of the concept.

The above scheme enables mismatch error shaping across the segmentation boundary and works in the binary domain. Unsegmented DACs require a 2× growth in the shaper size for each binary bit added. By using the segmentation scheme above, it is possible to limit the growth of the mismatch error shaper to smaller power and area.

7.2 Calibration-Based Techniques

Calibration-based techniques have been used in SAR ADCs for quite some time [15]. Calibration can be analogue or digitally based, but in the case of SARs have tended to be mainly digitally based. It involves determining the capacitor weights for each significant DAC element within the SAR and applying a correction digitally. These techniques make use of the stability of capacitor structures to do foreground calibration and store the capacitor weights in memory.

More recently, mismatch error shaping and calibration have been combined [16, 17]. The mismatch error shaping can be zero or higher order and in conjunction with calibration can achieve sub-1 ppm INL. It is binary based to consume low area and power. An example of this can be seen in Fig. 13.

8 Dynamic Amplifiers

Dynamic amplifiers have been used in NS SAR in recent years [18]. The amplifiers can be as simple as inverters running from a battery capacitor to give gain. The amplifier works from a fixed amount of charge, driven by a non-overlapping clock generator. The inverter shown in Fig. 14 is powered from a capacitor driven from a non-overlapping clock generator. The inverter acts as an amplifier and discharges the battery capacitor.

As the noise depends on the device gm, either large device sizes or battery capacitors are required. If these get too large, a more standard amplifier may be used and the bias current turned on and off [19], as shown in Fig. 15. Constantly on devices that generate bias can cause flicker noise to leak through (e.g., MN4 (Fig. 15)).

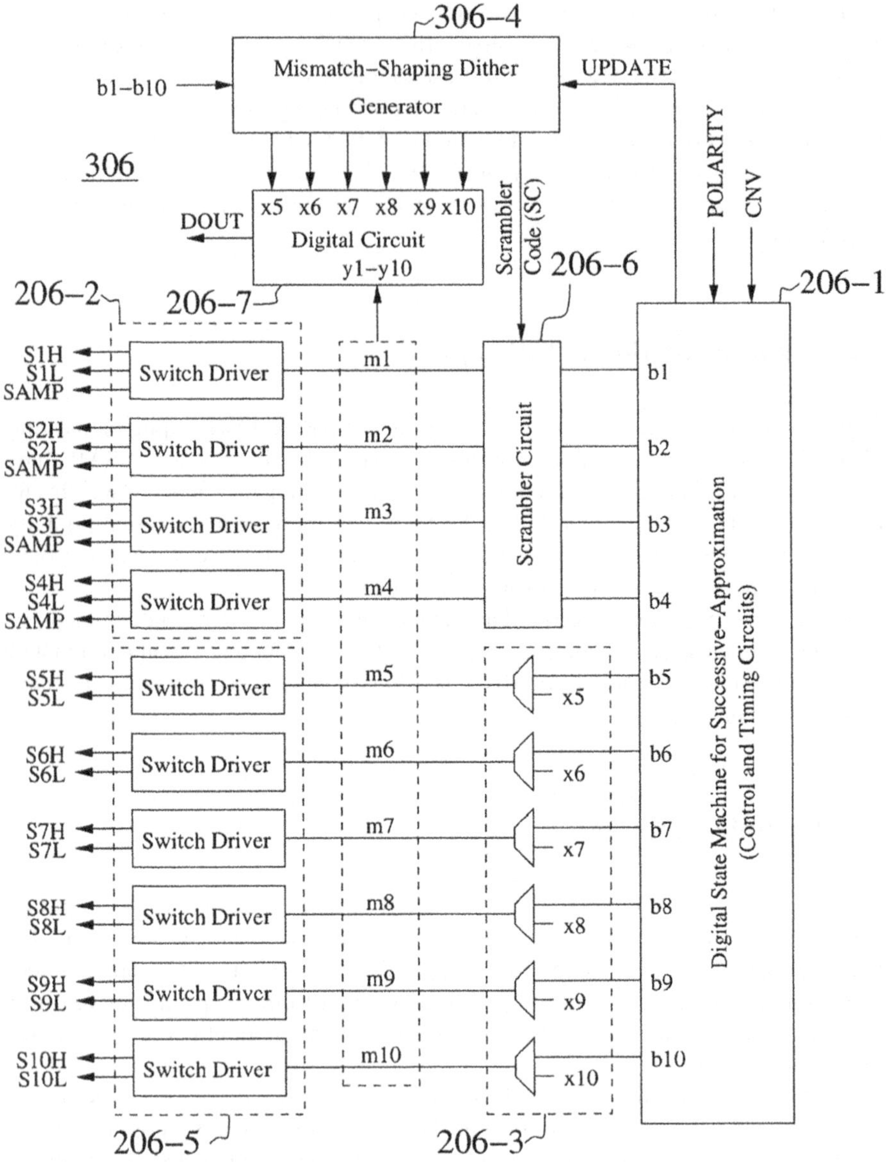

Fig. 13 Mismatch error shaping topology – Steensgaard

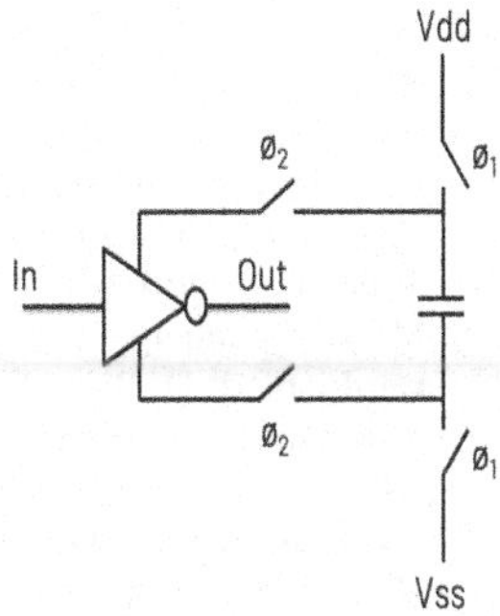

Fig. 14 Dynamic amplifier

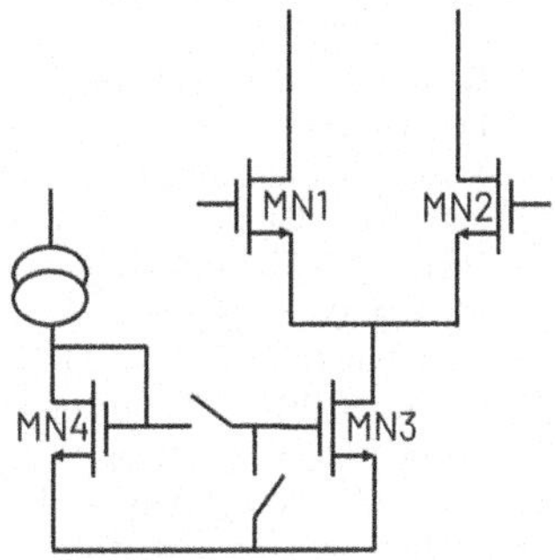

Fig. 15 Dynamic differential pair

9 Conclusions

This chapter has reviewed the key blocks that enable an NS SAR, i.e., a loop filter and summation blocks, highlighting some of the key trends in the area such as multi-sampling and kT/C noise suppression, combinations of which cause potential problems with trade-offs all pushing the unit capacitors to reduce and mismatch to increase. Some techniques to help address these new issues have been highlighted. Finally, dynamic amplifier has been briefly reviewed, and the potential of flicker noise to be of concern has been raised.

References

1. J. Fredenburg, M. Flynn, A 90MS/S 11MHz bandwidth 62dB SNDR noise-shaping SAR ADC, in *ISSCC*, (2012)
2. S. Li, B. Qiao, M. Gandara, N. Sun, A 13-ENOB 2nd-order noise-shaping SAR ADC realizing optimized NTF zeros using an error-feedback structure, in *ISSCC*, (2018)
3. Y.-Z. Lin, C.-Y. Lin, S.-C. Tsou, C.-H. Tsai, L. Chao-Hsin, A 40MHz-BW 320MS/s passive noise-shaping SAR ADC with passive signal-residue summation in 14nm FinFET, in *ISSCC*, (2019)
4. S. Li, B. Qiao, M. Gandara, D.Z. Pan, N. Sun, A 13-ENOB second-order SAR ADC realizing optimized NTF zeros using the error-feedback structure. IEEE JSSC **53**(12), 3484–3496 (2018)

5. J. Silva, U. Moon, J. Steensgaard, G.C. Ternes, Wideband low-distortion delta-sigma ADC topology. Electron. Lett. **37**(12), 737–738 (2001)
6. B. Murmann, ADC Performance Survey 1997–2022 [Online]. Available http://web.stanford.edu/~murmann/adcsurvey.html. Rev: 20220719
7. Y.-S. Shu, L.-T. Kuo, T.-Y. Lo, An oversampling SAR ADC with DAC mismatch error shaping achieving 105dB SFDR and 101dB SNDR over 1kHz BW in 55nm CMOS, in *ISSCC*, (2016)
8. M. Konijnenburg, R. van Wegberg, S. Song, H. Ha, W. Sijbers, X. Jiawei, S. Stanzione, C. van Liempd, D. Biswas, A. Breeschoten, P. Vis, C. Van Hoof, N. Van Helleputte, A 769μW battery-powered single-chip SoC with BLE for multi-modal vital sign health patches, in *ISSCC*, (2019)
9. E. Eland, S. Karmakar, B. Gönen, R. van Veldhoven, K. Makinwa, A 440μW, 109.8dB DR, 106.5dB SNDR discrete-time zoom ADC with a 20kHz BW, in *VLSI*, (2020)
10. R. Kapusta, H. Zhu, C. Lyden, Sampling circuits that break the KT/C thermal noise limits, in *JSSCC*, (2014)
11. R. Kapusta, J. Shen, S. Decker, H. Li, E. Ibaragi, A 14b 80MS/s SAR ADC with 73.6dB SNDR in 65nm CMOS, in *ISSCC*, (2013)
12. J. Liu, X. Tang, W. Zhao, L. Shen, N. Sun, A 13b 0.005 mm^2 40MS/s SAR ADC with kT/C noise cancellation, in *ISSCC*, (2020)
13. US10333543B1 – Analog-to-digital converter with noise-shaped dither
14. R. Adams, K. Nguyen, A 113-dB SNR oversampling DAC with segmented noise-shaped scrambling. IEEE JSSC **33**(12), 1871–1878 (1998)
15. US7129879B1 – Method of and apparatus for characterizing an analog to digital converter
16. US8810443B2 – Analog-to-digital converter system and method
17. Analog Devices AD4630-24 Datasheet
18. X. Tang, X. Yang, W. Zhao, C.-K. Hsu, J. Liu, L. Shen, A. Mukherjee, W. Shi, D.Z. Pan, N. Sun, A 13.5b-ENOB second-order noise-shaping SAR with PVT-robust closed-loop dynamic amplifier, in *ISSCC*, (2020)
19. S. Ma, L. Liu, T. Fang, J. Liu, W. Nanjian, A discrete-time audio delta sigma modulator using dynamic amplifier with speed enhancement and flicker noise reduction techniques. IEEE JSSC **55**(2), 333–343 (2020)

Noise-Shaping SAR ADCs: From Discrete Time to Continuous Time

Hanyue Li, Yuting Shen, Eugenio Cantatore, and Pieter Harpe

Abstract Noise-shaping (NS) SAR ADCs become popular recently, thanks to their low-power and high-resolution features. This article first summarizes and benchmarks different discrete-time (DT) NS-SAR implementations in literature. An open-loop duty-cycled residue amplifier is selected as a power-efficient solution to realize high residue gain. Then, a digital-predicted mismatch error shaping technique is introduced to improve the DAC linearity. The proposed DT NS-SAR ADC achieves 80 dB SNDR and 98 dB SFDR in a 31.25 kHz bandwidth while consuming 7.3 μW. Next, the NS-SAR architecture is extended from DT operation to continuous-time (CT) operation. The ADC sampling switch is removed, and the loop filter is duty cycled to realize the CT NS-SAR operation. Compared to DT designs, the CT NS-SAR ADC is easy to drive and has an inherent anti-aliasing function. As a proof of concept, the proposed CT NS-SAR ADC achieves 77 dB SNDR and 86 dB SFDR in a 62.5 kHz bandwidth with a power consumption of 13.5 μW.

1 Introduction

Noise-shaping successive approximation register (NS-SAR) analog-to-digital converters (ADCs) have emerged in recent years as a promising architecture to realize high-resolution and low-power data converters [1]. It combines the merits from a SAR ADC and a $\Delta\Sigma$ ADC and thus reaches an optimization between resolution and power consumption. Its application ranges from sensor readout [2] to wireless communication [3], and its research remains active in both academia and industry.

This article focuses on two types of NS-SAR ADCs: discrete-time (DT) NS-SAR ADCs and continuous-time (CT) NS-SAR ADCs. For DT NS-SAR ADCs, different loop filter design choices will be benchmarked in terms of their power

H. Li (✉) · Y. Shen · E. Cantatore · P. Harpe
Eindhoven University of Technology, Eindhoven, The Netherlands
e-mail: h.li2@tue.nl

P. Harpe et al. (eds.), *Biomedical Electronics, Noise Shaping ADCs, and Frequency References*, https://doi.org/10.1007/978-3-031-28912-5_9

efficiency. The mismatch error shaping (MES) technique will be discussed as an effective solution to enhance the linearity. For CT NS-SAR ADCs, its architecture will be explained in detail, and its design considerations will be elaborated. Design examples will be presented for each type of ADC, and conclusions will be drawn at last.

2 Discrete-Time Noise-Shaping SAR ADC

The diagram of a DT NS-SAR ADC [4] is shown in Fig. 1. Its basic principle is as follows: After the normal SAR conversion is finished, the residue voltage V_{RES} is directly obtained at the output of its digital-to-analog converter (DAC). This residue voltage is then integrated by the loop filter $L(z)$ and added to the next input signal. In this way, the quantization noise and comparator noise can be high-pass shaped by using a low-pass $L(z)$ function.

The NS-SAR ADC has the same noise-shaping principle as the $\Delta\Sigma$ ADC, but the hardware implementation is different. First, the NS-SAR ADC contains only one DAC in the system, unlike a SAR-assisted $\Delta\Sigma$ ADC [5] where one DAC is needed in the SAR quantizer and another DAC is needed in the noise-shaping feedback loop. Second, there exists one feedforward path from the ADC input to the comparator input in the signal diagram. This feedforward path does not alter the noise transfer function (NTF), but makes sure that the integrator only sees the noise signal, and its output swing and linearity requirements are relaxed. These two features provide more possibilities in the loop filter choice, as will be discussed later.

This section is based on [6], and it will discuss the following topics in the DT NS-SAR ADC: choice of loop filter structure and circuit topology and mitigation of the DAC mismatch error.

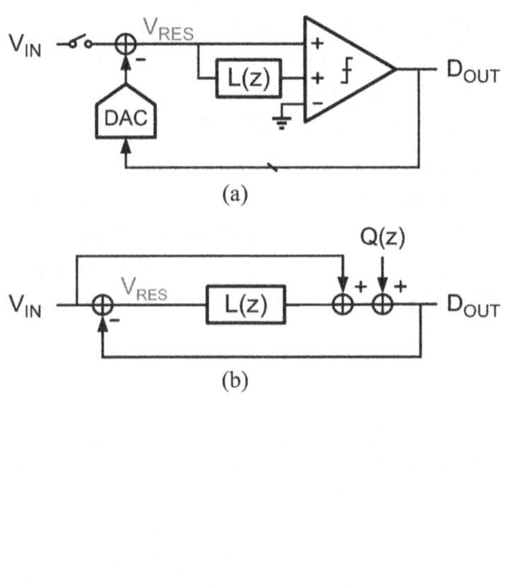

Fig. 1 (**a**) Block diagram and (**b**) signal diagram of a DT NS-SAR ADC

2.1 Loop Filter

Because the residue information is directly available on the DAC output after the SAR conversion, an NS-SAR ADC does not require an additional feedback DAC to subtract the ADC output from the input. Therefore, the loop filter does not necessarily have to be a switched-capacitor integrator as in a conventional DT ΣΔ ADC. Figure 2 shows the three commonly used loop filter structures:

1. A passive integrator followed by a gain stage [7], referred to as "passive loop filter (PLF)."
2. An active open-loop amplifier followed by a passive integrator [8], referred to as "open-loop amplifier (OLA)."

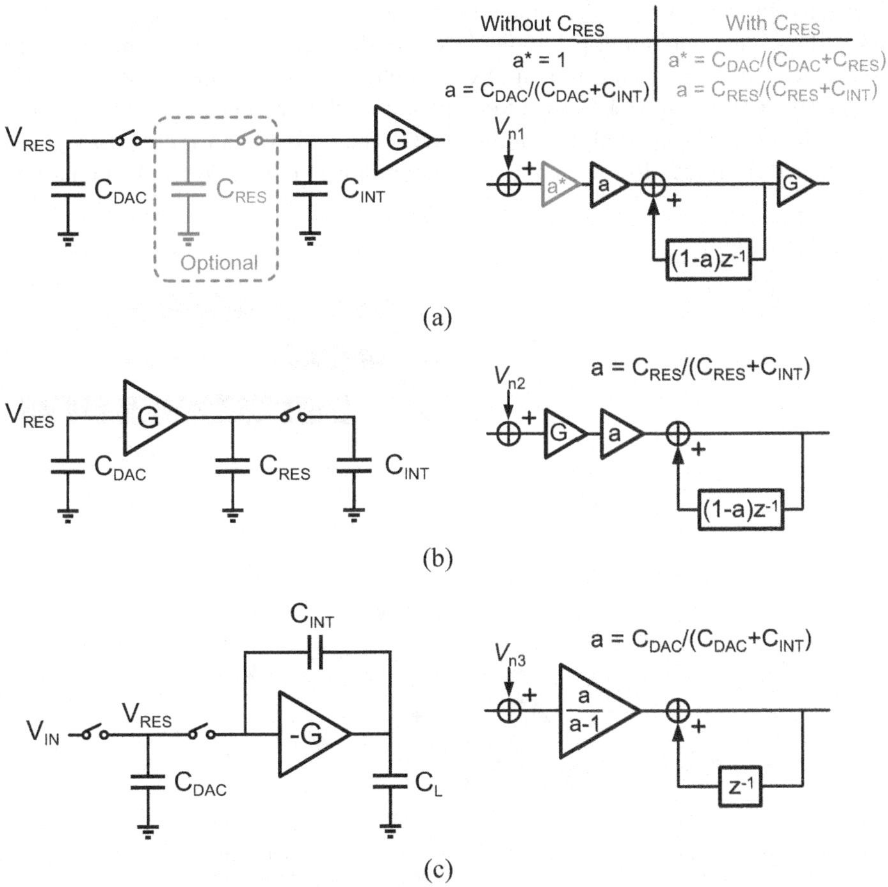

Fig. 2 Different loop filter structures in NS-SAR ADCs: (**a**) PLF, (**b**) OLA, and (**c**) CLI

3. A closed-loop switched-capacitor integrator [9], referred to as "closed-loop integrator (CLI)."

Because the majority of the noise in a NS-SAR ADC comes from the sampling kT/C noise and the loop filter noise, the power efficiency of the DAC and loop filter can be used as an indicator to reflect the power efficiency of the whole converter. We can use the figure-of-merit (FoM) definition from [10] to benchmark the power efficiency of different loop filter structures. This FoM is defined as:

$$\text{FoM} = (\text{Energy consumption}) \cdot (\text{Noise power}) \tag{1}$$

where "energy consumption" means the total energy consumed by the DAC (E_{DAC}) and the loop filter (E_{LF}) and "noise power" includes the noise from DAC sampling ($\overline{V_{n,\mathrm{DAC}}^2}$) and loop filter ($\overline{V_{n,\mathrm{LF}}^2}$) as well. A smaller FoM value indicates that lower energy is needed for the same noise target. The FoM expressions for the three loop filters can be found in [6]. Figure 3 shows the theoretical FoM range for each loop filter structure when different DAC switching schemes and circuit topologies are considered. For the CLI-based loop filter, both floating inverter amplifier (FIA) [11] and conventional amplifiers are included. The FoM value for closed-loop FIA is obtained from [12] following the same definition as in (1). Figure 4 summarizes

Fig. 3 Theoretical FoM of different loop filters

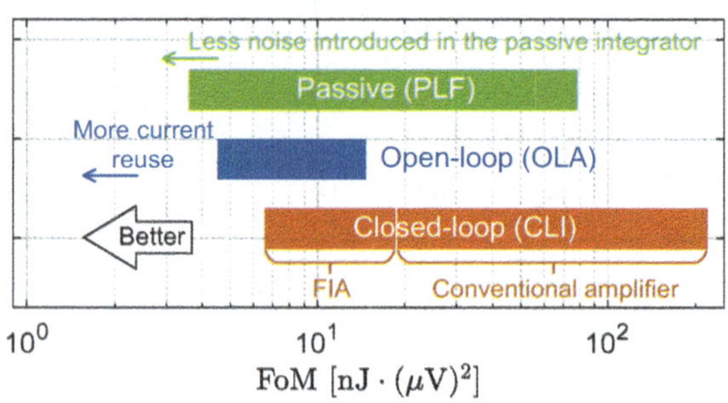

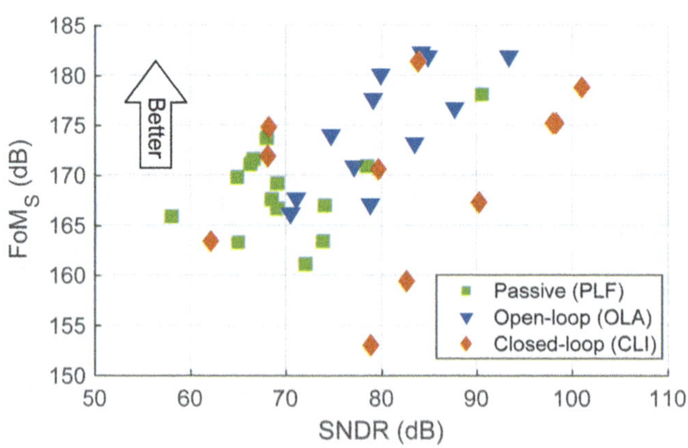

Fig. 4 NS-SAR ADC survey

the performance of the prior NS-SAR ADC designs categorized by their loop filter structures from [1] and three recent works [13–15]. Note that in Fig. 3, a lower FoM means a higher power efficiency, while in Fig. 4 a higher Schreier FoM (FoM_S) is better. By comparing Figs. 3 and 4, we can have the following observations:

1. The PLF has the best theoretical FoM defined in (1). The FoM improvement in PLF mainly comes from smart charge-sharing schemes among capacitors to reduce the additional sampling noise [16]. However, this power-efficient PLF in [16] has only been demonstrated in a first-order NS-SAR ADC and may not be easy to extend to higher order. Therefore, its noise-shaping ability can be limited when compared to OLA- and CLI-based designs.
2. The OLA-based designs show excellent power efficiency, because the noise from the passive integrator is attenuated by the amplifier gain and the amplifier usually consumes only dynamic power. The key to achieve a better FoM is to use a low-power amplifier topology. Inverter-based amplifiers [6, 10] and multiphase settling technique [17] can be considered for the purpose of saving power. The cascade of passive integrators [6] or active stages [17] can be used to realize high-order noise-shaping with minimal hardware overhead.
3. The CLI-based designs exhibit a wide range of FoM values, which is caused by the different choices on the amplifier topology. A conventional amplifier with a static bias current could be power hungry, while the emerging FIA-based closed-loop integrator [11] has demonstrated a significant power reduction benefit.

Overall, the calculated FoM from Fig. 3 matches well with the performance of actual designs listed in Fig. 4. Based on the above benchmark results, this work chooses the OLA-based loop filter for its high power efficiency and its ability to achieve higher-order noise-shaping.

Figure 5 shows the proposed DT NS-SAR ADC. Its loop filter consists of one active amplifier and two cascaded passive integrators. The whole converter operates as follows: After the SAR quantization, the residue voltage V_{RES} is amplified during Φ_{AMP}, and the result is stored on the amplifier load capacitors C_{RES}. Then, the first integration takes place between the two sets of C_{RES} and C_{INT1} (see Fig. 5c). After that, the second integration continues between C_{RES} and C_{INT2} (see Fig. 5d). Therefore, a second-order noise-shaping is achieved. With an amplifier gain of 18× and a capacitor ratio of 3 ($C_{INT1}/C_{RES} = C_{INT2}/C_{RES} = 3$), the in-band noise suppression reaches 30 dB at an oversampling ratio (OSR) of 16.

To achieve the required 18× gain, a duty-cycled amplifier is used in this design, as shown in Fig. 6. This amplifier is enabled by the control clock Φ_{AMP}. By designing the power-on time shorter than the amplifier's time constant, the power efficiency of this amplifier is comparable to an FIA [6]. In this work, the dc gain of the amplifier is 40×, and the active time is designed as 0.58× of the amplifier's time constant to achieve the desired 18× gain. The bias current generation circuit is also duty cycled. The bias voltage V_B is stored on a large capacitor C_B (2 pF). The voltage V_B is refreshed in every 16 sampling clock cycles to avoid leakage drift. Even though a lower current in the bias circuit can be used together with a high-ratio current mirror to provide the same current for the main amplifier, this

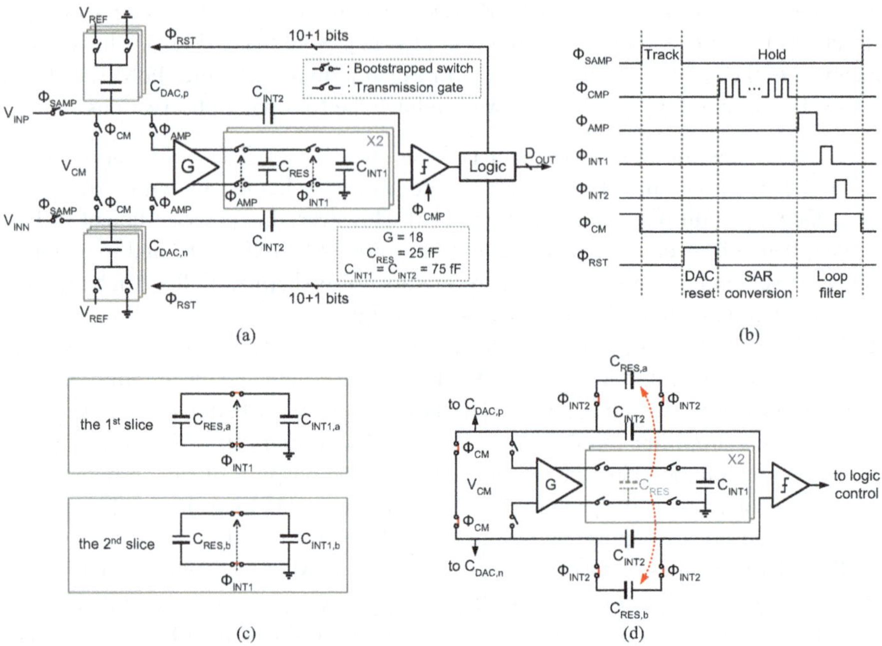

Fig. 5 (**a**) Proposed DT NS-SAR ADC. (**b**) Timing diagram. (**c**) The first integration phase. (**d**) The second integration phase

duty-cycled operation ensures that the amplifier and its bias generation circuit both consume only dynamic power and the power consumption of the whole converter can thus scale with its sampling frequency.

2.2 *Mismatch Error Shaping*

The mismatch errors in the DAC array can limit the linearity of the NS-SAR ADC. Mismatch error shaping (MES) has been proposed in [2] to shape the DAC mismatch errors out of band. Its principle is illustrated in Fig. 7. After the SAR conversion of the previous input signal is done, only the most significant bit (MSB) capacitor is reset, while the control signals for the other bits are held. Before the SAR conversion for the current input starts, all the least significant bits (LSB) are reset (Φ_{RST} in Fig. 5b), which adds the mismatch error E(n-1) contained in the previous LSB voltage ($z^{-1}V_{LSBs}$) to the current input. This additional LSB signal is then subtracted in the digital domain, and a first-order mismatch error shaping can be realized. However, the extra LSB voltage added at the input may cause the ADC to saturate. To avoid this problem, a digital prediction scheme has been proposed in [18]. By using the current ADC output to predict the next sample, we can know

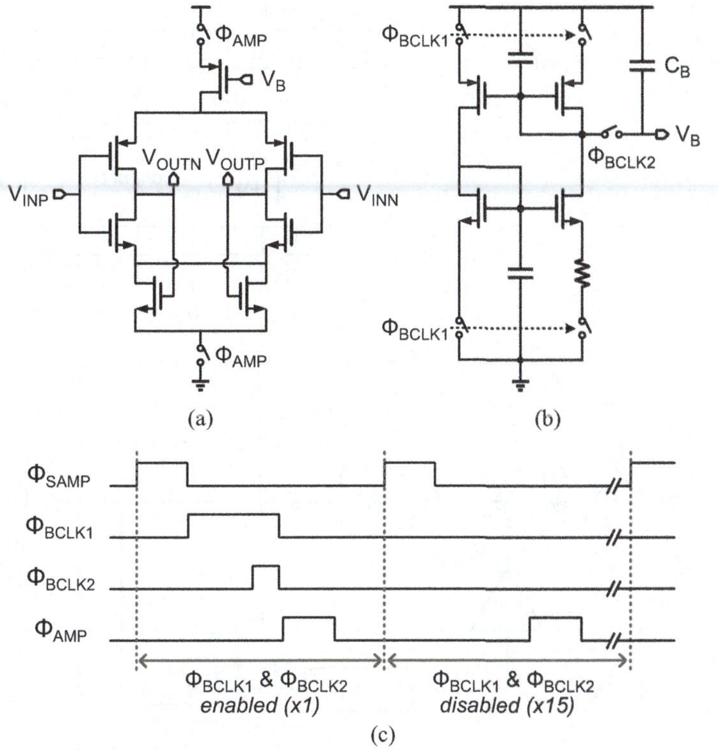

Fig. 6 (**a**) Duty-cycled amplifier. (**b**) Duty-cycled bias circuit. (**c**) Timing diagram for the amplifier and bias circuit

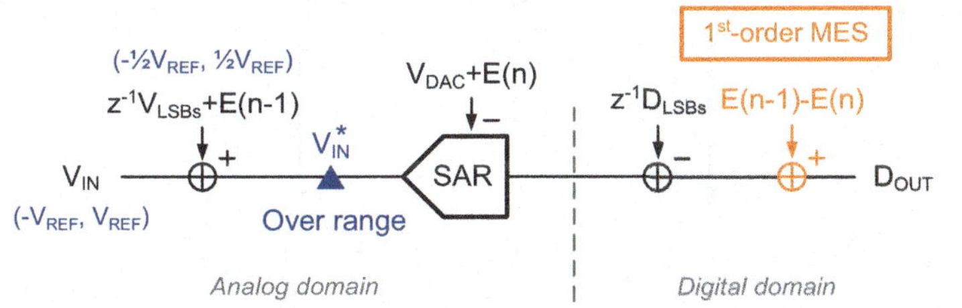

Fig. 7 Principle of MES

if overrange will occur and then toggle the MSB capacitor accordingly to prevent overrange. In this way, the ADC full input range can be kept.

The digital prediction scheme in [18] uses a tri-level decision criterion, which corresponds for compensating positively, compensating negatively, or no compensation, dependent on the predicted signal level. However, this tri-level prediction cannot deal with the mismatch between the two MSB capacitors (C_L and C_R in Fig. 8) when a split switching scheme is used. Therefore, in this work, we adopt a

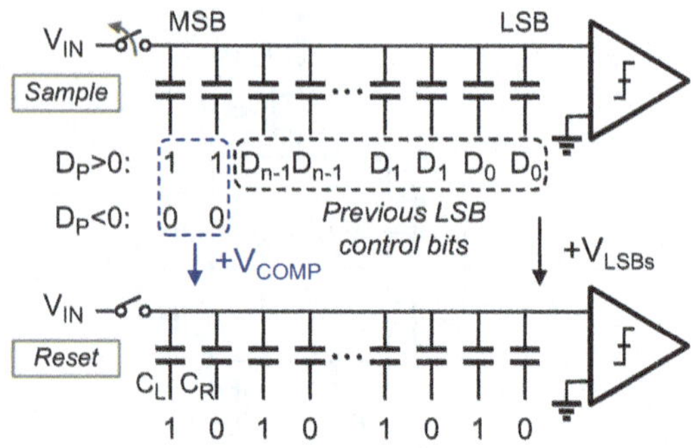

Fig. 8 DAC switching sequence in the proposed MES with a two-level digital prediction

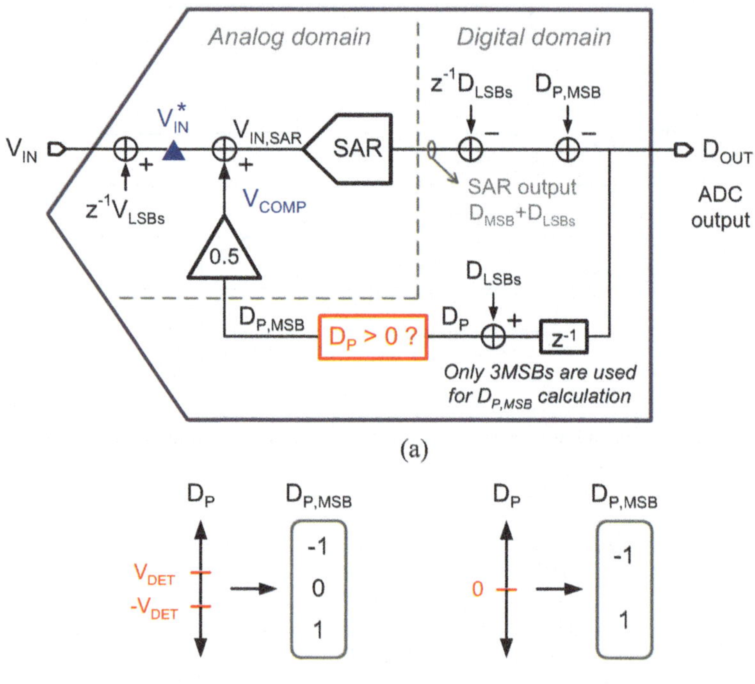

Fig. 9 (**a**) Principle of the MES with digital prediction. (**b**) Tri-level prediction scheme proposed in [18]. (**c**) Two-level prediction scheme in this work

two-level prediction scheme where the compensation is done based on the polarity of the prediction result, as shown in Fig. 9. Since a two-level switching ensures an inherently linear two-point compensation voltage (V_{COMP}), the mismatch between C_L and C_R will not cause distortion in the spectrum.

The simplification from the tri-level prediction to the two-level prediction leads to a reduction in the maximum tolerable prediction error. Figure 10 illustrates the prediction errors in the tri-level scheme. The possible ranges of the voltage before

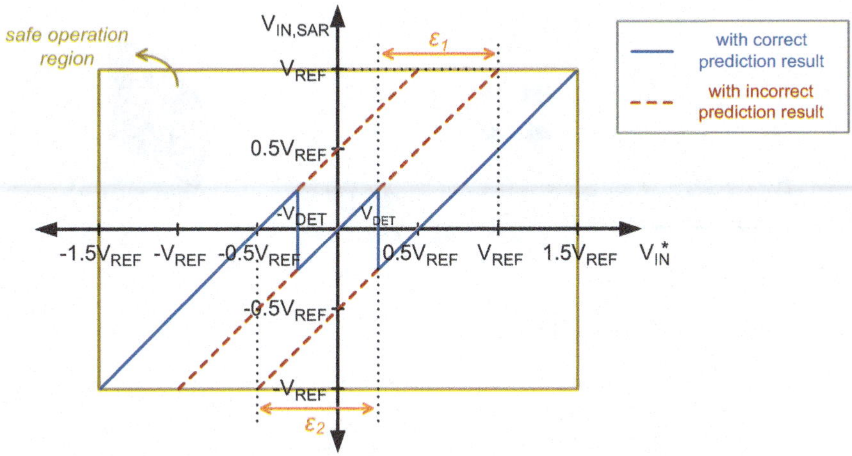

Fig. 10 Illustration of the tolerable prediction error

compensation (V_{IN}^*) and after compensation ($V_{\text{IN,SAR}}$) are shown, and the goal of the prediction scheme is to make the combination of V_{IN}^* and $V_{\text{IN,SAR}}$ locate in the yellow area ("safe operation region"). When the prediction error occurs (dashed red line), $V_{\text{IN,SAR}}$ with the wrong compensation may go beyond the yellow box. The horizontal distance of the extended red line indicates the prediction error that can be accepted. Therefore, the maximum tolerable prediction error (ε_{tol}) can be expressed as the minimum value between ε_1 and ε_2, namely:

$$\varepsilon_{\text{tol}} = \min(\varepsilon_1, \varepsilon_2) = \min(V_{\text{REF}} - V_{\text{DET}}, 0.5V_{\text{REF}} + V_{\text{DET}}). \tag{2}$$

In [18], V_{DET} is $0.25V_{\text{REF}}$, while in our design V_{DET} is reduced to 0. This reduces ε_{tol} from $0.75V_{\text{REF}}$ to $0.5V_{\text{REF}}$, which can be translated to a tighter OSR requirement from 3.8 to 5.6 [6].

In summary, the two-level prediction approach can avoid the input range loss and solve the mismatch problem between the two MSB capacitors in the split DAC switching scheme but at the cost of a stricter OSR requirement.

2.3 *Measurement Results*

The proposed DT NS-SAR ADC is fabricated in a 65 nm CMOS technology, and the die photo is shown in Fig. 11a. It operates at 1 MHz sampling rate with an OSR of 16. Under a 1.2 V supply, it consumes 7.3 μW power with a power breakdown as shown in Fig. 11b. Thanks to the MES, the DAC capacitance in this design is limited by the noise requirement, and it is only 1 pF per side with a power consumption of

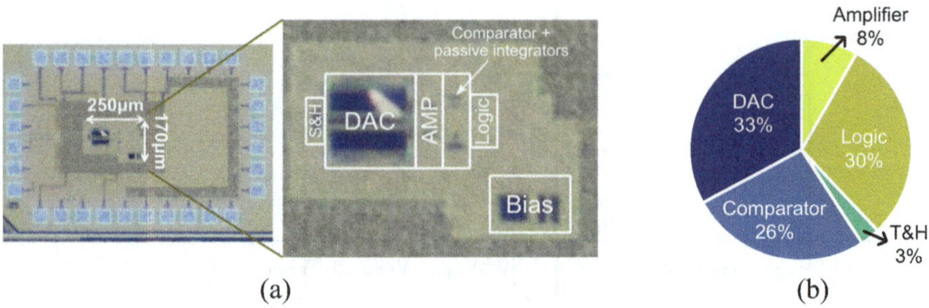

Fig. 11 (**a**) Die photo. (**b**) Power breakdown

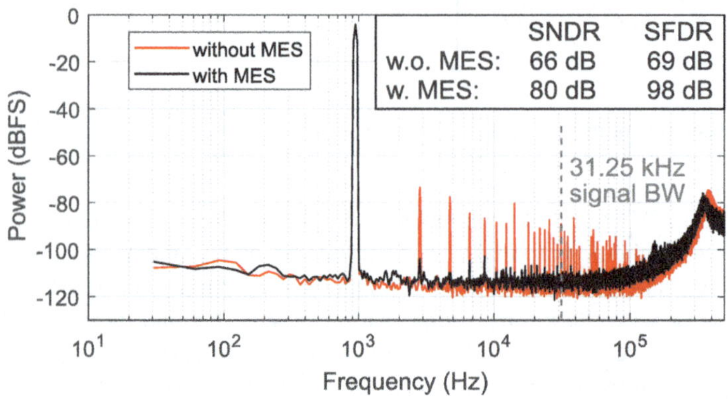

Fig. 12 Measured spectrum without and with MES

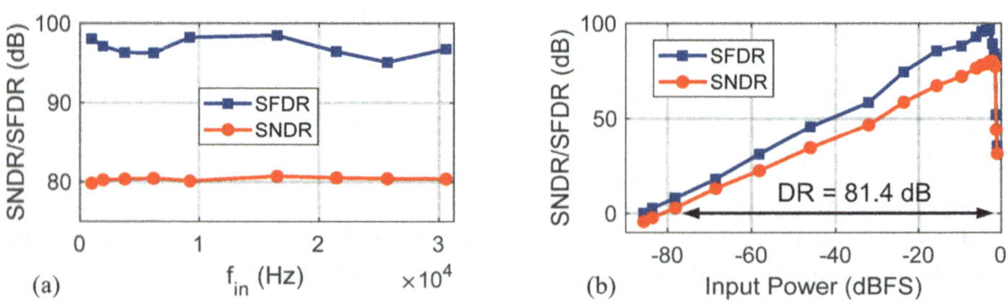

Fig. 13 (**a**) Measured performance over different input frequencies and (**b**) different input amplitudes

33% of the total power. The amplifier consumes only 8% of the total power, thanks to its duty-cycled operation.

Figure 12 shows the measured spectrum without and with MES. When MES is enabled, the measured signal-to-noise and distortion ratio (SNDR) is 80 dB and the spurious-free dynamic range (SFDR) is 98 dB. As can be seen, MES facilitates 29 dB improvement in SFDR. As can be seen from Fig. 13, over the whole

31.25 kHz bandwidth, the measured SNDR is above 80 dB, and SFDR remains above 95 dB. The dynamic range (DR) is 81.4 dB.

3 Continuous-Time Noise-Shaping SAR ADC

As a hybrid architecture of SAR ADC and $\Delta\Sigma$ ADC, NS-SAR ADCs are expected to have the main features from the original ADC architectures. As we know, $\Delta\Sigma$ ADCs can be implemented as DT converters as well as CT converters. However, NS-SAR ADCs are conventionally designed as a DT converter. In this section, the NS-SAR architecture is extended to the CT domain, so that the NS-SAR architecture can have the advantages from CT designs, such as easy to drive and inherent anti-aliasing function. This section is based on [19], and it will introduce the concept of a CT NS-SAR ADC and provides its circuit implementation.

3.1 Architecture

Because the NS-SAR ADC is derived from a SAR ADC which typically operates in a DT manner, most NS-SAR ADCs also work as a DT converter by nature. On the other hand, a CT noise-shaping ADC would need to perform a continuous integration of the residue voltage, and this integration would be interrupted in an NS-SAR architecture due the DT SAR operation. Therefore, the fundamental challenge to develop a CT NS-SAR is to deal with the conflict between the DT-operated SAR conversion and the desired CT residue integration.

This challenge can be illustrated from two aspects. First, the residue voltage V_{RES} only becomes available on the DAC after the SAR conversion is completed, which makes a continuous residue integration impossible during the SAR conversion phase. Second, the residue information on the DAC is lost in the ADC tracking phase, because the DAC is reconnected to the ADC input. In our design, two techniques are proposed to solve these two problems: First, the loop filter is duty cycled to leave sufficient time for the SAR conversion without significantly degrading the noise-shaping performance. Second, the sampling switch is removed to keep V_{RES} also in the tracking phase, and the input signal is ac coupled to the comparator, similar to the first stage of a CT two-step SAR ADC [20].

Figure 14 shows the diagram of the proposed CT NS-SAR ADC. The loop filter is enabled by the sampling clock Φ_{S}, and this duty-cycling operation is modeled as a clock signal multiplying the loop filter input. Here, the duty-cycle rate α is defined as the SAR conversion time T_{SAR} over the sampling period T_{S}, namely:

$$\alpha = T_{\mathrm{SAR}}/T_{\mathrm{S}}. \tag{3}$$

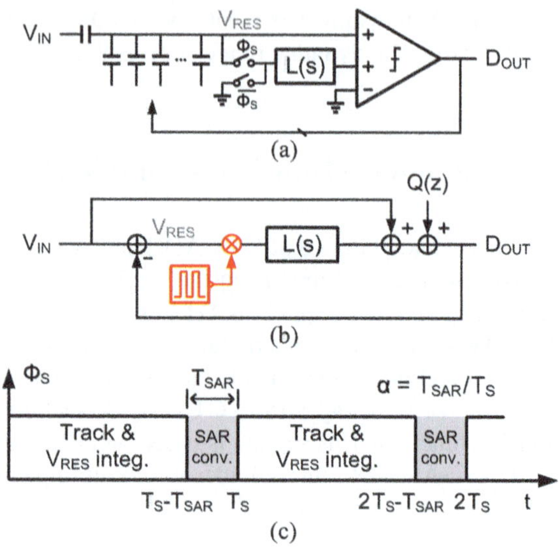

Fig. 14 (**a**) Block diagram, (**b**) signal diagram, and (**c**) timing diagram of the proposed CT NS-SAR ADC

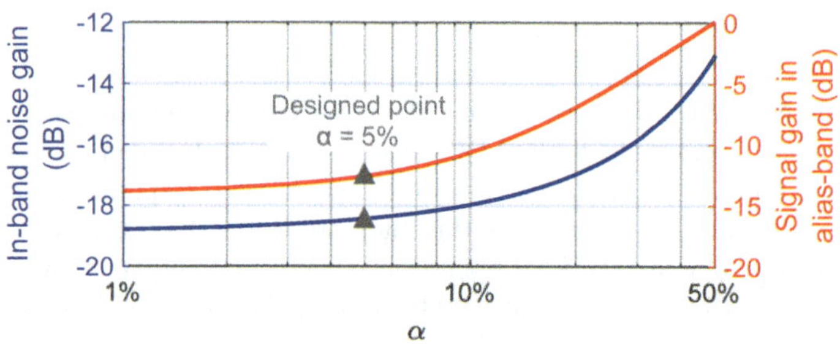

Fig. 15 In-band noise gain and alias-band signal gain with different α values

For a first-order loop filter ($L(s) = 1/s$), assuming the sampling frequency is 1 Hz, the noise transfer function (NTF) and the signal transfer function (STF) can be derived as [19]:

$$\mathrm{NTF}(z) = \frac{1 - z^{-1}}{1 - \alpha z^{-1}}. \tag{4}$$

$$\mathrm{STF}(f) = \frac{1 - e^{-j2\pi f}}{1 - \alpha e^{-j2\pi f}} \cdot \left[\frac{1}{j2\pi f} \frac{1}{1 - e^{-j2\pi f}} \left(1 - e^{-j2\pi f(1-\alpha)} \right) + 1 \right]. \tag{5}$$

Figure 15 shows the in-band noise gain and the alias signal suppression with different α values. As can be seen, when α is small, the noise attenuation and the anti-aliasing function are hardly affected. In this design, the sampling frequency is

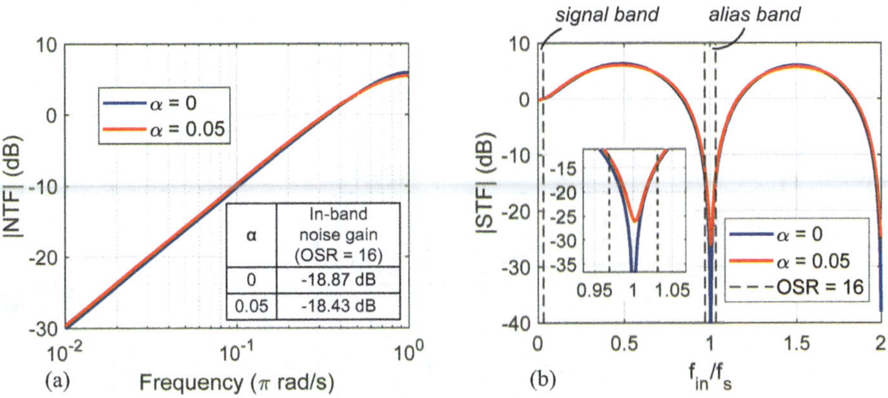

Fig. 16 (**a**) NTF with different α values. (**b**) STF with different α values

2 MHz, and α is 5% to leave enough time for the SAR conversion while minimizing the degradation on NTF and STF. Figure 16 compares the NTF and STF magnitudes when α is 0 and α is 5%. As expected, the in-band noise gain is only degraded by 0.44 dB, and the anti-aliasing property is well preserved.

Figure 17 shows the overall ADC architecture. The input signal is ac coupled through capacitor C_C to the DAC and comparator, and a pseudo-resistor R_B provides its dc bias. The bias resistor R_B and the capacitors C_C and $C_{DAC,p(n)}$ form a high-pass filter which blocks the input dc information [20]. The integrator is on in the 95% of the sampling period, and it is switched off in the remaining 5% time for the SAR conversion. During the SAR conversion phase, the DAC is first reset, and then the converter performs the binary search quantization. Because the input signal is not a static voltage sampled on the DAC, two redundancy bits are added in the DAC to compensate for the quantization error occurring from the varying input.

Since the integrator only integrates the small residue voltage, its output swing is relaxed, and thus a Gm-C integrator is used in this design. Instead of inserting switches as shown in Fig. 14a between the integrator input and the DAC, this work uses two input pairs inside the integrator which are operated in a current-steering manner, as shown in Fig. 18. The purpose of this implementation is to avoid any DAC output disturbance brought by the enable switches from Fig. 14a and the integrator input parasitic capacitance. When Φ_S is high, the blue branch is enabled, and the residue integration takes place. When Φ_S is low, the red branch is activated to stop the integration, and the SAR conversion starts. The bias current of the integrator is 200 nA, which leads to a transconductance of 4.5 μS to meet the noise requirement.

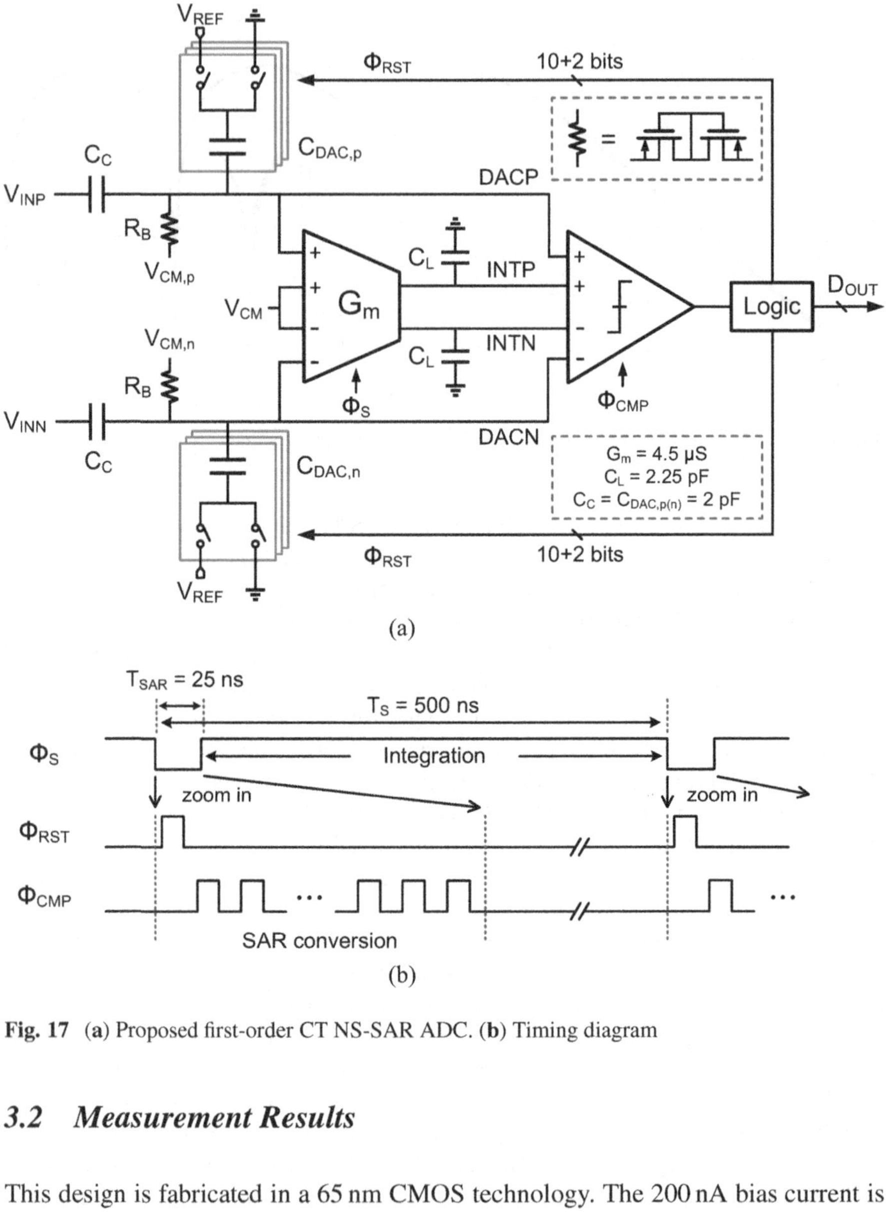

Fig. 17 (**a**) Proposed first-order CT NS-SAR ADC. (**b**) Timing diagram

3.2 Measurement Results

This design is fabricated in a 65 nm CMOS technology. The 200 nA bias current is provided off-chip, and the offsets from the comparator and integrator are calibrated off-chip by tuning the DAC bias voltages $V_{CM,p}$ and $V_{CM,n}$. The supply voltage is 1 V, and the total power consumption is 13.5 μW. The DAC, comparator, and logic consume similar power, while the integrator consumes only 4% of the total power (Fig. 19).

Segmented data-weighted averaging (DWA) is applied to the first five MSBs and three middle bits in the DAC array to improve the linearity. At a sampling rate of

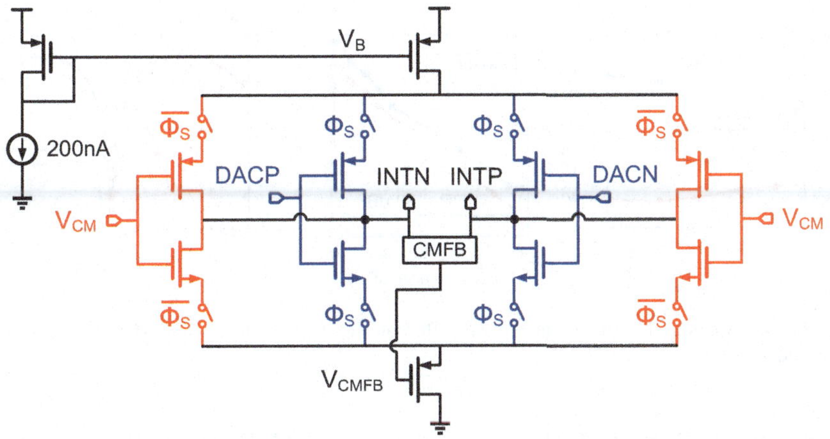

Fig. 18 Duty-cycled integrator

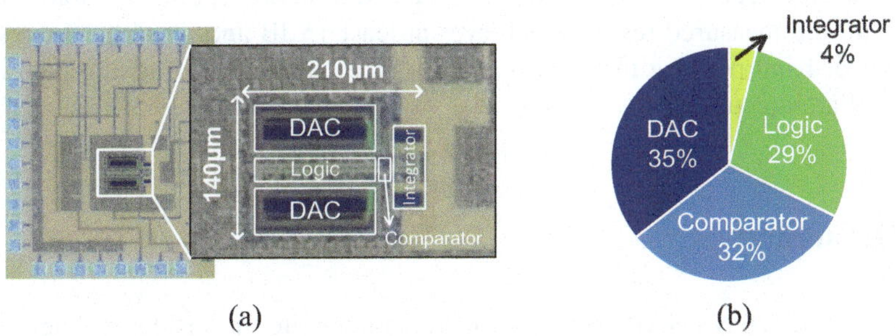

Fig. 19 (**a**) Die photo. (**b**) Power breakdown

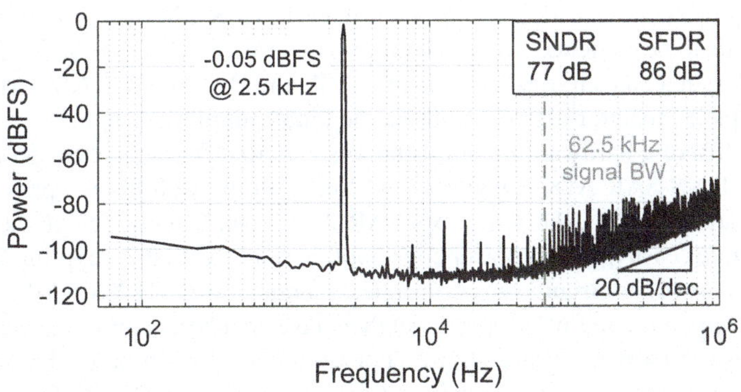

Fig. 20 Measured spectrum

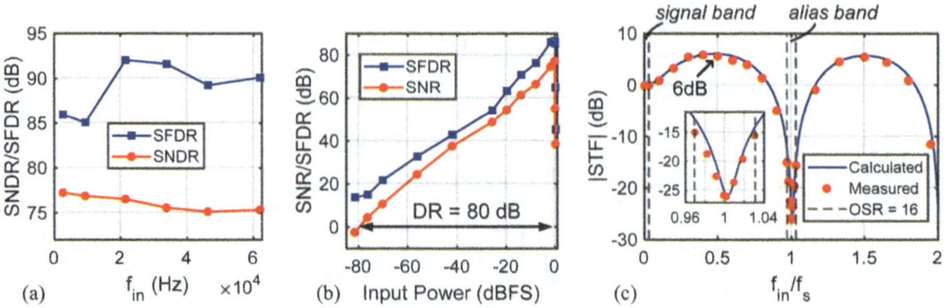

Fig. 21 (**a**) Measured performance over different input frequencies and (**b**) different input amplitudes. (**c**) Measured STF

2 MHz and an OSR of 16, the measured SNDR is 77 dB, and SFDR is 86 dB, as shown in Fig. 20. The SNDR remains above 75 dB, and the SFDR remains above 85 dB in the whole bandwidth. The measured DR is 80 dB, as shown in Fig. 21b. The measured STF is shown in Fig. 21c, and the theoretical calculation matches well with the measured result. It achieves at least 15 dB anti-aliasing suppression. The out-of-band peaking in STF is caused by the inherent feedforward path in the NS-SAR architecture.

4 Discussion

Both DT and CT NS-SAR ADC architectures integrate the residue voltage on the capacitive DAC after the SAR conversion to achieve a noise-shaping function, but the main difference is the implementation of the loop filter and DAC array. The loop filter in a DT NS-SAR ADC is composed of DT circuits, such as passive integrators and DT amplifiers. Instead, the loop filter in a CT NS-SAR ADC is a CT integrator which needs to be switched off periodically. The DAC in a DT NS-SAR ADC is the same as that in a conventional SAR ADC, while the DAC in a CT NS-SAR ADC is an ac-coupled capacitor network without the input sampling switch.

These different hardware implementations lead to different system properties of DT and CT NS-SAR ADCs, which are in line with the well-known pros and cons from DT and CT $\Delta\Sigma$ ADCs. DT NS-SAR ADCs can usually operate at different sampling rates easily, and its power consumption scales with the sampling frequency because (most of) its circuits only consume dynamic power. CT NS-SAR ADCs lose this flexibility in the sampling speed, but they have an inherent anti-aliasing function and are easier to drive. Because they do not require a high current to charge the load capacitor to a certain level in a given tracking time as in DT ADCs, the power consumption of the input driver could be potentially reduced.

When compared to conventional $\Delta\Sigma$ ADCs, NS-SAR ADCs are more hardware-efficient because they reuse the DAC in the SAR quantizer to perform the feedback voltage subtraction as well. Moreover, the loop filter in NS-SAR ADCs only deals with the small residue voltage, which relaxes the integrator output swing and eases the system-level design. For example, the proposed CT NS-SAR ADC can avoid the static power consumption from the resistive or current DAC found in typical CT $\Delta\Sigma$ ADCs, and it does not need an extra feedforward amplifier to reduce the integrator output swing as in [21]. However, due to the recursive SAR conversion steps and the constraint on the duty-cycle rate α, the CT NS-SAR ADC is only suitable for low-speed applications.

5 Conclusion

This paper presents NS-SAR ADC designs in both DT and CT domains. For the DT NS-SAR ADC, three loop filter structures are benchmarked in terms of their power efficiency. A duty-cycled amplifier is proposed to achieve high gain and dynamic power consumption together with its duty-cycled bias generation circuits. A two-level digital-predicted MES technique is used to improve the linearity without saturating the ADC. For the CT NS-SAR ADC, a 5% duty-cycling is applied to the loop filter to accommodate for the SAR conversion without significant degradation on the noise-shaping performance. As other CT ADC architectures, the proposed CT NS-SAR ADC has an implicit anti-aliasing function and is easy to drive.

Acknowledgments This work with project number 16594 is financed by the Dutch Research Council (NWO). We kindly acknowledge EUROPRACTICE for its MPW and design tool support.

References

1. L. Jie et al., An overview of noise-shaping SAR ADC: from fundamentals to the frontier. IEEE Open J. Solid State Circuits Soc. **1**, 149–161 (2021)
2. Y.-S. Shu, L.-T. Kuo, T.-Y. Lo, An oversampling SAR ADC with DAC mismatch error shaping achieving 105 dB SFDR and 101 dB SNDR over 1 kHz BW in 55 nm CMOS. IEEE J. Solid State Circuits **51**(12), 2928–2940 (2016)
3. C.-Y. Lin, Y.-Z. Lin, C.-H. Tsai, C.-H. Lu, An 80MHz-BW 640MS/s time-interleaved passive noise-shaping SAR ADC in 22nm FDSOI process, in *IEEE International Solid-State Circuits Conference (ISSCC)*, vol. 64 (2021), pp. 378–380
4. J.A. Fredenburg, M.P. Flynn, A 90-MS/s 11-MHz-bandwidth 62-dB SNDR noise-shaping SAR ADC. IEEE J. Solid State Circuits **47**(12), 2898–2904 (2012)
5. C.C. Lee, E. Alpman, S. Weaver, C.-Y. Lu, J. Rizk, A 66dB SNDR 15MHz BW SAR assisted $\Delta\Sigma$ ADC in 22nm tri-gate CMOS, in *IEEE Symposium VLSI Circuits* (2013), pp. C64–C65
6. H. Li, Y. Shen, H. Xin, E. Cantatore, P. Harpe, A 7.3-μW 13-ENOB 98-dB SFDR noise-shaping SAR ADC with duty-cycled amplifier and mismatch error shaping. IEEE J. Solid State Circuits **57**(7), 2078–2089 (2022)
7. W. Guo, N. Sun, A 12b-ENOB 61μW noise-shaping SAR ADC with a passive integrator, in *IEEE European Solid-State Circuits Conference (ESSCIRC)* (2016), pp. 405–408

8. C.-C. Liu, M.-C. Huang, A 0.46 mW 5 MHz-BW 79.7 dB-SNDR noise-shaping SAR ADC with dynamic-amplifier-based FIR-IIR filter, in *IEEE International Solid-State Circuits Conference (ISSCC)* (2017), pp. 466–467
9. K. Obata, K. Matsukawa, T. Miki, Y. Tsukamoto, K. Sushihara, A 97.99 dB SNDR, 2 kHz BW, 37.1 μW noise-shaping SAR ADC with dynamic element matching and modulation dither effect, in *IEEE Symposium VLSI Circuits* (2016), pp. 1–2
10. X. Tang et al., An energy-efficient comparator with dynamic floating inverter amplifier. IEEE J. Solid State Circuits **55**(4), 1011–1022 (2020)
11. X. Tang et al., A 13.5-ENOB, 107-μW noise-shaping SAR ADC with PVT-robust closed-loop dynamic amplifier. IEEE J. Solid State Circuits **55**(12), 3248–3259 (2020)
12. R.S.A. Kumar, N. Krishnapura, P. Banerjee, Analysis and design of a discrete-time delta-sigma modulator using a cascoded floating-inverter-based dynamic amplifier. IEEE J. Solid State Circuits **57**(11), 3384–3395 (2022)
13. C. Yang et al., An area-efficient SAR ADC with mismatch error shaping technique achieving 102-dB SFDR 90.2-dB SNDR over 20-kHz bandwidth. IEEE Trans. Very Large Scale Integr. (VLSI) Syst. **29**(8), 1575–1585 (2021)
14. T. Wang, T. Xie, Z. Liu, S. Li, An 84 dB-SNDR low-OSR 4th-order noise-shaping SAR with an FIA-assisted EF-CRFF structure and noise-mitigated push-pull buffer-in-loop technique, in *IEEE Int. Solid-State Circuits Conf. (ISSCC)*, vol. 65 (2022), pp. 418–420
15. K. Hasebe et al., A 100kHz-bandwidth 98.3dB-SNDR noise-shaping SAR ADC with improved mismatch error shaping and speed-up techniques, in *IEEE Symp. VLSI Tech. Circuits* (2022), pp. 56–57
16. J. Liu et al., A 90-dB-SNDR calibration-free fully passive noise-shaping SAR ADC with 4$\times$ passive gain and second-order DAC mismatch error shaping. IEEE J. Solid-State Circuits **56**(11), 3412–3423 (2021)
17. L. Jie, B. Zheng, H.-W. Chen, M. P. Flynn, A cascaded noise-shaping SAR architecture for robust order extension. IEEE J. Solid-State Circuits **55**(12), 3236–3247 (2020)
18. J. Liu, C.-K. Hsu, X. Tang, S. Li, G. Wen, N. Sun, Error-feedback mismatch error shaping for high-resolution data converters. IEEE Trans. Circuits Syst. I Reg. Papers **66**(4), 1342–1354 (2019)
19. H. Li, Y. Shen, E. Cantatore, P. Harpe, A first-order continuous-time noise-shaping SAR ADC with duty-cycled integrator, in *IEEE Symp. VLSI Tech. Circuits* (2022), pp. 58–59
20. L. Shen et al., A two-step ADC with a continuous-time SAR-based first stage. IEEE J. Solid-State Circuits **54**(12), 3375–3385 (2019)
21. J.-S. Bang, H. Jeon, M. Je, G.-H. Cho, 6.5 μW 92.3dB-DR biopotential-recording front-end with 360mVpp linear input range, in *IEEE Symp. VLSI Circuits* (2018), pp. 239–240

The Zoom ADC: An Evolving Architecture

Efraïm Eland, Shubham Mehrotra, Shoubhik Karmakar, Robert van Veldhoven, and Kofi A. A. Makinwa

Abstract Zoom ADCs combine a coarse SAR ADC with a fine delta-sigma modulator ($\Delta\Sigma$M) to efficiently obtain high energy efficiency and high dynamic range. This makes them well suited for use in various instrumentation and audio applications. However, zoom ADCs also have drawbacks. The use of over-ranging in their fine modulators may limit SNDR, large out-of-band interferers may cause slope overload, and the quantization noise of their coarse ADC may leak into the baseband. This chapter presents an overview of recent advances in zoom ADCs that tackle these challenges while maintaining high energy efficiency. Prototypes designed in standard 0.16 μm technology achieve SNDRs over 100 dB in bandwidths ranging from 1 to 24 kHz while consuming only hundreds of μWs.

1 Introduction

Audio applications often require analog-to-digital converters (ADCs) with high dynamic range (DR), high energy efficiency, and low area [1–4]. By combining a low-power successive-approximation register (SAR) ADC with a high-resolution delta-sigma modulator ($\Delta\Sigma$M), zoom ADCs can meet all these requirements [1–6]. The SAR ADC coarsely determines the references of the fine $\Delta\Sigma$M, drastically reducing loop filter swing and enabling energy-efficient design. The overall digital output is then obtained by simply summing the outputs of both converters.

E. Eland (✉) · S. Mehrotra
Delft University of Technology, Delft, the Netherlands

NXP Semiconductors, Eindhoven, the Netherlands
e-mail: efraim.eland@nxp.com

S. Karmakar · K. A. A. Makinwa
Delft University of Technology, Delft, the Netherlands

R. van Veldhoven
NXP Semiconductors, Eindhoven, the Netherlands

P. Harpe et al. (eds.), *Biomedical Electronics, Noise Shaping ADCs, and Frequency References*, https://doi.org/10.1007/978-3-031-28912-5_10

Proposed ΔΣMs with finite impulse response (FIR) DACs and negative R-assisted integrators are also capable of satisfying the requirements of audio applications [7–10]. An FIR DAC essentially filters out the fed-back quantization noise and thereby also relaxes loop filter swing. However, it introduces an extra delay in the feedback path, which requires an additional compensation path to maintain stability and restore NTF [7, 10]. This delay also limits the extent to which the loop filter's input swing can be reduced. Similarly, the swing at the virtual ground of an active integrator can be reduced by connecting it to a negative resistance [8, 9]. This effectively increases the integrator's gain and improves its linearity. However, since the negative resistance is realized by an active circuit, it also produces noise and consumes power. Furthermore, foreground calibration is required to ensure good matching between the negative resistance and the integrator's equivalent input resistance. In comparison, zoom ADCs seem to present a good tradeoff between design complexity, energy efficiency, resolution, and no need for calibration.

However, zoom ADCs also have drawbacks. In order to absorb SAR ADC non-idealities, their fine ΔΣMs are usually designed to provide at least ±1 LSB of over-ranging [2, 4–6]. In the case of a 1-bit ΔΣM, this means that the modulator's DAC must span at least three SAR LSBs, leading to a significant loss of SQNR. Another issue is the leakage of the SAR ADC's quantization noise, to which zoom ADCs, like other MASH ADCs, are susceptible. This is because summing the outputs of the SAR ADC and the ΔΣM tacitly assumes that the signal transfer function (STF) of the latter is exactly unity, which will usually not be the case, especially at high frequencies [6]. These issues can be mitigated by increasing the modulator's OSR or by using a digital noise cancellation filter. But both approaches inevitably increase power consumption [2, 6]. Furthermore, previous zoom ADCs also suffered from limited robustness to out-of-band interferers.

This chapter gives an overview of the evolution of zoom ADC architectures for instrumentation and audio applications that tackle these issues and is organized as follows: first, the system-level design for zoom ADCs is discussed (Sect. 10.2). Different techniques used in recent zoom ADCs for instrumentation and audio applications (1–24 kHz BW) are explained to solve the zoom ADC's drawbacks such as susceptibility to out-of-band interferers (Sect. 10.2.1), loss of SQNR due to over-ranging (Sect. 10.2.2), and SAR quantization noise leakage (Sect. 10.2.3). This is followed by the discussion of challenges in the amplifier design for DT zoom ADCs (Sect. 10.3). Recent CT zoom ADCs are discussed next (Sect. 10.4), with a focus on amplifier nonlinearity (Sect. 10.4.1) and DAC drivers (Sect. 10.4.2), followed by a CT zoom ADC design example (Sect. 10.4.3). This chapter closes with a conclusion and comparison with state-of-the-art ADCs (Sect. 10.5).

2 System-Level Design of the Zoom ADC

A dynamic zoom ADC [4], as shown in Fig. 1a, consists of an N-bit SAR ADC, which performs a coarse conversion and outputs an N-bit code k. This digital value

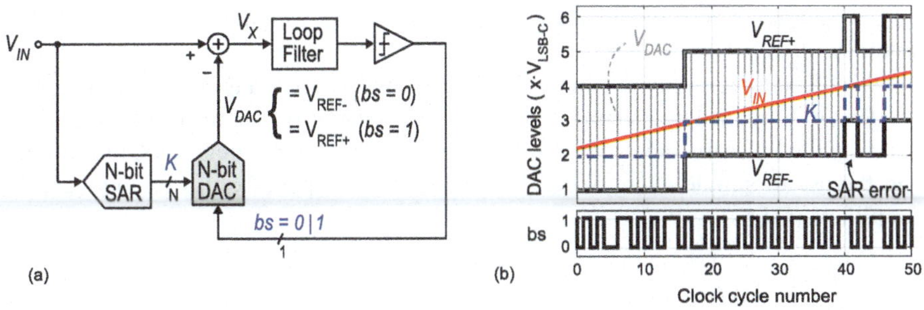

Fig. 1 (**a**) Simplified block diagram of the dynamic zoom ADC. (**b**) Time-domain waveforms of different signals in the zoom ADC with an over-ranging of $M = 1$

k is then used to determine the high and low references of a fine $\Delta\Sigma$M, respectively, as

$$V_{\mathrm{REF+}} = (k + 1 + M) \bullet V_{\mathrm{LSB,C}} \tag{1}$$

$$V_{\mathrm{REF-}} = (k - M) \bullet V_{\mathrm{LSB,C}} \tag{2}$$

where $V_{\mathrm{LSB,C}}$ is the quantization step size corresponding to the N-bit SAR and M is an over-ranging factor. The fine $\Delta\Sigma$M DAC toggles between these references depending on the bitstream output of the comparator (bs), essentially zooming in on the signal, and operating as a conventional 1-bit $\Delta\Sigma$M, but achieving a significantly higher signal-to-quantization noise ratio (SQNR) due to the small step size. Figure 1b shows the resulting signals in the case of $M = 1$.

As compared to one of the first incremental zoom ADCs proposed in [5], the parallel operation of the SAR and fine $\Delta\Sigma$M in a dynamic zoom ADC effectively allows a much higher bandwidth. If $M = 0$, no error can be tolerated in the coarse SAR conversion since a conversion error would lead to the $\Delta\Sigma$M references not straddling the input signal, thus leading to $\Delta\Sigma$M overload. Over-ranging, i.e., making $M > 0$, is used to relax the accuracy requirements of the SAR ADC [4].

As the SAR ADC uses a separate capacitive DAC, its quantization levels will also exhibit some mismatch with respect to those of the main DAC used by the $\Delta\Sigma$M to set the fine references. Any error made by the SAR ADC due to its noise, linearity, and offset will result in an error in the coarse code k. Without over-ranging, the overall accuracy of the zoom ADC would, therefore, be limited by both the SAR ADC and $\Delta\Sigma$M DAC. Over-ranging ensures that the fine references of the $\Delta\Sigma$M are still valid for a given input as long as the error in the SAR conversion is below M LSBs. Thus, the SAR ADC does not limit the overall accuracy. This is illustrated in Fig. 1b, where despite the error in k, the input remains bounded by the fine references. It must be noted that although over-ranging relaxes the SAR ADC constraints, the main N-bit $\Delta\Sigma$ DAC must still be designed to achieve the intended target linearity. This is enabled by implementing data-weighted averaging (DWA) in the N-bit $\Delta\Sigma$ DAC.

The relaxed requirements on the SAR ADC due to over-ranging greatly simplify its design. Furthermore, zooming reduces the swing at the input of the loop filter, relaxing the linearity and driving requirements of the $\Delta\Sigma$M integrators, thus allowing the use of simple energy-efficient inverter-based operational transconductance amplifiers (OTAs).

2.1 Asynchronous SAR ADC

The time-domain operation of a dynamic zoom ADC is shown in Fig. 2a, b. The coarse ADC is an N-bit synchronous SAR ADC, whose conversion time takes N clock cycles of the DT$\Delta\Sigma$M ($N = 5$ in Fig. 2). The references are set by k, which represents the signal's value at the moment of the coarse ADC's sampling $V_{in}(t_{s,C})$. This is used to compare it with the sampled input of the DT$\Delta\Sigma$M at its sampling moment ($t_{s,F}$), which is $V_{in}(t_{s,F})$. In the case of an N-bit coarse SAR ADC, the input is sampled at $t_{s,C}$, and the corresponding k is available N cycles later. Assuming k sets the references immediately, the minimum difference $\Delta t = t_{s,F} - t_{s,C}$ is N clock periods. However, this k value will be used for the next N cycles. Thus, the maximum Δt would be $2N$ clock cycles. This requires more over-ranging, i.e., $M > 1$, to achieve proper operation, as shown in Fig. 2b. A higher N is desirable to reduce the $\Delta\Sigma$M input range and increase energy efficiency. However, it raises Δt. As shown in Fig. 2a, when the input signal changes too fast to be tracked by

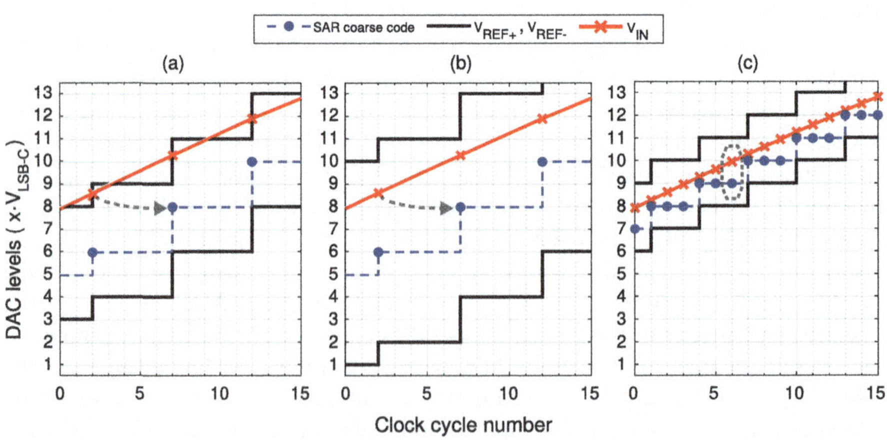

Fig. 2 Time-domain operation of the dynamic zoom ADC for a fast-changing input. (**a**) Coarse code and corresponding fine reference are updated at every five cycles by a 5-bit SAR ADC for (**a**) $M = 2$ and for (**b**) $M = 4$. (**c**) Fine reference is updated every cycle by a 5-bit asynchronous SAR, with an over-ranging of $M = 1$ [6]

the coarse SAR ADC, it can extend beyond the modulator's stable input range, as in the case of large out-of-band interferers. The duration of the coarse conversion also puts a limit on the maximum full-scale input frequency ($f_{\text{in,max}}$), thereby making the $\Delta\Sigma$M susceptible to overload in the presence of large out-of-band interferer signals. This is because the references of the DT$\Delta\Sigma$M are only updated once every N clock cycles, while the $\Delta\Sigma$M assumes that the signal rests in between these set reference levels between two reference update moments. The stable input range of $\Delta\Sigma$M can be expressed as [4]:

$$f_{\text{in,max}} < \alpha \frac{2M+1}{2N\left(2^N-1\right)\pi} f_s \tag{3}$$

Equation (3) represents the relation of $f_{\text{in,max}}$ to f_s, M, and N, which are parameters of the coarse conversion, and α, which defines the topology-dependent stable input range of the $\Delta\Sigma$M and is $\leq$1. It can be seen that a higher N or a lower f_s reduces $f_{\text{in,max}}$.

An asynchronous SAR ADC is a better alternative, because after being triggered by a clock edge, synchronous to f_s, its internal execution of the binary search algorithm is self-timed. In [6], the asynchronous SAR ADC's total conversion time was much less than half a clock cycle, allowing it to update the $\Delta\Sigma$M references every clock cycle as shown in Fig. 2c; hence, Δt was $0.5/f_s$. We can express $f_{\text{in,max}}$ in this case as [6]:

$$f_{\text{in,max}} < 2\alpha \frac{2M+1}{\left(2^N-1\right)\pi} f_s \tag{4}$$

The dependency of $f_{\text{in,max}}$ on N is less drastic in (4) compared to (3). It is seen that a synchronous SAR ADC limits $f_{\text{in,max}}$ dramatically. Although this can be alleviated by increasing M, or reducing N, both would result in increased $\Delta\Sigma$M input swing and quantization noise and therefore degraded energy efficiency. An asynchronous SAR ADC offers higher $f_{\text{in,max}}$ and thus improves the robustness to out-of-band interferers. This allows the use of $M = 1$, as reported in [1–3, 6] and shown in Fig. 2c. This directly reduces the input swing of the $\Delta\Sigma$M and improves its linearity, and thus its energy efficiency, which is usually limited by nonlinearity.

2.2 *SQNR Recovery*

An example waveform for $M = 1$ (minimum over-ranging) is shown in Fig. 3a, when the fine references $V_{\text{REF+}}$ and $V_{\text{REF-}}$ are updated at the sampling rate. For $M = 1$, the DAC of a 1-bit $\Delta\Sigma$M will then span 3-$V_{\text{LSB,C}}$. So even with this minimum over-ranging, the modulator's quantization error will increase by 3×, reducing its SQNR by ~9.5 dB compared to the case with no over-ranging. Although this can be restored by increasing the OSR, it comes at the expense of increased power consumption.

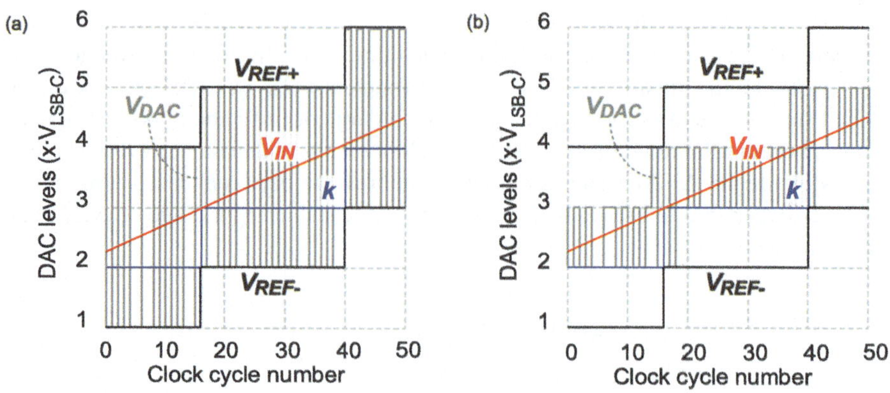

Fig. 3 SAR output (k) and ΔΣM DAC swings with $M = 1$ for (**a**) a 1b quantizer and (**b**) a 2b quantizer

To recover the lost SQNR and to effectively utilize the two intermediate levels, a 2-bit quantizer can be used, as shown in Fig. 3b [1]. It should be noted that the DAC itself remains unchanged, and so does the data-weighted averaging (DWA) scheme required to obtain high linearity [11]. The resulting reduction in quantization noise enables a corresponding decrease in OSR to achieve the same SQNR, which, in turn, leads to reduced analog and digital power consumption, a significant reduction in comparison to the slightly larger power consumption of the two additional comparators of the 2-bit flash quantizer.

2.3 *SAR Quantization Noise Leakage*

As shown in Fig. 4, a zoom ADC can be modeled as a 0-N MASH ADC by splitting its DAC into two halves, one driven by the SAR ADC and the other driven by the ΔΣM. The overall digital output $Y_{OUT} = k + Y_{\Delta\Sigma M}$ can then be expressed as [1]:

$$Y_{OUT} = V_{IN}(z) + Q_{SAR}(z) \bullet (\mathrm{STF} - 1) + Q_{2-\mathrm{bit}} \bullet \mathrm{NTF} \quad (5)$$

where Q_{SAR} and $Q_{2\text{-bit}}$ represent the quantization noise of the SAR and the ΔΣM quantizer, respectively. As expected, $Q_{2\text{-bit}}$ is shaped by the NTF. However, the cancellation of Q_{SAR} is limited by *STF-1*, which is equal to *-NTF* for a feedforward DT loop filter. Since the quantization noise of a 5-bit ADC is quite tonal, it leaks into the output spectrum of the zoom ADC, referred to as "fuzz" in the literature [6]. The in-band fuzz degrades the SNDR of the ADC and limits its bandwidth.

SAR ADC quantization noise can be reduced by increasing OSR, at the expense of higher power consumption, especially in a switched-capacitor implementation [4]. Alternatively, a digital noise cancellation filter can be used to process *k* before

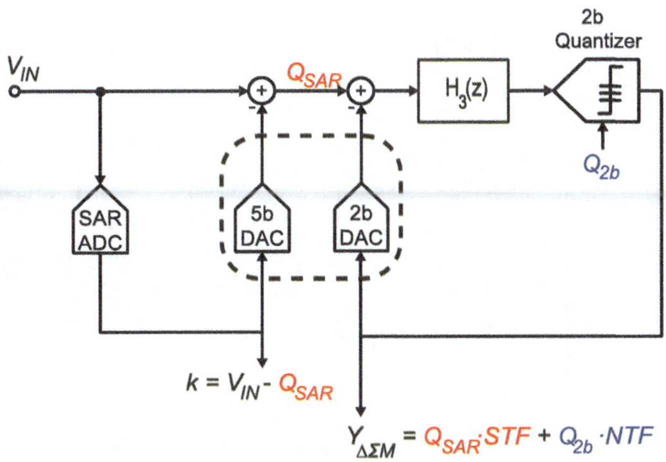

Fig. 4 Intuitive block diagram of the coarse-fine operation in the N-bit DAC

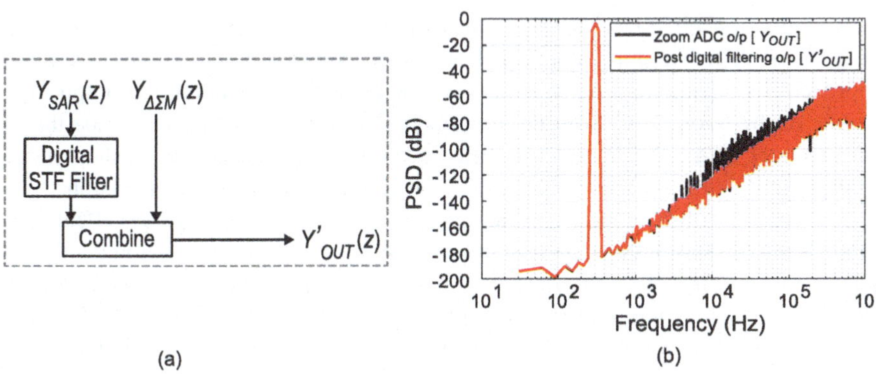

Fig. 5 (**a**) Fuzz filtering using a digitally matched STF filter. (**b**) The resulting zoom ADC output spectrum

combining it with $Y_{\Delta\Sigma M}$ [6], as in MASH architectures. As shown in Fig. 5a, this involves passing the output code of the SAR ADC (k) through a digital filter matched to the STF before combining it with the bitstream output. This results in almost perfect fuzz suppression, as shown in Fig. 5b. However, the required digital STF filter increases the complexity and power consumption. Furthermore, any mismatch between the analog STF and the reconstruction filter will degrade the fuzz suppression performance, therefore likely requiring calibration.

In [1], a low-power fuzz cancellation technique is proposed. It is based on the observation that, from (5), Q_{SAR} leakage can be prevented by ensuring that the modulator has a unity STF. One way of doing this is by implementing an input feedforward path [12]. As shown in Fig. 4, the modulator's input is the residue of the SAR ADC. Thus, this can be extracted and added to the input of

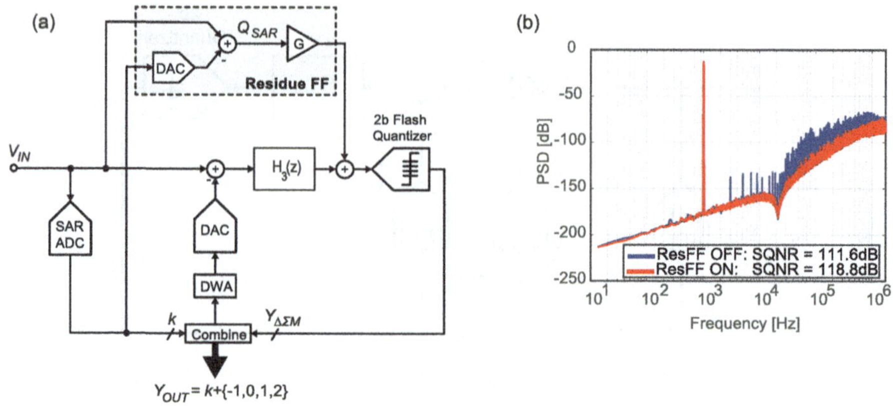

Fig. 6 (**a**) Block diagram of a zoom ADC with third-order $\Delta\Sigma$M with 2-bit quantizer, 5-bit SAR ADC, and residue feedforward [1]; (**b**) Output spectrum of the zoom ADC

the modulator's quantizer. Rather than extracting the small residue at the input of the SAR ADC's comparator, which would require complex circuitry, a simpler approach is to generate a replica. This can be done by subtracting the output of a replica of the SAR DAC from the input signal, as shown in Fig. 6a. Since the 2-bit quantizer of the $\Delta\Sigma$M has a well-defined gain (K_q), the output of the residue feedforward path is scaled by a factor $G = 1/K_q$ for optimal fuzz suppression.

Although the fuzz reduction techniques proposed in [1, 6] show excellent fuzz suppression in simulations, measurement results show that the amount of fuzz suppression is limited by the mismatch between the primary and the cancellation paths. Another technique proposed in [3] uses a 4-bit passive noise-shaping SAR (NS SAR) [13] as a coarse ADC, which effectively decorrelates and shapes the tonal quantization noise of the basic SAR ADC, and thus reduces the fuzz enough to obtain a thermal noise-limited output spectrum.

At the end of the SAR conversion cycle, the residue signal which is readily available on SAR DAC is extracted with a gain of 2 and then passively integrated with a gain of 0.4 to achieve first-order noise shaping (NS) in the SAR ADC. Passive integration preserves the energy efficiency of the SAR ADC while improving its SNDR significantly. This modest NS is sufficient to decorrelate and suppress the tonal quantization noise of the 4-bit SAR ADC to enable a thermal noise-limited output spectrum for the zoom ADC. The simulated output spectrum of the 4-bit noise-shaping SAR ADC is shown in Fig. 7b; it shows the suppression of the tonal quantization noise of the SAR ADC (~9 dB). The output spectrum of a zoom ADC with a 4-bit NS SAR and a third-order CIFF loop filter with a notch at 19 kHz is shown in Fig. 8 [3]. It is shown that the improvement in SQNR is ~7 dB while only using four additional capacitors, a few switches, and digital logic, without requiring any calibration.

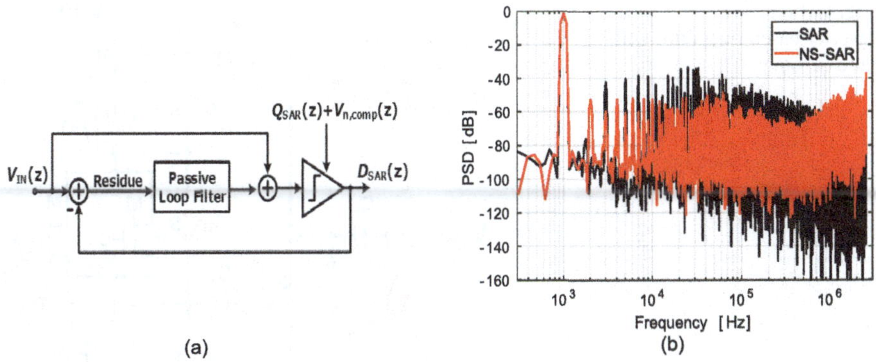

Fig. 7 (**a**) Block diagram of 4-bit noise-shaping SAR ADC [3] and (**b**) output spectrum of the noise-shaping SAR ADC

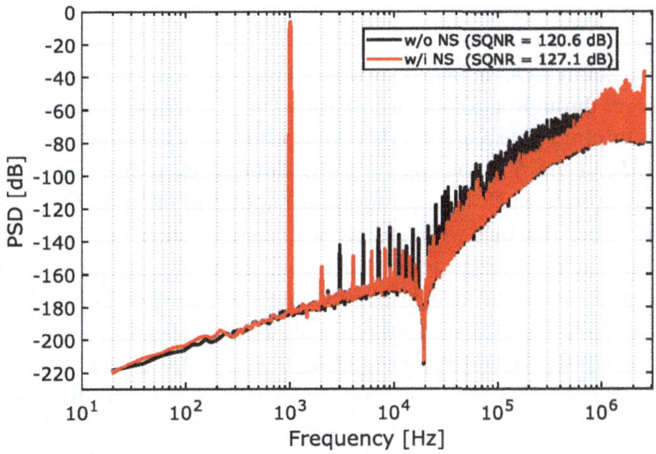

Fig. 8 Output spectrum of a zoom ADC with third-order loop filter and 4-bit SAR ADC w/i and w/o noise shaping [3]

3 Amplifiers in DT Zoom ADCs

Amplifiers in switched-capacitor integrators for high-resolution ADCs require fast settling, low thermal and $1/f$ noise, and excellent energy efficiency. This sub-section discusses the different amplifiers that have been used in DT zoom ADCs as well as the circuit techniques (auto-zero, CDS, and chopping) needed to meet these requirements.

A pseudo-differential inverter-based OTA is used in [4] for its energy efficiency. Auto-zeroing is used to reduce the effect of offset and 1/f noise, whereas differential sampling is used to improve the overall CMRR. To mitigate the effect of PVT variations, a dynamic biasing scheme for inverter-based OTAs is used for the first

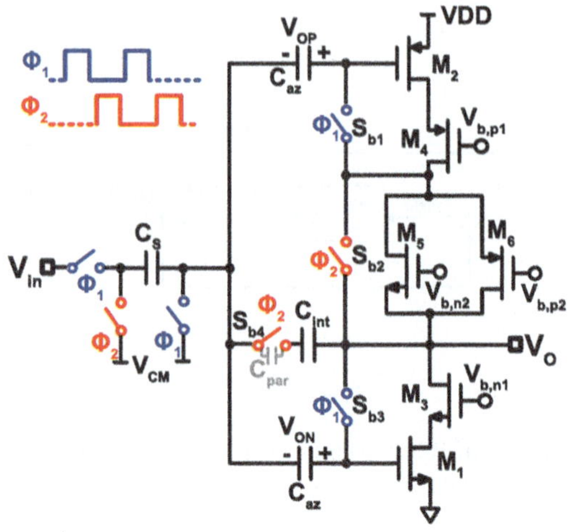

Fig. 9 Proposed inverter-based integrator with auto-zeroing

integrator of a zoom ADC in [4], as shown in Fig. 9. Instead of switching the floating current source through cascode transistors [5], switches $S_{b1\text{–}3}$ are introduced. During the sampling phase φ_1, diode connections are established around the input transistors ($M_{1\text{–}2}$) via S_{b1} and S_{b3}, and the floating current source ($M_{5\text{–}6}$) forces the same bias current (125 μA) through the input and cascode ($M_{3\text{–}4}$) transistors. At the same time, the bias voltages as well as the offset and the 1/*f* noise are sampled on the auto-zeroing capacitors C_{az} (2 pF each). In the integration phase φ_2, diode connections are broken by opening the switches S_{b1} and S_{b3}, and the floating current source consisting of M_5 and M_6 is simply bypassed by S_{b2}. Since there is no switching capacitive load to the biasing circuit, its power consumption can be minimized. Furthermore, the proposed biasing scheme results in a much more compact design by eliminating two large cascode transistors. A simple SC common-mode feedback (CMFB) circuit as in [14] is adequate to avoid output common-mode drift in the pseudo-differential implementation. Though very effective for *1/f* noise and offset suppression, this technique uses relatively large auto-zero capacitors to fulfill the noise requirements.

In [6], a correlated double sampling (CDS) scheme is implemented in a second-order zoom ADC to suppress the offset of OTA_1 [15], as shown in Fig. 10. A simple current-starved OTA with cascodes is used as the first integrator for its high energy efficiency, DC gain, and PSRR, as shown in Fig. 10. While the input is shorted to the outer plate of C_S during phase φ_1, OTA_1 is connected in unity feedback and samples its offset and 1/*f* noise on the other plate. During φ_2, this offset is effectively canceled, while the input is integrated. Due to the finite DC gain of OTA_1, the offset sampled at the virtual ground node due to unity feedback is $V_{OFF} \cdot A/(1 + A)$. As a result, an input-referred offset of approximately V_{OFF}/A remains. A typical offset of a few millivolts will be suppressed to a few microvolts if the OTA gain is around 60 dB. Compared to [4], this architecture omits the use of large auto-zeroing caps while achieving similar levels of 1/*f* noise and offset suppression. However, due

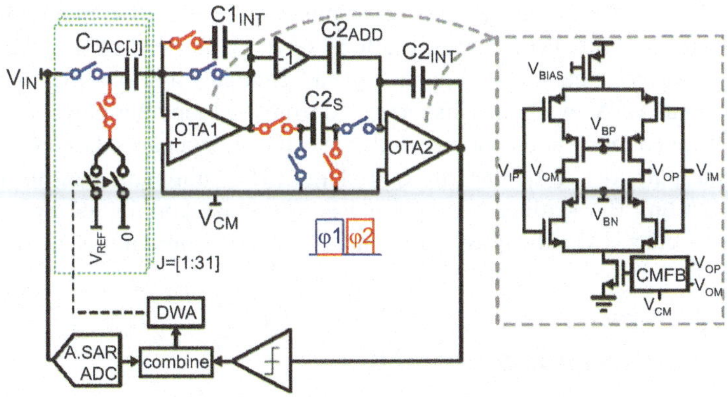

Fig. 10 Simplified single-ended block diagram of a second-order zoom ADC [6]

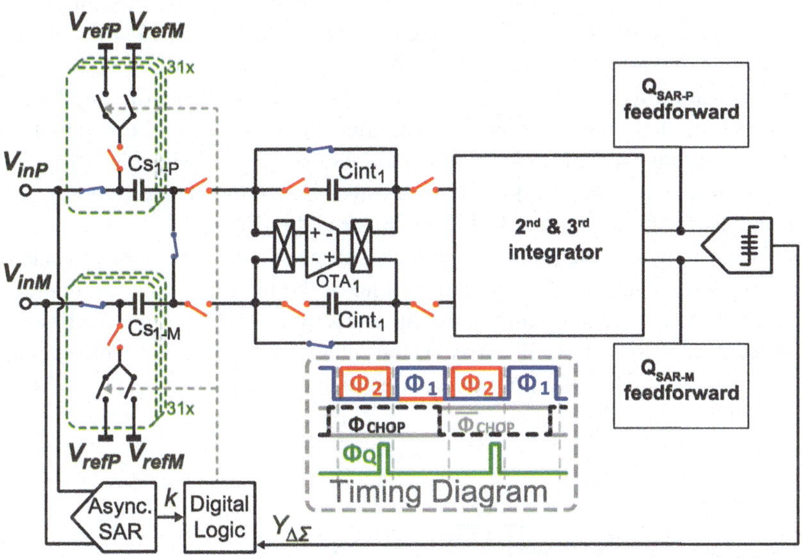

Fig. 11 The first integrator of a zoom ADC with third-order loop filter as proposed in [1]

to its pseudo-differential sampling operation, its CMRR is limited compared to the fully differential sampling used in [4].

To improve CMRR and reduce area, while still reducing $1/f$ noise and offset, [1] uses chopping and differential sampling, as shown in Fig. 11. The input is sampled using a fully differential sampling network. The sampling switches are bootstrapped to maintain high linearity [16], and thick oxide switches are used in the bootstrapping circuitry, thereby reducing complexity. The first-stage OTA is similar to the one used in Fig. 10 [6].

While the input is being tracked during phase φ_1, OTA_1 is configured in unity feedback and is disconnected from the loop filter. At the end of φ_1, the input is sampled onto C_S, whereas OTA_1, having had enough time to settle, is chopped [15]. Chopping OTA_1 while it is disconnected from the loop filter prevents chopping artifacts from coupling to the input signal. Since the chopper switches are connected to the input pairs of OTA_1, they are quite large (31× minimum size) to minimize their impact on OTA settling time and noise. The noise contributed by the output chopper switches is significantly lower, and so these are minimum-size devices.

4 Continuous-Time Zoom ADCs

Previously discussed zoom ADC designs employed switched-capacitor (SC) front-ends that required input and reference drivers capable of delivering large signal-dependent peak currents. For high linearity applications (>100 dB), the power dissipation of the drivers for DTΔΣMs will be higher than that of the ADC itself due to the large sampling capacitors (C_s), in some cases necessitating on-chip buffers, at the expense of chip area and power consumption [17]. It is well known that ADCs based on continuous-time delta-sigma modulators (CTΔΣMs) generally do not require anti-aliasing filters, while their resistive input impedance imposes relaxed requirements for the input driver [17], as illustrated in Fig. 12.

In contrast to SC integrators, in which charge is transferred in exponentially decaying pulses, and only the result at the end of the integration period matters, the charge transfer in a CT integrator is a continuous process. Therefore, the linearity of this process depends on the linearity of the integrator's amplifier. This subchapter will describe two different amplifier architectures that were used in zoom ADCs [2, 3] to reach high linearity.

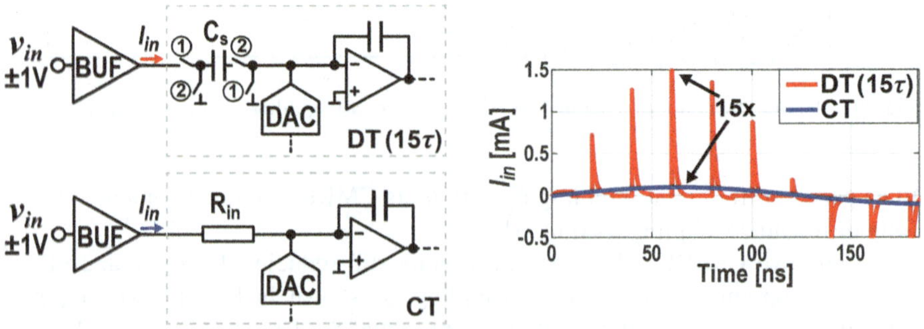

Fig. 12 Block diagram and buffer current output of a DT and CT zoom ADC

4.1 Amplifiers in CT Zoom ADCs

In a CT loop filter, the first integrator's linearity is the most critical. This is often realized with a fully differential amplifier (Fig. 13a). However, the linearity of a fully differential amplifier is worse than that of its pseudo-differential counterpart (Fig. 13b). This is because the fixed tail current makes the amplifier's transconductance (gm) compressive. Simulations were made to compare the linearity of the proposed amplifier with that of its fully differential counterpart. Both the amplifiers are biased in weak inversion, have the same I_{bias} and device sizing, and thus have the same power consumption and gm. In Fig. 13, the nonlinear components of their differential output currents are shown after being normalized to the tail current (I_{bias}). It can be seen that the proposed pseudo-differential amplifier is much more linear than its fully differential counterpart. It requires 2× less power for the same linearity. However, removing the tail current source makes a pseudo-differential amplifier difficult to bias robustly. The dynamic biasing techniques proposed for SC designs [4, 18, 19] are not suitable for CT operation. Furthermore, pseudo-differential amplifiers usually suffer from poor power supply rejection ratio (PSRR) and common mode rejection ratio (CMRR) [4].

The proposed pseudo-differential amplifier is shown in Fig. 14. Chopping is often employed to reduce 1/*f* noise in audio CTΔΣMs. In [2], a capacitively coupled inverter-based pseudo-differential amplifier incorporating chopping is proposed. As shown in Fig. 14, it uses ac coupling capacitances (C_c) and large resistors ($R_b = 3\ \text{M}\Omega$) to bias its input transistors at the desired current levels and simultaneously block input common-mode variations. The biasing voltages (V_{bni}, V_{bpi}, V_{bnc}, and V_{bpc}) are generated by a constant-gm biasing circuit.

The combination of R_b and C_c behaves like a high-pass input filter. Setting its corner frequency below the audio band (<20 Hz) would require extremely large resistors and/or capacitors making this approach difficult to integrate. Instead, choppers are used to up-modulate audio signals to f_{chop} before this filter and then demodulate them back into an output DC. In this way, the high-pass filter's corner frequency only has to be lower than f_{chop}. To avoid folding down the quantization

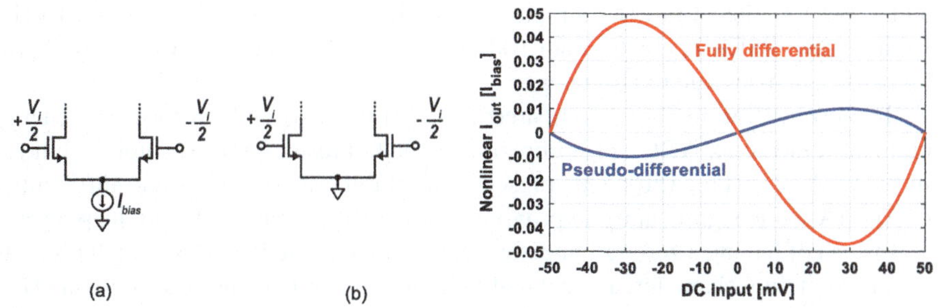

Fig. 13 Nonlinear components of I_{out} for a fully differential (**a**) and a pseudo-differential (**b**) amplifier

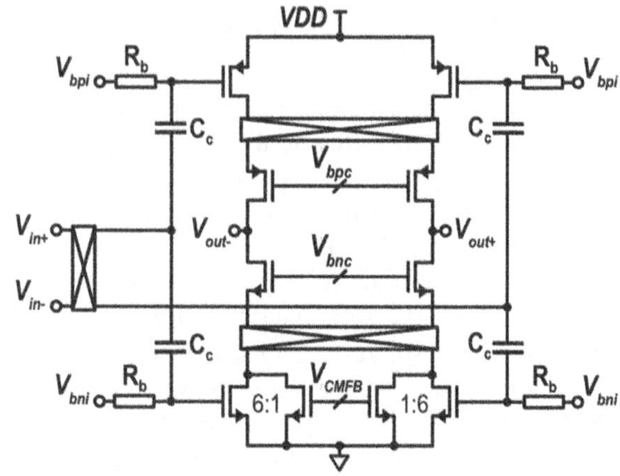

Fig. 14 Simplified schematic of the proposed amplifier in [2]

noise present at the virtual ground node, the choppers are driven at the sampling frequency ($f_{chop} = f_s$) [7, 20]. Since the output choppers are placed in a high bandwidth node between the input devices and the cascodes, the DC gain reduction due to these is negligible.

For linearity, the coupling capacitors ($C_c = 2$ pF) are implemented as metal fringe capacitors and designed to be much larger than the gate capacitances of the input transistors to minimize signal attenuation. The polysilicon biasing resistors ($Rb = 3$ MΩ) are chosen to ensure that the high-pass corner frequency is much less than f_s. In the layout, R_b is placed under C_c to reduce the total area of the four R_b–C_c pairs to 0.01 mm^2.

The NMOS input transistors are split in a 6:1 ratio, with the smaller branch being used for common-mode feedback (CMFB). A CT CMFB circuit is used to sense and stabilize the amplifier's output common-mode voltage [18]. The input and output choppers also chop the offset and low-frequency noise contributed by the CMFB loop itself.

The total power consumption of the amplifier is 205 μW, including the chopper drivers, biasing, and CMFB circuits. Its nominal and minimum DC gains are 60 and 55 dB, respectively, over process, voltage, and temperature (PVT) (−55 °C–150 °C and 1.6–2 V). The amplifier's simulated CMRR is greater than 70 dB up to 1 kHz. Its simulated PSRR is greater than 100 dB up to 1 kHz and greater than 50 dB for higher frequencies due to chopping.

To improve the linearity compared to [2] (Fig. 15a) and, in turn, the energy efficiency, the zoom ADC in [3] implements the tail resistor linearization technique in the OTA [21] (Fig. 15b). This preserves all the merits of the pseudo-differential (PD) OTA while significantly improving its linearity and thus reducing its power.

Instead of having a zero tail impedance, as shown in the PD OTA, which gives its *gm* an expanding characteristic and makes it more linear than the conventional OTA, the TRL OTA sets an optimum tail impedance (R_{tail}). With the optimum R_{tail}, the dominant third-order nonlinearity of the OTA (when the input transistors are biased

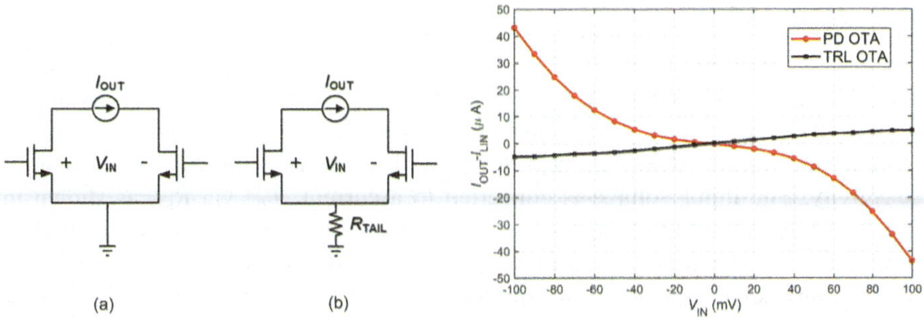

Fig. 15 Linearity comparison between pseudo-differential OTA and TRL OTA

Fig. 16 Tail resistor-linearized OTA

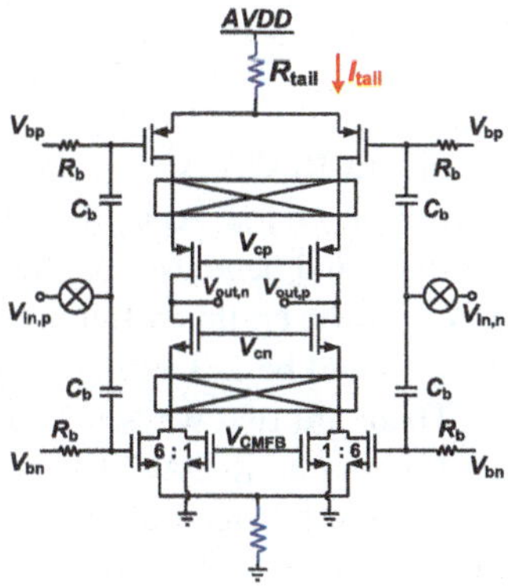

in weak or moderate inversion) can be canceled, thereby improving its linearity significantly.

To illustrate this, the output characteristics (I_{out}-I_{lin} vs. V_{in}) of the PD OTA and the TRL OTA are biased in weak inversion with the same transistor sizes, and I_{tail} (112 μA) is plotted in Fig. 15. The value of R_{tail} to maximize the linearity of the TRL OTA is calculated using (6) [21] and is further optimized based on the simulation results.

$$R_{\mathrm{tail}} = \frac{\eta V_T}{2I_{\mathrm{tail}}} \tag{6}$$

In Eq. (6), I_{tail} is the bias current of the amplifier, V_{T} is the thermal voltage, and η is a process-dependent ideality factor. For the same I_{tail} and g_m, the TRL OTA is ~17× more linear than the PD OTA as shown in Fig. 15. This allows for a significant reduction in I_{tail} and allows it to be determined solely by the thermal noise considerations. The schematic of the tail resistor-linearized (TRL) OTA is shown in Fig. 16.

4.2 DAC Drivers

The DAC of the CT zoom ADC is one of its most critical blocks, as it directly impacts its total input-referred noise, total harmonic distortion (THD), and clock jitter sensitivity. An NRZ DAC is preferred for high energy efficiency and low jitter sensitivity. The input voltage is converted to a current (I_{in}) via R_{in}, as shown in Fig. 17. After subtracting the DAC current (I_{DAC}), their difference (I_{OTA}) is then integrated. The maximum value of I_{OTA} defines the output current requirements of the OTA and hence its power consumption. The maximum input current ($I_{\mathrm{in,max}}$) for a sinusoidal input with amplitude $V_{\mathrm{in,max}}$ is:

$$I_{\mathrm{in,max}} = \frac{V_{\mathrm{in,max}}}{R_{\mathrm{in}}}$$

Figure 18 shows I_{OTA} and I_{DAC} for NRZ and RZ DACs for a zoom ADC based on a 3-bit coarse ADC. For an NRZ DAC, the difference between I_{in} and I_{DAC} is constant and decreases as the resolution of the coarse DAC is increased. For an RZ DAC, however, this difference is much larger, since I_{DAC} is sometimes zero, and so I_{OTA} should be as large as $I_{\mathrm{in,max}}$.

There are two ways to implement a two-level NRZ DAC: as a current DAC (I-DAC) or as an R-DAC. However, an I-DAC will generate extra distortion due to the interaction between the nonlinear output impedance of its current sources and the voltage swing at the virtual ground of the OTA. An R-DAC is not only more linear, but it also has lower thermal and 1/*f* noise [22]. Thus, an R-DAC is used in [2, 3].

ISI refers to the signal-dependent errors that occur at code transitions due to the mismatch between finite rise/fall times of the currents generated by the unit elements of the R-DAC. The use of DWA makes this problem even worse because it increases the number of unit element transitions in the DAC and introduces even-order distortion [23]. In these works [2, 3], a novel ISI reduction technique is proposed to solve this problem.

In the output of the differential R-DAC unit element shown in Fig. 19, there are four different transition edges: t_{rp}, t_{rn}, t_{fp}, and t_{fn}. If the total amount of positive and negative DAC output currents within one period would match, there would be no nonlinear ISI error [23]. One approach to achieve this is to match a rising edge

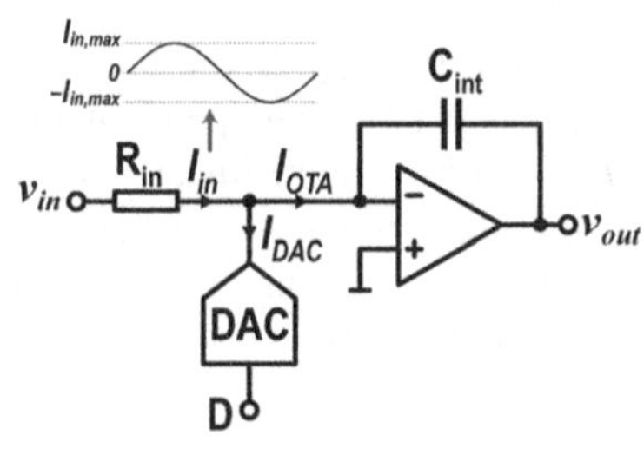

Fig. 17 I_{DAC} and I_{OTA} for NRZ and RZ DACs

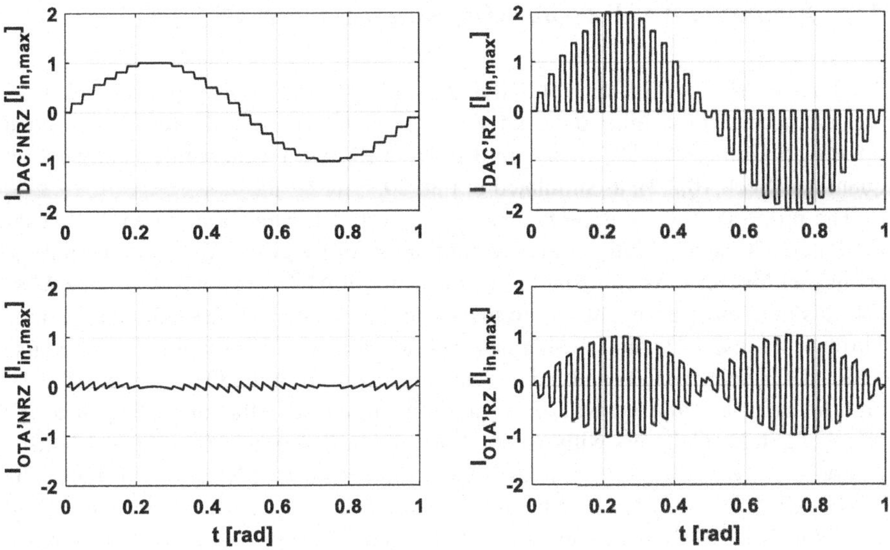

Fig. 18 I_{DAC} and I_{OTA} for NRZ and RZ DACs

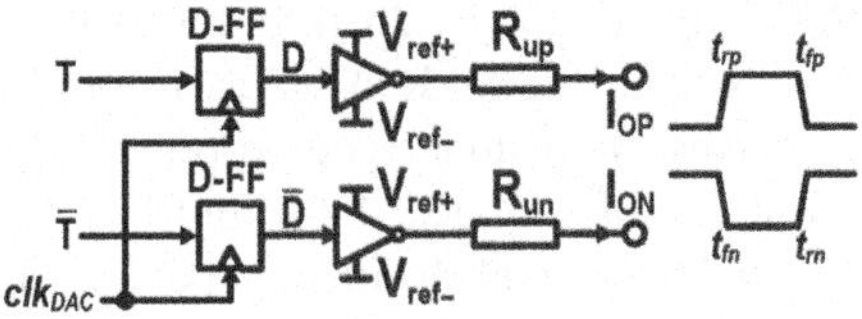

Fig. 19 Schematic of an R-DAC cell

with its corresponding falling edge (match t_{rp} and t_{fn}, and match t_{rn} and t_{fp}) [7, 24]. However, this is hard to guarantee in practice since the speed of the rising edges is set by PMOS drivers, while the speed of the falling edges is set by NMOS drivers. Thus, background calibration is often necessary for this approach [7, 24].

Alternatively, we note that to avoid ISI, it is only necessary to match the rising and falling edges of the positive and the negative half DACs (match t_{rp} and t_{rn}, and match t_{fp} and t_{fn}). This is comparatively easy to achieve because the edges that need to be matched are generated by the same type of devices. However, the positive and the negative DAC unit resistors also need to match, as they also influence the resulting rise and fall times. Simulations indicated that the 1% matching needed for low DWA in-band noise (IBN) is also more than enough to achieve < -120 dB HD2. The positive and negative half DACs should then be laid out next to each other. Noting that the ON resistances of the DAC switches are much smaller than R_{up} and R_{un}, the matching requirements on the driver inverters can be relaxed to 5%. The switch driving signal asymmetry, which is also a source of ISI error, is reduced by using two separate flip-flops to drive D and $\overline{D}$, as shown in Fig. 19.

4.3 Example: A CT Zoom ADC with a Coarse NS SAR

A CT zoom ADC with 106.6 dB $SNDR_{MAX}$ in a 24 kHz BW is proposed in [3]. To reach a thermal noise-limited design, the target SQNR of 120 dB SQNR is reached by using a 4-bit SAR ADC with a third-order $\Delta\Sigma$M with local feedback and a 2-bit quantizer (Sect. 10.2.2), as simulated in Fig. 20.

The proposed CT zoom ADC is shown in Fig. 21. Since this zoom ADC uses a 4-bit coarse ADC, the tonal quantization noise of the SAR ADC and its internal swings will increase as compared to previous zoom ADCs with a 5-bit coarse ADC [2]; this puts more stringent requirements on the linearity of the individual OTAs. Through the use of analog techniques such as passive noise shaping in the coarse SAR ADC (Sect. 10.2.3) and tail resistor linearization in the OTA_1 (Sect. 10.4.1), high linearity is achieved without compromising energy efficiency. To reduce the effect of ISI, DAC drivers with matched rise and fall times are used (Sect. 10.4.2).

For a 1 kHz, −0.35 dBFS input signal, the measured SNDR and THD of the 4-bit SAR ADC with NS "OFF" are 30.3 dB and − 30.4 dB, respectively (Fig. 22). With NS "ON," its in-band quantization noise is suppressed by about 9.5 dB, while its SNDR and THD improve by 14.5 dB and 14.6 dB, respectively. Turning NS "ON" effectively suppresses in-band fuzz and improves the SNDR and THD of the zoom ADC by 1.2 dB and 5.5 dB, respectively. The tone at $f_s/2$ with NS "ON" comes from the SAR ADC as it toggles predominately between two levels while resolving the residue. The proposed ADC achieves a competitive Schreier figure of merit (FOM_S) of 183.4 dB and FOM_{SNDR} of 182.7 dB, together with the lowest in-band noise and distortion spectral density (NDSD = −150.4 dBFS/Hz) among state-of-the-art audio ADCs.

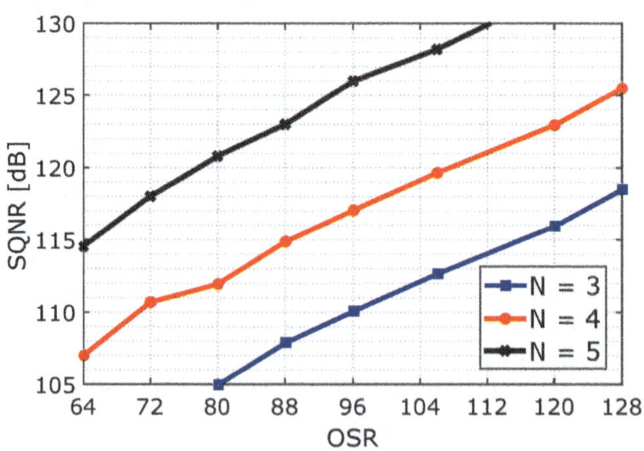

Fig. 20 OSR vs. SQNR for a third-order CT zoom ADC with notch and 2-bit quantizer

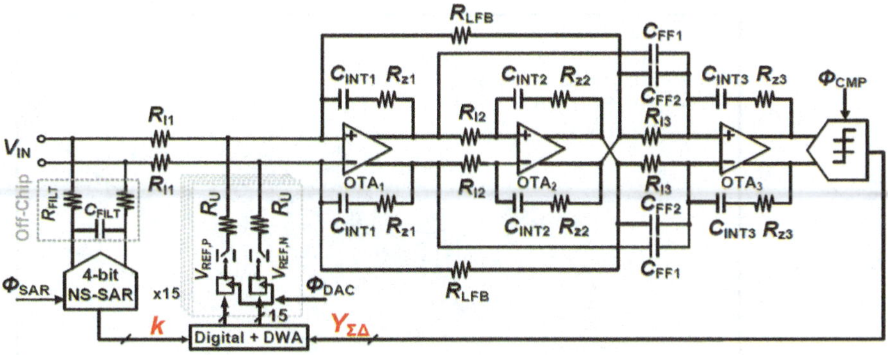

Fig. 21 Simplified schematic of the proposed CT zoom ADC

Fig. 22 Measured PSD of zoom ADC (2^{21} points, 8 averages, Blackman-Harris window)

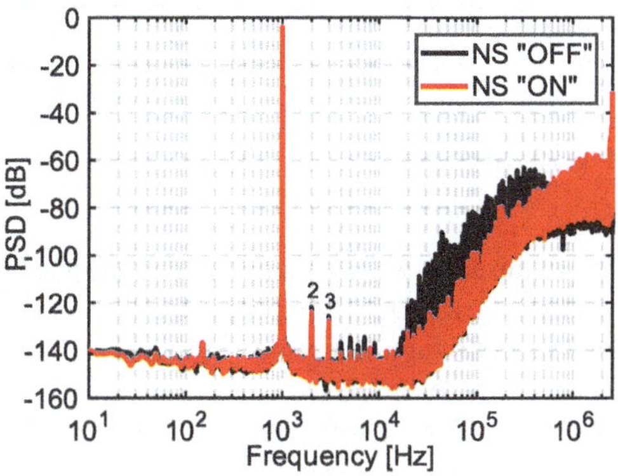

5 Conclusions

This book chapter describes the evolution of zoom ADCs for high-resolution instrumentation and audio applications. The use of an asynchronous SAR ADC as a coarse converter drastically improves the immunity to out-of-band interferers while keeping the over-ranging at ± 1 LSB. The use of a 2-bit flash quantizer in the fine $\Delta\Sigma$M fully recovers the SQNR loss (~9 dB) due to the use of 1 LSB over-ranging and improves the immunity to out-of-band interferers while keeping the benefits of over-ranging such as relaxed specifications for the SAR ADC intact. To improve the impact of fuzz on linearity, techniques that rely on the suppression of the SAR quantization noise, such as an NS SAR, are preferred over techniques that rely on cancellation, due to their low power, low area, and tolerance to circuit mismatch. For the DT zoom ADCs, differential sampling, together with chopping OTA_1, is preferred over auto-zeroing and CDS, due to its benefits in terms of CMRR, low offset, and low $1/f$ noise. CT loop filters for high-resolution zoom ADCs improve

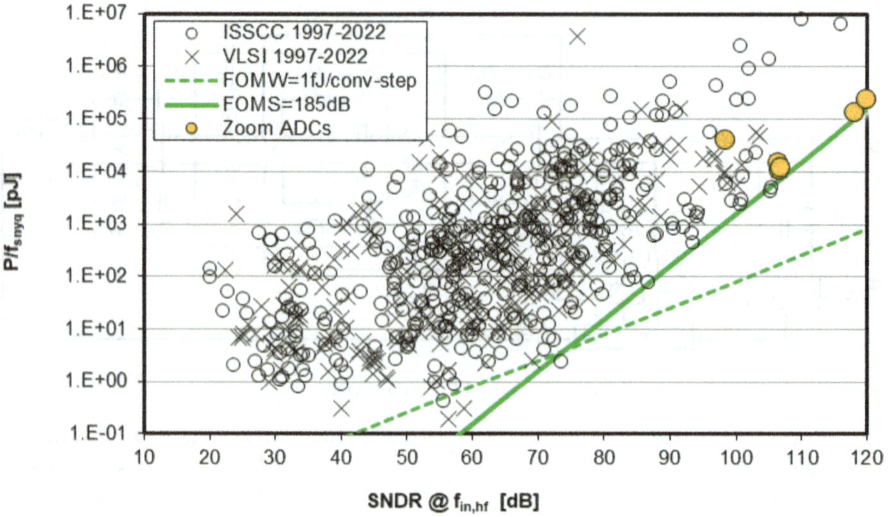

Fig. 23 ADC survey

the requirements for the input and reference buffers due to their resistive load and are preferred over DT loop filters. A tail resistor-linearized OTA improves the linearity of the first-stage integrator by nearly 17× as compared to a tail current source, allowing it to be biased solely for noise, and thus improves the energy efficiency of the ADC. Matching rise and fall times in the DAC drivers improves the overall linearity of the system.

These advances in zoom ADCs lead to high resolution at very high energy efficiency (Schreier DR and SNDR FoMs >180 dB) as shown in Fig. 23 and Table 1 and make them well suited for instrumentation and audio applications. In recent publications, however, other architectures reach better energy efficiency and lower area, at similar resolution and bandwidth. So, what future approaches can be taken to evolve the zoom ADC towards even higher performance? Is a combination of architectures the key, or could a more "digital" zoom ADC lead to improved performance in advanced technology nodes? Only silicon will tell us the answer to these questions.

Table 1 Table of comparison

	[3]	[25]	[26]	[27]	[1]	[8]	[2]	[6]	[4]
Architecture	CT Zoom	DT PPD	CT	CT FIRDAC	DT Zoom	CT 3-level	CT Zoom	DT Zoom	DT zoom
Tech (nm)	160	180	28	65	160	65	160	160	160
Area (mm^2)	0.36	0.03	0.07	0.39	0.27	0.28	0.27	0.25	0.16
Supply (V)	1.8	1.8/1.1	1.8/1.0	1.2	1.8	1.2	1.8	1.8	1.8
Power (μW)	590	203.5	116	139	440	134	618	280	1120
f_s (MHz)	5.12	5.80	6.144	7.2	3.5	8	5.12	2	11.29
BW (kHz)	24	20	24	24	20	24	20	1	20
SNR_{max} (dB)	107.2	106.7	100.7	102.0	107.5	101.0	108.1	119.1	106
$SNDR_{max}$ (dB)	106.6	105.4	100.6	100.9	106.5	99.4	106.4	118.1	103
DR (dB)	107.3	108.8	104.4	104.8	109.8	103.5	108.5	120.3	109
FoM_{SNDR}[a] (dB)	182.7	185.3	183.7	183.3	183.1	181.9	181.5	183.6	175.5
FoM_S[b] (dB)	183.4	188.7	187.5	187.2	186.4	186.0	183.6	185.6	181.5

[a]$FoM_{SNDR} = SNDR + 10\log_{10}(BW/Power)$; [b]$FoM_S = DR + 10\log_{10}(BW/Power)$

References

1. E. Eland, S. Karmakar, B. Gönen, V.R. Veldhoven, K. Makinwa, A 440-μW, 109.8-dB DR, 106.5-dB SNDR discrete-time zoom ADC with a 20-kHz BW. IEEE J. Solid State Circuits **56**(4), 1–2 (2021)
2. B. Gönen, S. Karmakar, V.R. Veldhoven, K. Makinwa, A continuous-time zoom ADC for low-power audio applications. *IEEE Journal of Solid-State Circuits* **55**(4) (2020)
3. S. Mehrotra, E. Eland, S. Karmakar, A. Liu, B. Gönen, V.R. Veldhoven, K. Makinwa, *A 590 μW, 106.6 dB SNDR, 24 kHz BW continuous-time zoom ADC with a noise-shaping 4-bit SAR ADC* (ESSCIRC Conference 2022 – 48th European Solid-State Circuits Conference (ESSCIRC), Milan, 2022)
4. B. Gönen, F. Sebastiano, R. Quan, V.R. Veldhoven, K. Makinwa, A dynamic zoom ADC with 109-dB DR for audio applications. IEEE J. Solid-State Circuits **52**(6) (2017)
5. Y. Chae, K. Souri, K. Makinwa, A 6.3 μW 20 bit incremental zoom-ADC with 6 ppm INL and 1 μV offset. IEEE J. Solid-State Circuits **48**(12) (2013)
6. S. Karmakar, B. Gönen, F. Sebastiano, V.R. Veldhoven, K. Makinwa, A 280 μW dynamic zoom ADC with 120 dB DR and 118 dB SNDR in 1 kHz BW. IEEE J. Solid-State Circuits **53**(12) (2018)
7. S. Billa, A. Sukumaran, S. Pavan, Analysis and design of continuous-time delta–sigma converters incorporating chopping. IEEE J. Solid State Circuits **52**(9), 2350–2361 (2017)
8. M. Jang, C. Lee, Y. Chae, *A 134 μW 24 kHz-BW 103.5db-dr ct $\Delta\Sigma$ modulator with chopped negative-R and tri-level fir DAC* (IEEE Int. Solid-State Circuits Conf. (ISSCC) Dig. Tech. Papers, San Francisco, 2020)
9. M. Jang, C. Lee, Y. Chae, Analysis and design of low-power continuous-time delta-sigma modulator using negative-R assisted integrator. IEEE J. Solid State Circuits **54**(1), 277–287 (2019)
10. A. Sukumaran, S. Pavan, Low power design techniques for single-bit audio continuous-time delta sigma ADCs using FIR feedback. IEEE J. Solid State Circuits **49**(11), 2515–2525 (2014)
11. R. Baird, T. Fiez, Linearity enhancement of multibit $\Delta\Sigma$ A/D and D/A converters using data weighted averaging. IEEE Trans. Circuits Syst. I, Reg. Papers **42**(12), 753–762 (1995)
12. J. Silva, U. Moon, J. Steensgaard, G.C. Temes, Wideband lowdistortion. IET Electron. Lett **37**(12), 737–738 (2001)
13. Y.-Z. Lin, C.-Y. Lin, S.-C. Tsou, C.-H. Tsai, C.-H. Lu, *20.2 A 40MHz-BW 320MS/s Passive Noise-Shaping SAR ADC With Passive Signal-Residue Summation in 14nm FinFET* (2019 IEEE International Solid-State Circuits Conference – (ISSCC), 2019)
14. Y. Chae, G. Han, Low voltage, low power, inverter-based switched-switched capacitor. IEEE J. Solid State Circuits **44**(2), 458–472 (2009)
15. C. Enz, G. Temes, Circuit techniques for reducing the effects of op-amp imperfections: Autozeroing, correlated double sampling, and chopper stabilization. Proc. IEEE **84**(11), 1584–1614 (1996)
16. A. Abo, P. Gray, A 1.5-V, 10-bit, 14.3-MS/s CMOS pipeline analog-to-digital converter. IEEE J. Solid State Circuits **34**(5), 599–606 (1999)
17. K. Nguyen, R. Adams, K. Sweetland, H. Chen, A 106-dB SNR hybrid oversampling analog-to-digital converter for digital audio. IEEE J. Solid State Circuits **40**(12), 2408–2415 (2005)
18. M.S. Akter, R. Sehgal, F. Goes, K.A.A. Makinwa, K. Bult, A 66-dB SNDR pipelined split-ADC in 40-nm CMOS using a class-AB residue amplifier. IEEE J. Solid State Circuits **53**(10), 2939–2950 (2018)
19. S. Lee, W. Jo, S. Song, Y. Chae, A 300-μW audio $\Delta\Sigma$ modulator with 100.5-dB DR using dynamic bias inverter. IEEE Trans. Circuits Syst. I Reg. Papers **63**(11), 1866–1875 (2016)
20. H. Jiang, B. Gönen, K.A.A. Makinwa, S. Nihitanov, *Chopping in continuous-time sigma-delta modulators* (Proc. IEEE Int. Symp. Circuits Syst. (ISCAS), 2017), pp. 1–4
21. S. Pan, K.A.A. Makinwa, A 10 fJ·K^2 Wheatstone bridge temperature sensor with a tail-resistor-linearized OTA. IEEE J. Solid State Circuits **56**(2), 501–510 (2021)

22. P. Shettigar, S. Pavan, Design techniques for wideband single-bit continuous-time $\Delta\Sigma$ modulators with FIR feedback DACs. IEEE J. Solid State Circuits **47**(12), 2865–2879 (2012)
23. L. Risbo, R. Hezar, B. Kelleci, H. Kiper, M. Fares, Digital approaches to ISI-mitigation in high-resolution oversampled multi-level D/a converters. IEEE J. Solid State Circuits **46**(12), 2892–2903 (2011)
24. T. He, M. Ashburn, S. Ho, Y. Zhang, G. Temes, A 50 MHz-BW continuous-time $\Delta\Sigma$ ADC with dynamic error correction achieving 79.8 dB SNDR and 95.2 dB SFDR. IEEE ISSCC Dig. Tech. Papers, 230–232 (2018)
25. C.Y. Lee, U.-K. Moon, *A 0.0375mm2 203.5μW 108.8dB DR DT single-loop DSM audio ADC using a single-ended ring-amplifier-based integrator in 180nm CMOS* (2022 IEEE International Solid-State Circuits Conference (ISSCC), 2022), pp. 412–414
26. C. Lo, J. Lee, Y. Lim, Y. Yoon, H. Hwang, J. Lee, M. Choi, M. Lee, S. Oh, J. Lee, *A 116μ W 104.4dB-DR 100.6dB-SNDR CT* $\Delta\sum$ *audio ADC using tri-level current-steering DAC with gate-leakage compensated off-transistor-based bias noise filter* (2021 IEEE International Solid-State Circuits Conference (ISSCC), 2021), pp. 164–166
27. S. Mondal, O. Ghadami, D.A. Hall, *A 139 μ W 104.8dB-DR 24kHz-BW CT* $\Delta\Sigma M$ *with chopped AC-coupled OTA-stacking and FIR DACs* (2021 IEEE International Solid-State Circuits Conference (ISSCC), 2021), pp. 166–168

Pushing the Limits of kT/C Noise in Delta-Sigma Modulators

Spyridon Kalogiros, Gerardo Salgado, Colin Lyden, Kevin McCarthy, and Ivan O'Connell

Abstract Thermal noise, which is sampled and aliases in-band in discrete-time systems, limits the achievable performance of switched-capacitor noise-shaping Analog-to-Digital Converters (ADCs). While the performance of such ADCs has advanced significantly over the last 20 years, as quantified by the Schreier figure of merit (FoM_S), the theoretical limit of 192 dB remains unchallenged. Over that period, the envelope of ADC performance has advanced from a FoM_S of 163 dB, 20 years ago, to 186 dB today, with a rate of advancement, corresponding to ADC performance, which is doubling every 1.6 years. However, this rate of advancement has started to slow in recent years. This chapter will review some of the recent advancements in relation to reducing the thermal noise in switched-capacitor Delta-Sigma Modulators. In addition, this chapter will address many of the challenges associated with breaking the 192 dB FoM_S performance barrier.

1 Introduction to kT/C Noise

Sampled thermal noise plays a crucial role in the performance of switched-capacitor ADCs, significantly degrading their Signal-to-Noise Ratio (SNR) performance, and it has been a field of research study for decades. To study its origins, the fundamental sampling circuit has to be considered, which consists of a MOSFET device biased

S. Kalogiros (✉)
5 Grande Rue, Sevres, France
e-mail: spyridon.kalogiros@umail.ucc.ie

G. Salgado
Tyndall National Institute, Cork, Ireland

C. Lyden · I. O'Connell
Analog Devices, Cork, Ireland

K. McCarthy
Department of Electrical and Electronic Engineering, University College Cork, Cork, Ireland

P. Harpe et al. (eds.), *Biomedical Electronics, Noise Shaping ADCs, and Frequency References*, https://doi.org/10.1007/978-3-031-28912-5_11

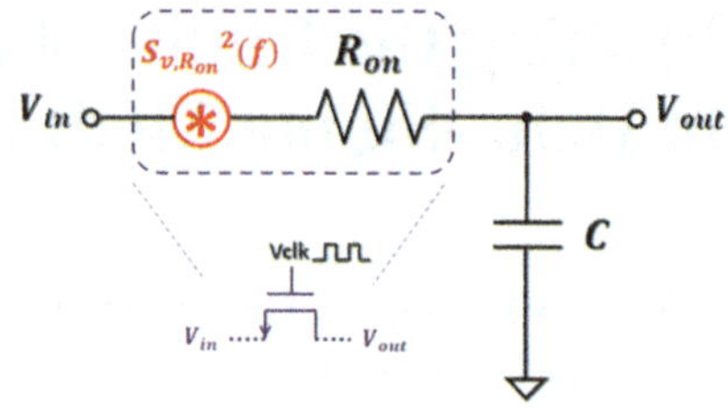

Fig. 1 Sampling equivalent circuit to study sampled thermal noise

in the deep triode region operating as a switch, and thus, it can be replaced by an equivalent resistor, R_{on}, driving a sampling capacitor, C, in series:

Illustrated in Fig. 1, $S_{v,Ron}$ is the power spectral density of the switch equivalent resistance's thermal noise source. Noise that arises from the random motion of carriers within a conductor is typically called thermal noise due to its temperature dependence. Its spectral power density is given by:

$$S_{v,R_{on}}{}^2(f) = 4kTR_{on}, \tag{1}$$

where k is the Boltzmann constant (1.380649×10^{-23} m^2 kg s^{-2} K^{-1}); T is the absolute temperature, in kelvin; and R_{on} is the equivalent on-resistance of the MOS device. In order to calculate the total mean squared noise voltage, which in switched-capacitor circuits is determined by the sampling capacitor, Parseval's theorem can be used [1]. The total mean squared noise voltage is the product of $S_{v,Ron}$ and the equivalent noise bandwidth (ENBW) of the circuit, and is given by:

$$\overline{v_n{}^2} = S_{v,R_{on}}{}^2(f) * \text{ENBW}, \tag{2}$$

where:

$$\text{ENBW} = \int_0^{\infty} \left| \frac{1}{1 + j2\pi f R_{on} C} \right|^2 df. \tag{3}$$

For a single-pole system, Eq. (3) becomes as follows [1]:

$$\text{ENBW} = \frac{1}{4R_{on}C}. \tag{4}$$

Substituting Equation (4) into (2) results in:

$$\overline{v_n{}^2} = 4kTR_{on} * \frac{1}{4R_{on}C} \tag{5}$$

The resulting noise voltage from (5) is the following:

$$\sqrt{\overline{v_n{}^2}} = \sqrt{\frac{kT}{C}}. \tag{6}$$

The existence of kT/C noise fundamentally limits the SNR performance of a switched-capacitor circuit. This limitation can be evident in the SNR equation below, considering the signal power of a sinusoidal-type input signal, with amplitude V (in volts), and the power of kT/C noise [2]:

$$\mathrm{SNR} = \frac{V^2/2}{kT/C}. \tag{7}$$

By introducing values to the capacitor logarithmically from 10 fF to 10 pF, the resulting kT/C noise voltages can be calculated, as shown in Table 1 below.

If an input signal with an amplitude of 1 V is considered, by sweeping the SNR from 20 dB to 140 dB in steps of 20 dB, the capacitor values that are needed to achieve each value of SNR are shown in Table 2 below [2]:

While it's relatively easy to achieve performances at the range of 60–80 dB, very high capacitor values are required to achieve performances from 100 dB and beyond, which is very difficult to be integrated in silicon.

From (6), it is evident that the resistance R_{on} has been eliminated from the calculation of kT/C noise voltage value, but when its power spectral density (PSD) is plotted in the frequency domain, R_{on} impacts both the bandwidth and the magnitude. A lower resistance increases the bandwidth (BW) and decreases the maximum noise voltage. The total integrated noise in any case remains the same. This is visualized in Fig. 2.

Table 1 kT/C noise voltage values for 10 fF–10 pF capacitor sizes

Capacitor	$\sqrt{kT/C}$ noise
10 fF	640 μV
100 fF	200 μV
1 pF	64 μV
10 pF	20 μV

Table 2 Capacitor values for a given SNR performance

SNR (dB)	C (pF)
20	0.00000083
40	0.000083
60	0.0083
80	0.83
100	83
120	8300
140	830000

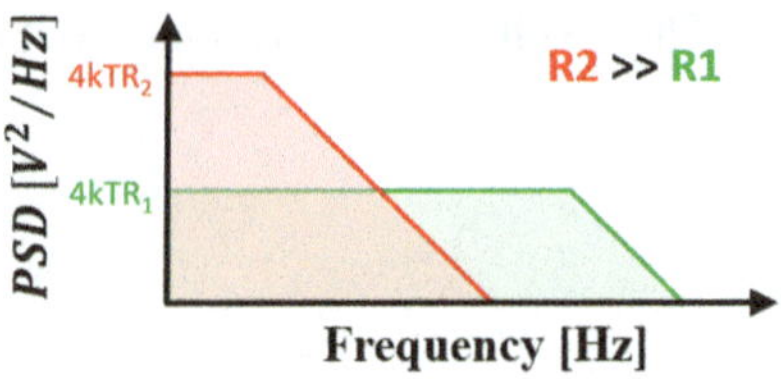

Fig. 2 Power spectral density of kT/C noise vs R_{on} resistance

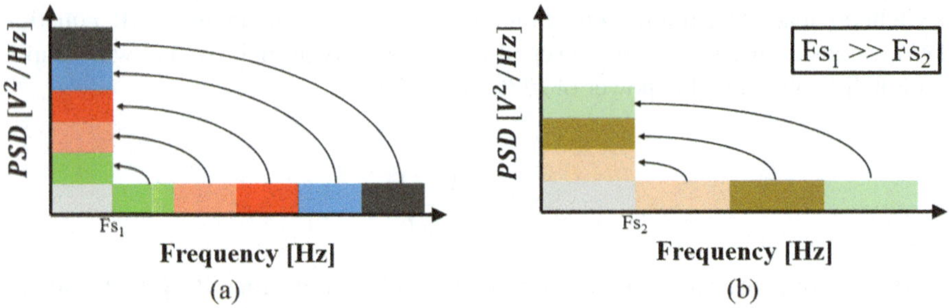

Fig. 3 In-band aliasing of thermal noise due to sampling

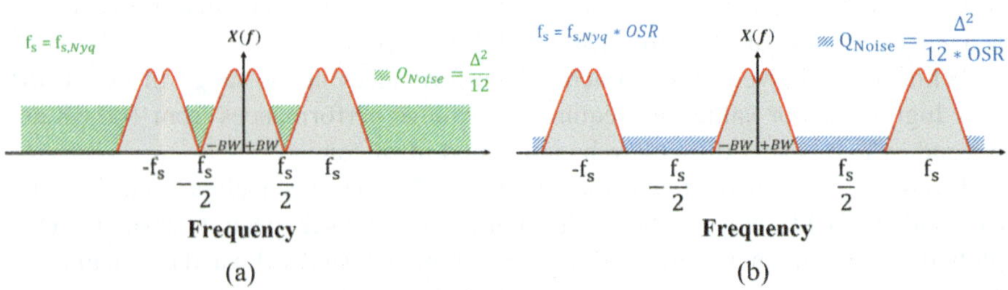

Fig. 4 (**a**) Noise floor in Nyquist-rate sampling. (**b**) Noise floor with oversampling with an oversampling ratio of OSR

Thermal noise when sampled in switched-capacitor circuits folds back and aliases from higher to lower frequencies, increasing the resultant in-band noise. This amount depends on the existing total integrated noise, which is set primarily by the capacitor size. As the sampling frequency becomes smaller, more in-band noise is generated. Figure 3 visualizes the noise aliasing.

The oversampling technique is a key aspect of oversampling ADCs, like delta-sigma modulators, to further improve their SNR performance, compared to their Nyquist-rate counterparts. Sampling at rates higher than two times the Nyquist frequency, $f_{s,\mathrm{Nyq}}$ (where $f_{s,\mathrm{Nyq}} = 2*\mathrm{BW}$), moves the signal aliases higher in frequency and significantly reduces the overall quantization noise (Q_{Noise}) floor [2]. Figure 4 illustrates the case of noise reduction with oversampling. On this figure, parameter Δ represents the quantization step size.

2 Introduction to the FoM_S and Current State-of-the-Art Trends

Delta-Sigma Modulators are characterized as noise-limited systems. Their performance is broadly benchmarked (in dB) by the Schreier Figure of Merit (FoM_S), which is the most appropriate figure of merit to be taken into account in the case of benchmarking noise-limited ADCs. Firstly, it includes a noise quantity in its calculation, and secondly, it captures most accurately the quadruplication of the consumed power per added bit [3]. The mathematical expression of FoM_S is given by [3]:

$$\mathrm{FoM_s\,(dB)} = \mathrm{SNDR\,(dB)} + 10\log\left(\frac{\mathrm{BW}}{P}\right), \tag{8}$$

where SNDR is the Signal-to-Noise and Distortion, BW is the bandwidth of the input signal (in Hz), and P is the total power consumption (in W). The FoM_S can be visualized in the energy per conversion vs SNDR plot of the state of the art, in Fig. 5 [4]. The red line on the plot indicates the FoM_S frontier and is associated with an increase by four of the power per added bit in noise-limited converters, whereas the green dashed line illustrates the frontier for technology-limited converters, benchmarked using the Walden FoM (FoM_W) [3] and associated with a doubling of the power per added bit.

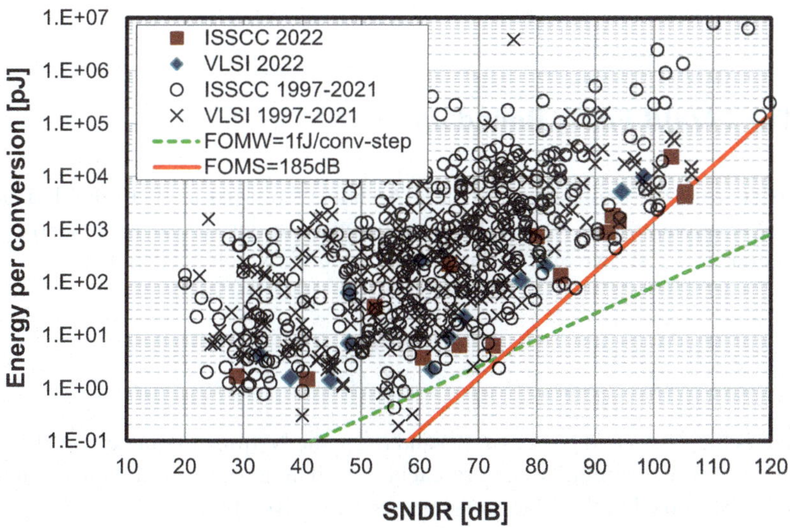

Fig. 5 Energy per conversion vs SNDR state-of-the-art plot [4]

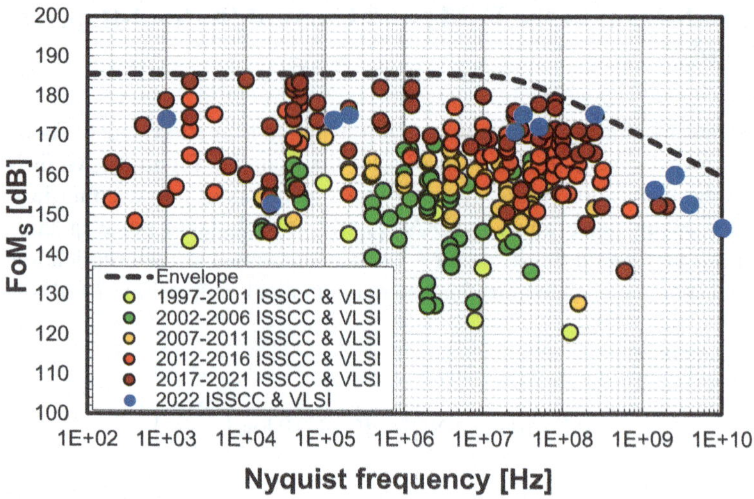

Fig. 6 FoM_S vs Nyquist frequency state-of-the-art plot [4]

Another representation of the FoM_S is shown in Fig. 6, where the FoM_S for the State-of-the-Art is plotted versus the Nyquist frequency, categorized in groups of 5 years advancement, apart from those of 2022 [4].

The rate of improvement indicates that there is an advancement of 1 dB in the FoM_S per year [3, 5, 6]. However, there is a fundamental limitation for this advancement, which is going to be analyzed in the next section of this chapter [3, 5].

3 The 192 dB FoM_S Boundary

Alongside with the expression in (8), kT/C noise from (6) leads to the derivation of a FoM_S boundary for ADCs. The first quantity to derive this is the energy per conversion, E_{conv}, shown also in the Y-axis of Fig. 5, which is given by:

$$E_{\text{conv}} = {}^{P}\Big/_{f_{\text{s,Nyq}}}, \tag{9}$$

where P is the dissipated power (in W) and $f_{\text{s,Nyq}}$ the Nyquist frequency (in Hz). The minimum power for circuits that are capable to handle rail-to-rail signal voltages, which is also linked to the kT/C noise presence, is given by [8]:

$$P_{\min} = 8kT * f_{\text{s,Nyq}} * \text{SNR} . \tag{10}$$

Combining Eqs. (9) and (10), the minimum energy it takes to drive a sampling capacitor is obtained as follows [3, 7–10]:

$$E_{\text{conv,min}} = 8kT * \text{SNR}. \tag{11}$$

Based on the Nyquist frequency, $f_{\text{s,Nyq}}$, which is equal to 2*BW, the FoM_S from (8) can be rewritten as:

$$\text{FoM}_\text{s}\,(\text{dB}) = \text{SNDR}\,(\text{dB}) + 10\log\left(\frac{f_{\text{s,Nyq}}}{2\text{P}}\right). \tag{12}$$

By assuming an ideal situation where SNDR = SNR with no distortion [3], expression (12) can be rewritten using (9) as:

$$\text{FoM}_\text{s}\,(\text{dB}) = \text{SNR}\,(\text{dB}) + 10\log\left(\frac{1}{2E_{\text{conv,min}}}\right). \tag{13}$$

By substituting the minimum energy from (11) to (13), the FoM_S now becomes:

$$\text{FoM}_\text{s} = \text{SNR}\,(\text{dB}) + 10\log\frac{1}{16kT * \text{SNR}}. \tag{14}$$

By applying logarithmic rules, (14) can be rewritten as:

$$\text{FoM}_\text{s} = \text{SNR}\,(\text{dB}) + 10\log 1 - 10\log\,(16kT) - 10\log\,(\text{SNR})\,, \tag{15}$$

which can be further simplified to:

$$\text{FoM}_\text{s} = -10\log\,(16kT)\,, \tag{16}$$

or, equivalently, 192 dB in approximation, at room temperature (T = 300° K). A more realistic approach to this boundary could be a number between 186 and 188 dB, since the 192 dB limitation considers only sampling energy [3] and ideal circuits. The 192 dB limitation can be shown on the State-of-the-Art FoM_S of Fig. 7 [4].

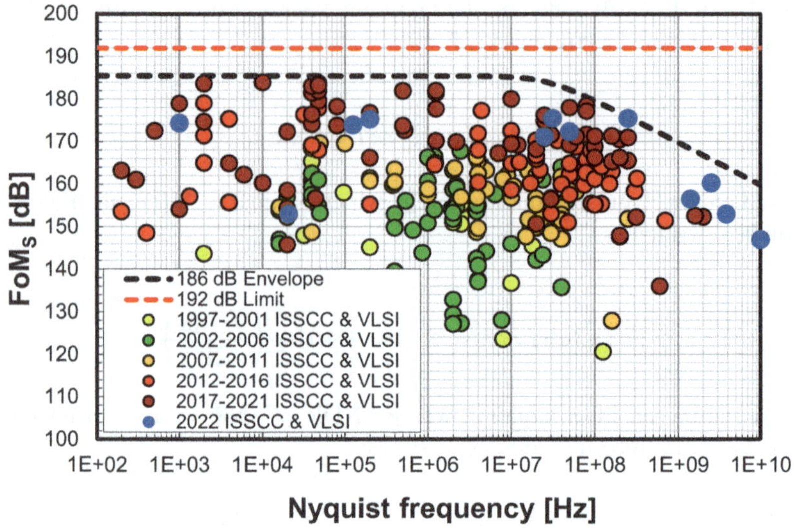

Fig. 7 FoM_S vs Nyquist frequency state-of-the-art plot with the 192 dB FoM_S limit [4]

4 *kT/C* Noise Cancellation: Previous Work

A lot of research has been conducted in order to reduce *kT/C* noise in switched-capacitor circuits. The work in [1] proposes a sampling circuit for image sensors that breaks the relationship between the dominant noise source and the impedance which limits the noise bandwidth, in order to reduce sampled thermal noise. This circuit is illustrated in Fig. 8.

The feedback configuration with a transconductance amplifier, G_m, and a relatively high resistor, R_{FB}, performs the noise reduction by decoupling thermal noise density and bandwidth. Resistors R_{FB} and R_L, alongside with the transconductance amplifier, form a new system with an effective transconductance $G_{m,eff}$, which is the G_m of the amplifier multiplied with the resistive divider formed by the two resistors. Eventually, the total sampled *kT/C* noise is attenuated by $1/(G_{m,\,eff}{*}R_{FB})$. However, the need for active elements in order to beat *kT/C* noise substantially increases the power consumption, and the FoM_S is not improved.

Another useful circuit for the purpose of beating *kT/C* noise in successive approximation (SAR) ADCs is proposed in [11]. In conventional SAR ADCs, the sampled *kT/C* noise is directly added on the input signal, significantly degrading the SNR performance. The work in [11] proposes the addition of a capacitor C_2 and a switch controlled by phase φ_2, placed after the pre-amplifier. Figure 9 illustrates this idea.

This added circuitry, alongside with the clocking scheme of phases φ_1 and φ_2, practically cancels the *kT/C* noise imposed by capacitor C_1 and introduces kT/C

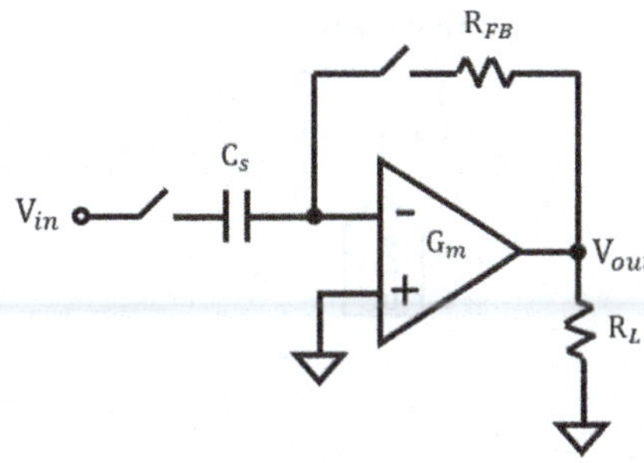

Fig. 8 Circuit from [1] for the reduction of sampled thermal noise

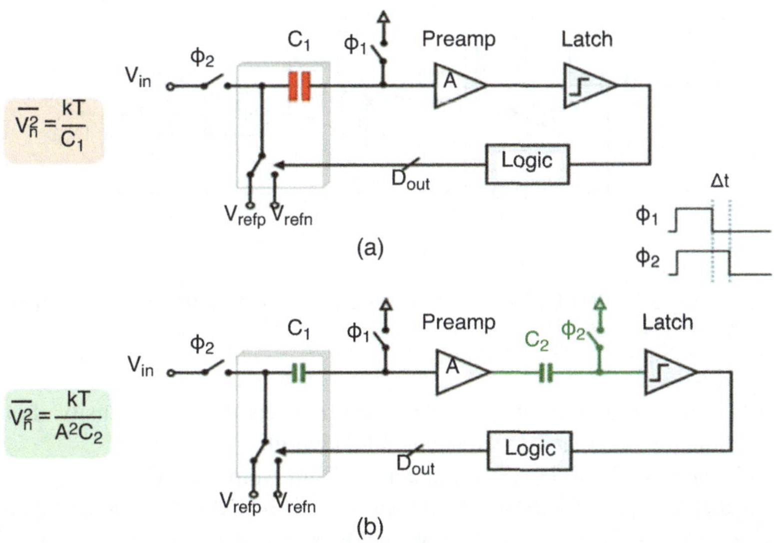

Fig. 9 (**a**) Conventional SAR ADC with bottom-plate sampling and comparator pre-amplifier; (**b**) SAR ADC with *kT/C* noise cancellation [11]

noise only from capacitor C_2, which is attenuated by the gain of the pre-amplifier, since the capacitor is placed after this block.

However, the additional pre-amplifier here has to drive a large capacitive load and a large capacitor C_2, as it has to fully settle for each bit conversion. This configuration significantly reduces sampled thermal noise, but due to increased power consumption, the FoM_S is not advanced.

An interesting approach on cancelling *kT/C* noise has been introduced in [12] and is depicted in Fig. 10. Traditionally, in Continuous-Time Delta-Sigma Modulators (CTDSMs), the power required to drive the input and the Digital-to-Analog Converter (DAC) is larger than the power of the first integrator's operational amplifier (opamp). Consequently, a large amount of power is burnt at the resistance R_{in} of the first integrator, as shown in Fig. 10a. The loading of the first integrator

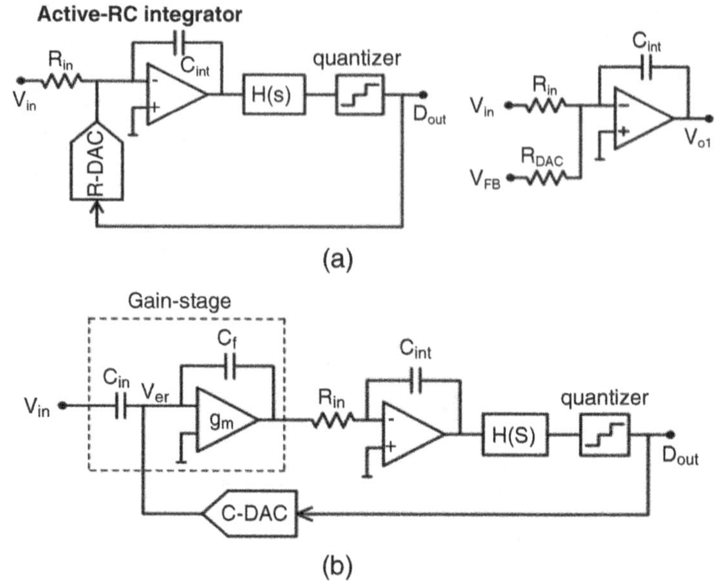

Fig. 10 (**a**) CTDSM with RC integrator front-end; (**b**) CTDSM with capacitive-gain amplifier (CGA) front-end [12]

should be reduced, and the transconductance that sets the noise density should be different than the one that sets the signal's bandwidth.

To alleviate the above issues, a Capacitive-Gain Amplifier (CGA) stage, with an input capacitor, C_{in}, and a feedback capacitor, C_f, is added before the first integrator of the CTDSM, as shown in Fig. 10b. With the existence of this, the "power x noise" tradeoff is significantly reduced by a factor of 18.1, from 29.5 kT to 1.62 kT. Due to the absence of kT/C noise from the input, the capacitors of that stage can be made small, and the power consumption is therefore dominated by the opamp used for the CGA stage. The closed-loop gain of the CGA block is given by C_{in}/C_f, and it is selected to a value that ensures a proper error signal (V_{in}-V_{DAC}) swing for the rest of the blocks in the loop-filter of the modulator.

The CGA block dramatically changes the noise performance of such a modulator, and it can be adapted to modulators which consist of switched-capacitor filters also. To operate properly, the feedback signal has to precisely track the input signal, and that enables larger gains to be applied with better noise performances. It not only improves the kT/C noise performance but advances the FoM_S. In the following section, the operation of this block and its benefits will be discussed.

5 Delta-Sigma Modulators: General Overview

Before introducing the CGA stage and analyzing its benefits, this section revises the basics of Delta-Sigma Modulators. Such a system is considered as a closed-loop system like the one in Fig. 11. It contains the input X(s), a loop-filter with a number of integrators according to its order which has a transfer function H(s), the output signal W(s), and the additive quantization noise source Q(s) to approximate the behavior of a quantizer. Finally, the output signal Y(s), which is fed back to the system, is subtracted from the input signal providing to the loop-filter a quantization error signal as an input. By applying basic algebra rules, the output Y(s) is given by [13–15]:

$$Y(\mathrm{s}) = W(\mathrm{s}) + Q(\mathrm{s}). \tag{17}$$

Moreover, the output of the loop-filter, W(s), can be further derived as:

$$W(\mathrm{s}) = H(\mathrm{s}) * (X(\mathrm{s}) - Y(\mathrm{s})). \tag{18}$$

By substituting W(s) from (18) into (17), the output Y(s) becomes:

$$Y(\mathrm{s}) = H(\mathrm{s}) * (X(\mathrm{s}) - Y(\mathrm{s})) + Q(\mathrm{s}). \tag{19}$$

This equation can be further derived as follows:

$$Y(\mathrm{s}) = \frac{H(\mathrm{s})}{1 + H(\mathrm{s})} X(\mathrm{s}) + \frac{1}{1 + H(\mathrm{s})} Q(\mathrm{s}). \tag{20}$$

Equation (20) contains two important quantities, the "Signal Transfer Function" (STF), multiplied by the input X(s), and the Noise Transfer Function (NTF), multiplied by the quantization noise Q(s), the two main transfer functions of a Delta-Sigma Modulator loop. The former has a low-pass characteristic for the input signal, and the latter a high-pass characteristic for the quantization noise, introducing the noise-shaping feature of the modulator. If an integrator is used, with a transfer

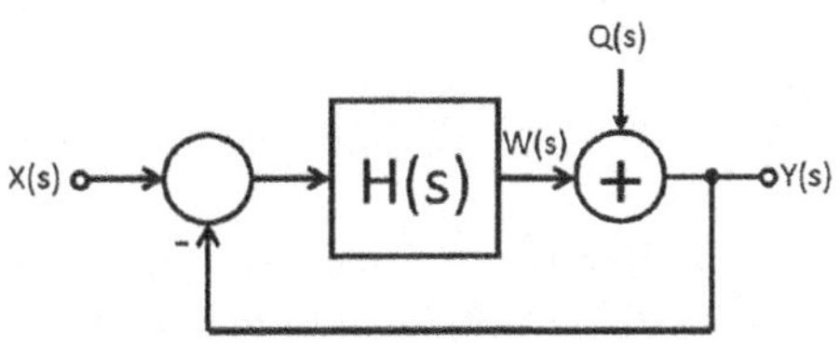

Fig. 11 s-domain model of a delta-sigma modulator

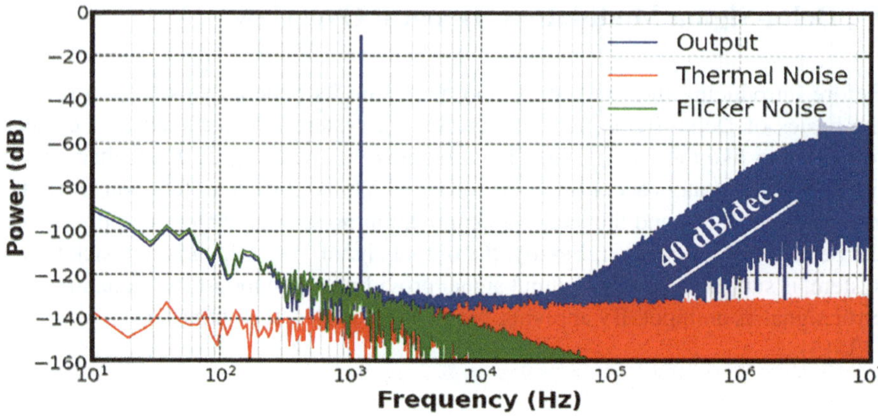

Fig. 12 PSDs of output signal, flicker, and thermal noise of the second-order delta-sigma modulator in [16]

function of $H(s) = 1/s$, the STF becomes equal to $1/(s + 1)$, and the NTF becomes equal to $s/(s + 1)$ [13–15].

The noise-shaping feature can be observed when looking at the output of the modulator, $Y(s)$, in the frequency domain. The work in [16] presents the output spectrum of a synthesized single-bit second-order Delta-Sigma Modulator, with the existence of flicker and thermal noise. Here, the two types of noise form a noise corner. This introduces a roll-off characteristic to the output spectrum at low frequencies due to the flicker noise and a noise floor characteristic at mid-range frequencies due to the thermal noise (Fig. 12). At higher frequencies, the noise-shaping takes place, depending on the order of the modulator. A first-order modulator introduces 20 dB/dec. noise-shaping. Then, for every increase by one in the order of the modulator, the noise shaping increases by 20 dB/dec. [13–16].

6 Introducing CGA Stage to Delta-Sigma Modulators

In a generic Delta-Sigma Modulator, like the one in Fig. 13a, the dominant noise sources, seen as input-referred, are the in-band thermal noise voltage from the DAC and the in-band thermal noise voltage from the first integrator of the loop-filter. The input thermal noise, thus, can be derived as follows:

$$V_{\text{th,in}} = \sqrt{V^2{}_{\text{th,DAC,In-Band}} + V^2{}_{\text{th,Int1,In-Band.}}} \qquad (21)$$

The input-referred noise voltages from the subsequent integrators of the loop-filter are divided by the product of gains of the preceding integrators and thus, are

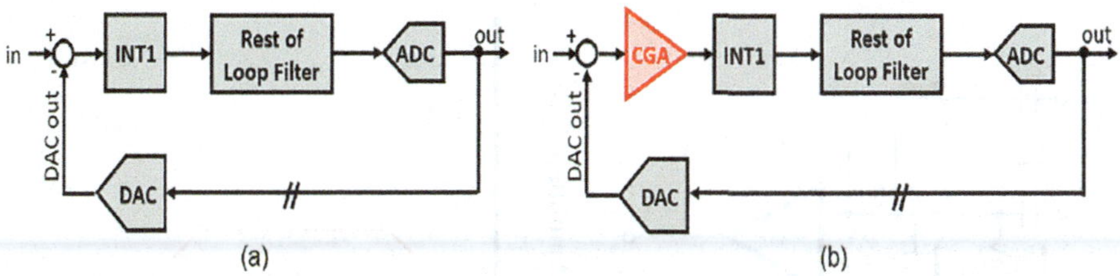

Fig. 13 (**a**) Delta-sigma modulator with no CGA. (**b**) Delta-sigma modulator with CGA to provide gain and attenuate the input-referred noise of the subsequent circuits

significantly attenuated. The in-band noise voltage of the first integrator, in the case of a switched-capacitor circuitry implementation, is the following:

$$V_{\mathrm{th,Int1,In-Band}} = \sqrt{\frac{kT}{C_{\mathrm{in,Int1}} * \mathrm{OSR}}}, \tag{22}$$

where $C_{\mathrm{in,Int1}}$ is the integrator's input capacitance and OSR the oversampling ratio.

By applying a gain stage upfront, before the first integrator and directly after the formation of the quantization error signal, as shown in Fig. 13b, the dominant noise sources change [12]. Thermal noise voltage from the CGA block becomes now the dominant noise source. However, the noise of the first integrator, as seen input-referred, is attenuated by a factor equal to the gain of the added CGA block. The input-referred noise voltage in this case can be derived as follows:

$$V_{\mathrm{th,in}} = \sqrt{V^2{}_{\mathrm{th,CGA,In-Band}} + V^2{}_{\mathrm{th,DAC,In-Band}} + \left(\frac{1}{G_{\mathrm{CGA}}} * V_{\mathrm{th,Int1,In-Band}}\right)^2}. \tag{23}$$

The addition of this gain stage upfront plays a critical role for the SNR and $\mathrm{FoM_S}$ improvement. That will be analytically explained in the following sections.

7 Theoretical Analysis of the CGA Stage

A CGA stage, like the one proposed in [12], is depicted in Fig. 14a. A second branch for a DAC input has been added to illustrate the feedback signal's input and also a parasitic capacitor, C_{par}, related to parasitics such as metal in real silicon implementations. Also, a resistance R_{B} is added which sets the bias voltage of the opamp. The ideal transfer function of this block is given by the following [12]:

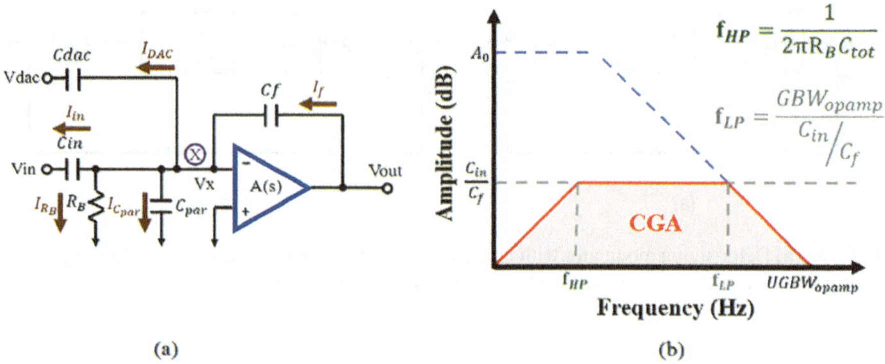

Fig. 14 (**a**) CGA block. (**b**) Associated transfer function characteristic

$$V_{\text{out}} = -\left(V_{\text{in}}\frac{C_{\text{in}}}{C_{\text{f}}} + V_{\text{dac}}\frac{C_{\text{dac}}}{C_{\text{f}}}\right). \tag{24}$$

However, a more realistic form can be derived, considering non-idealities like the finite DC gain and the dominant pole of the opamp. The opamp can be modeled as a single-pole system with a transfer function, A(s), given by [17, 18]:

$$A(s) = \frac{A_0}{1 + {}^{s}\!/_{\omega_0}}, \tag{25}$$

where A_0 is the DC gain of the opamp and ω_0 its dominant pole, in radian frequency.

By applying Kirchhoff's current law (KCL) at node X of Fig. 14a, the current, I_f, is equal to the sum of currents I_{in}, I_{DAC}, I_{RB}, and I_{Cpar} (or equivalently $I_f = I_{in} + I_{DAC} + I_{RB} + I_{Cpar}$), thus:

$$[V_{\text{out}}(s) - V_X]\, sC_f = [V_X - V_{\text{in}}(s)]\, sC_{\text{in}} + [V_X - V_{\text{DAC}}(s)]\, sC_{\text{DAC}} + \frac{V_X}{R_B} + V_X sC_{\text{par}}. \tag{26}$$

The output voltage, V_{out}, can be further derived as $V_X * A(S)$, and thus, the voltage at node X is given by:

$$V_X = \frac{V_{\text{out}}(s)}{A(s)} = \frac{V_{\text{out}}(s)}{\frac{A_0}{1 + {}^{S}\!/_{\omega_0}}}. \tag{27}$$

By using Eq. (27) to replace V_X, Eq. (26) now becomes:

$$\left[V_{out}(s)-\frac{V_{out}(s)}{\frac{A_0}{1+s/\omega_0}}\right]sC_f=\left[\frac{V_{out}(s)}{\frac{A_0}{1+s/\omega_0}}-V_{in}(s)\right]sC_{in}+\left[\frac{V_{out}(s)}{\frac{A_0}{1+s/\omega_0}}-V_{DAC}(s)\right]$$

$$\times sC_{DAC}+\frac{\frac{V_{out}(s)}{\frac{A_0}{1+s/\omega_0}}}{R_B}+\frac{V_{out}(s)}{\frac{A_0}{1+s/\omega_0}}sC_{par}. \tag{28}$$

By further rearranging Eq. (28) in order to derive the transfer function, $V_{out}(s)/V_{in}(s)$, it becomes as follows:

$$V_{out}(s)=-\frac{sA_0V_{dac}C_{dac}+sA_0V_{in}C_{in}}{s^2\left(\frac{C_f}{\omega_0}+\frac{C_{dac}}{\omega_0}+\frac{C_{in}}{\omega_0}+\frac{C_{par}}{\omega_0}\right)+s\left(A_0C_f+C_f+C_{dac}+C_{in}+C_{par}+\frac{1}{\omega_0R_B}\right)+\frac{1}{R_B}} \tag{29}$$

The transfer function characteristic of (29) is shown in Fig. 14b and has three main spectral areas of interest. Initially, it has a 20 dB/dec. high-pass characteristic due to the formation of a high-pass filter from the existence of C_{in} and R_B. A high-pass filter pole is therefore formed at f_{HP}, and the response thereafter becomes flat. At higher frequencies, the opamp with a DC gain of A_0 rolls off the CGA gain until its Unity-Gain Bandwidth (UGBW) frequency. At f_{LP}, the point where the opamp response crosses the CGA's characteristic, another CGA pole f_{LP} (low-pass) can be calculated, and it is shown in Fig. 14b.

A slight modification of the CGA block can be also implemented without the existence of the resistor R_B, since there are circuitry mechanisms that can provide a stable bias voltage for the opamp, like the case of a switched-capacitor DAC. In this case, the transfer function of (29) can be further simplified as:

$$V_{out}=-\frac{A_0V_{dac}C_{dac}+A_0V_{in}C_{in}}{s\left(\frac{C_f}{\omega_0}+\frac{C_{dac}}{\omega_0}+\frac{C_{in}}{\omega_0}+\frac{C_{par}}{\omega_0}\right)+\left(A_0C_f+C_f+C_{dac}+C_{in}+C_{par}\right)}. \tag{30}$$

The poles f_{HP} and f_{LP} of the CGA transfer function characteristic in Fig. 14b can be precisely calculated by solving the second-order equation of the denominator in

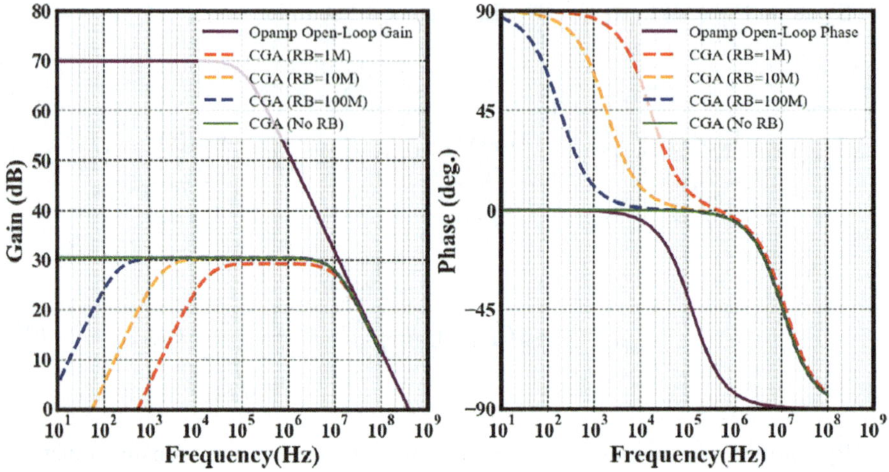

Fig. 15 Behavioral simulation of CGA transfer function with varying R_B. (**a**) CGA gain bode plot; (**b**) CGA phase bode plot

Eq. (29). By applying basic algebra in order to find the roots of this equation, they are extracted as follows:

$$f_{HP} = \left| \frac{-A_0 C_f \omega_0 - C_{tot}\omega_0 - \frac{1}{R_B} + \omega_0 \sqrt{A_0{}^2 C_f{}^2 + C_{tot}{}^2 + \frac{1}{\omega_0{}^2 R_B{}^2} + 2A_0 C_f C_{tot} + \frac{2A_0 C_f - 2C_{tot}}{\omega_0 R_B}}}{2C_{tot}} \right|, \tag{31}$$

$$f_{LP} = \left| \frac{-A_0 C_f \omega_0 - C_{tot}\omega_0 - \frac{1}{R_B} - \omega_0 \sqrt{A_0{}^2 C_f{}^2 + C_{tot}{}^2 + \frac{1}{\omega_0{}^2 R_B{}^2} + 2A_0 C_f C_{tot} + \frac{2A_0 C_f - 2C_{tot}}{\omega_0 R_B}}}{2C_{tot}} \right|. \tag{32}$$

Figure 15 shows the behavioral simulation with gain and phase bode plots of the CGA block's transfer function from (29), while the R_B resistance is varied logarithmically, from 1 MΩ to 100 MΩ, which in turn alters the pole f_{HP}. The bigger the resistor R_B, the lower the pole f_{HP} becomes. Also, the case of a zero R_B is plotted, where there is no pole formed. The opamp bode and phase plots are also shown to illustrate how opamp contributes to the CGA gain roll-off at higher frequencies. Table 3 summarizes the parameter values for the setup of this behavioral simulation.

Figure 16 illustrates the simulated poles f_{HP} and f_{LP}, given the parameters of Table 3, as they are calculated using Eqs. (31) and (32). They appear on the

Table 3 Parameter values for the behavioral simulation of CGA transfer function

Parameter	Description	Value
f_s	Sampling frequency	6.25 MHz
V_{in}	Input voltage	1 V
V_{DAC}	DAC input voltage	1 V
C_{in}	Input capacitor	500 fF
C_{DAC}	DAC capacitor	60 fF
G_{CGA}	CGA closed-loop gain (C_{in}/C_f)	30
C_f	Feedback capacitor	C_{in}/G_{CGA}
C_{par}	Parasitic capacitor	20 fF
C_{tot}	Total capacitance	$C_{in} + C_{DAC} + C_f + C_{par}$
GBW	CGA opamp gain bandwidth	10*f_s
$A_{0,\text{ min}}$	CGA opamp minimum DC gain	70 dB
$f_{\text{pole, Hz}}$	CGA opamp dominant pole frequency	GBW/$A_{0,\text{ min}}$
$f_{\text{pole, rad}}$	CGA opamp dominant pole radial frequency	$2\pi f_{\text{pole, Hz}}$

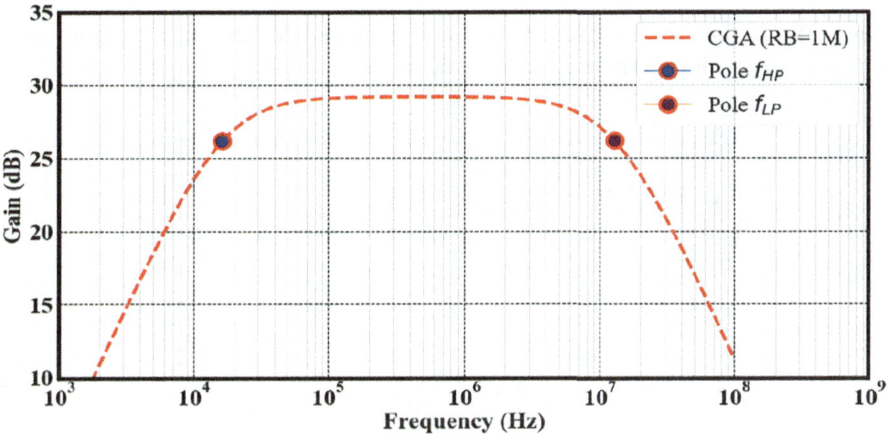

Fig. 16 Verification of CGA fHP and fLP poles

spectrum very precisely, at - 3 dB from the maximum closed-loop gain of the CGA characteristic, with $R_B = 1\ \text{M}\Omega$.

8 Impact of the CGA on the SNR Performance of Delta-Sigma Modulators

The SNR for a Delta-Sigma Modulator can be generally calculated (in dB) as follows:

$$\mathrm{SNR\,(dB)} = P_{\mathrm{sig.}}\,(\mathrm{dBFS}) - V_{\mathrm{n,tot}}\,(\mathrm{dB})\,, \tag{33}$$

where $P_{\mathrm{sig.}}$ is the signal power, in dBFS, and $V_{\mathrm{n,tot}}$ the total input-referred noise voltage, in dB. The signal power, $P_{\mathrm{sig.}}$, can be calculated (in dBFS) as follows:

$$P_{\mathrm{sig.}}\,(\mathrm{dBFS}) = 10\log\left(\frac{A^2/2}{{A^2}_{\mathrm{FS}}/2}\right), \tag{34}$$

where A is the amplitude of the input signal (in Volts) and A_{FS} the full-scale amplitude (in Volts).

The total input-referred noise voltage, $V_{\mathrm{n,tot}}$, is given by:

$$V_{\mathrm{n,tot}}\,(\mathrm{dB}) = 20\log\left(\sqrt{{N_{\mathrm{Q}}}^2 + {N_{\mathrm{th}}}^2}\right), \tag{35}$$

where N_{Q} is the total quantization noise voltage and N_{th} the total input-referred thermal noise voltage. Those are calculated as follows [13–15]:

$$N_{\mathrm{Th}} = \sqrt{\left(\frac{\sqrt{\frac{4kTC_{\mathrm{dac}}}{\mathrm{OSR}}}}{C_{\mathrm{in}}}\right)^2 + \left(\sqrt{\frac{4kT\gamma\mathrm{BW}}{G_{\mathrm{m,CGA}}}}\right)^2 + \frac{1}{G_{\mathrm{CGA}}}\left(\sqrt{\frac{4kT}{C_{\mathrm{in,Int1}} * \mathrm{OSR}}}\right)^2}, \tag{36}$$

$$N_{\mathrm{Q}} = \frac{10^{\frac{N_{\mathrm{Q}}(\mathrm{dB})}{10}}}{\sqrt{\mathrm{BW}}}, \tag{37}$$

where:

$$N_{\mathrm{Q}}\,(\mathrm{dB}) = 10\log 10\left(\frac{\mathrm{BW}}{\mathrm{NBW}}\right) + 10\log 10\left(\sqrt{\frac{\Delta^2}{12\pi\,(2L+1)}\left(\frac{\pi}{\mathrm{OSR}}\right)^{2\mathrm{L}+1}}\right). \tag{38}$$

Thermal noise voltage from the DAC is the first quantity in the square root of Eq. (36). It is calculated as the thermal noise charge of the DAC capacitor, divided by the oversampling ratio, OSR, to keep the in-band quantity, and this result is divided by the input capacitor of the CGA, $\mathrm{C_{in}}$, in order to get the equivalent kT/C noise voltage, multiplied by the $\mathrm{C_{DAC}}$ capacitance, which can provide noise scaling according to its size compared to $\mathrm{C_{in}}$. The second quantity in the square root of (36) is the thermal noise voltage from the opamp [18]. The third quantity is the in-band

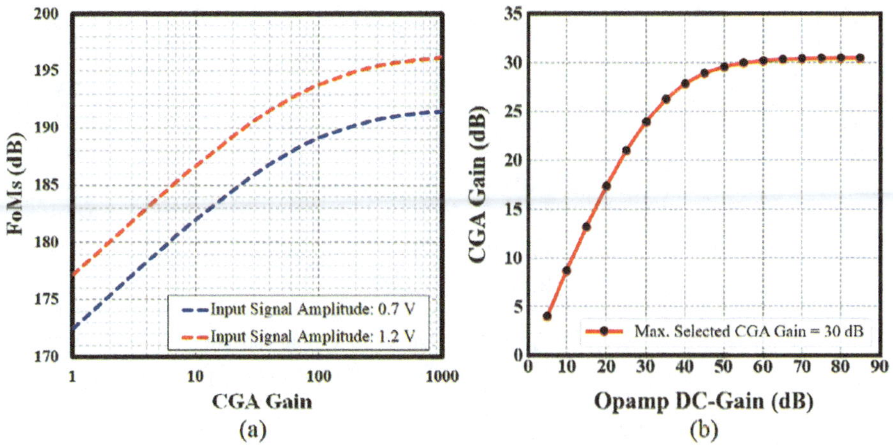

Fig. 17 (**a**) FoM_S vs CGA gain. (**b**) CGA gain vs opamp DC gain

thermal noise voltage of the first integrator in the loop-filter [13–15], divided by the gain of the CGA, as seen in-band.

The effect of CGA block on the overall SNR and FoM_S performance is shown in Fig. 17a. By sweeping the CGA closed-loop gain logarithmically from 1 to 1000, the improvement on the SNR and FoM_S is observed. Further improvement occurs when input signals with larger amplitudes are applied. The best performance improvement occurs when the CGA gain is in the range between 2 and 200. Further increase beyond 200 does not contribute to better SNR and FoM_S performances, as the SNR stabilizes. However, for very high CGA gain numbers beyond 200, a very large C_{in} capacitance is required to provide a reasonable size of C_f, provided that the CGA closed-loop gain is given by C_{in}/C_f. Thus, the difficulty to integrate large capacitances in real silicon implementations sets a limit to the maximum usable CGA gain.

Moreover, by sweeping the DC gain of the opamp in the CGA, and by keeping the CGA closed-loop gain to 30 dB, the minimum DC gain of the opamp for which the maximum closed-loop gain is achieved can be found, and in the present behavioral simulation, it is 70 dB, as shown in Fig. 17b. This is critical for optimum design purposes. The values of the parameters for the Eqs. (33, 34, 35, 36, 37 and 38), and with which the results of Fig. 17 were extracted, are listed in Table 4, in addition to the parameters which are already shown in Table 3.

Shown in Eq. (36), the existence of CGA block significantly reduces the thermal noise voltage from the first integrator, seen as input-referred. Thus, the CGA block becomes the dominant thermal noise voltage source. That explains the SNR and FoM_S improvement, observed in Fig. 17a. In order to approach FoM_S performances of 186 dB and higher, a quantization error amplification by at least 10, according to Fig. 17a, has to be introduced before the loop-filter. The amplification requirement, for a targeted performance, gets lower when input signals with higher amplitudes are

Table 4 The rest of parameter values for the behavioral simulation of total input-referred noise voltage in a Delta-Sigma Modulator with CGA block

Parameter	Description	Value
A	Input signal amplitude	0.7 V
A_{FS}	Full-scale amplitude	1.2 V
BW	Input signal bandwidth	100 kHz
Nfft	Number of total frequency bins	4096016
NBW	Noise bandwidth	1/Nfft
OSR	Oversampling ratio	32
Δ	Quantization steps	64
L	Modulator order	3
$G_{M,CGA}$	CGA opamp transconductance	10 μS
$C_{in,Int1}$	Integrator 1 input capacitor	500 fF
γ	Transistor technological parameter	2/3

applied. Due to this amplification process, the DAC feedback signal has to precisely track the input signal, providing in that way a very small quantization error to the CGA stage.

9 Conclusions

The ongoing demand for power-efficient ADCs is even more present, driven by rapidly increasing energy costs and the need to minimize self-heating in deep sub-micron nodes. The efficiency of noise-limited ADCs in particular, like Delta-Sigma Modulators, is benchmarked by the FoM_S, and its rate of advancement has slowed in the recent years. While the theoretical FoM_S limit of 192 dB will most likely result in a practical of 186–188 dB, recent publications have begun to approach it and in two cases match these numbers [19, 20]. However, to exceed these numbers, new architectures will be required, whereby the quantization error signal is amplified prior to sampling. This approach is necessary to reduce the impact that sampling has on wide-band thermal noise sources and can pave the way for future ADCs to achieve FoM_S of 190 dB and greater.

References

1. R. Kapusta, H. Zhu, C. Lyden, Sampling circuits that break the kT/C thermal noise limit. IEEE J. Solid-State Circuits **49**(8), 1694–1701 (2014)
2. M. Pelgrom, *Analog-to-Digital Conversion*, 3rd edn. (Springer, 2017)
3. B. Murmann, The race for the extra decibel: A brief review of current ADC performance trajectories. IEEE L. Solid-State Circuit Magazine **7**(3), 58–66 (2015)
4. Murmann, ADC performance survey 1997–2022 [Online]. http://web.stanford.edu/~murmann/adcsurvey.html

5. M. Keller, B. Murmann, Y. Manoli, *Analog-digital interfaces, NANO-CHIPS 2020* (Springer, 2010), pp. 93–116
6. B. Murmann, M. Verhelst, Y. Manoli, *Analog-to-information conversion, NANO-CHIPS 2030* (Springer, 2020), pp. 275–292
7. B. Murmann, *A/D converter trends: Power dissipation, scaling and digitally assisted architectures* (Proc. IEEE Custom Integrated Circuits Conf, 2008), pp. 105–112
8. C.C. Enz, E.A. Vittoz, *CMOS Low-Power Analog Circuit Design* (Emerging Technologies: Designing Low Power Digital Systems, 1996), pp. 81–82
9. E.A. Vittoz, *Future of analog in the VLSI environment*, vol 2 (IEEE International Symposium on Circuits and Systems, 1990), pp. 1372–1375
10. B.J. Hosticka, Performance comparison of analog and digital circuits. Proc. IEEE **73**(1), 25–29 (1985)
11. J. Liu, X. Tang, W. Zhao, L. Shen, N. Sun, A 13-bit 0.005-mm2 40-MS/s SAR ADC with kT/C noise cancellation. IEEE J. Solid-State Circuits **55**(12), 3260–3270 (2020)
12. H. Chandrakumar, D. Marković, A 15.2-ENOB 5-kHz BW 4.5- μ W chopped CT ΔΣ -ADC for artifact-tolerant neural recording front ends. IEEE Journal of Solid-State Circuits **53**(12), 3470–3483 (2018)
13. S. Pavan, R. Schreier, G.C. Temes, *Understanding delta-sigma data converters*, 2nd edn. (IEEE Press, Wiley, 2017)
14. J. de la Rosa, *Sigma-delta converters: practical design guide*, 2nd edn. (IEEE Press, Wiley, 2018)
15. S. Norsworthy, R. Schreier, G. Temes, *Delta-sigma data converters: Theory, design and simulation* (IEEE Press, Wiley, 1997)
16. S. Kalogiros, G. Salgado, K. McCarthy, I. O'Connell, *Behavioral modeling of low-frequency noise in switched-capacitor circuits using python* (2022 20th IEEE Interregional NEWCAS Conference (NEWCAS), 2022), pp. 446–449
17. K. Martin, A. Sedra, Effects of the op amp finite gain and bandwidth on the performance of switched-capacitor filters. IEEE Trans. Circuit System **28**(8), 822–829 (1981)
18. B. Razavi, *Design of analog CMOS integrated circuits*, 2nd edn. (McGraw Hill, New York, 2017)
19. S.-E. Hsieh, C.-C. Hsieh, *A 0.4V 13b 270kS/S SAR-ISDM ADC with an opamp-less time-domain integrator* (2018 IEEE International Solid – State Circuits Conference – (ISSCC), 2018), pp. 240–242
20. J. Steensgaard, R. Reay, R. Perry, D. Thomas, G. Tu, G. Reitsma, *A 24b 2MS/s SAR ADC with 0.03ppm INL and 106.3dB DR in 180nm CMOS* (2022 IEEE International Solid-State Circuits Conference (ISSCC), 2022), pp. 168–170

A Second-Order 5bit Hybrid CT/DT Delta-Sigma ADC Implementing Novel Techniques for ELD Compensation and Coefficient Trimming

Roberto Modaffari, Paolo Pesenti, and Germano Nicollini

Abstract A general and well-known trend of the consumer electronic sensor market provides the increasing of accuracy together with the reduction of the power consumption in every new product generation. Further, the semiconductor shortage we are facing in this moment has stressed the already critical role played by both die size and chip testing time in the product definition process, representing themselves a serious limit to the industrial capability. In this context, for a given analog sensing chain, the ADC is usually considered a key block whose proper choice and design can address all these aspects, resulting in a whole system optimization. In this chapter, the design of a second-order 5bit hybrid CT/DT delta-sigma modulator is presented together with the architectural solutions used to limit the impact of the flicker noise on ADC accuracy, save current consumption, and reduce testing time and area occupation.

1 Introduction

MEMS-based sensors became very common in the last 15 years [1]. Their capillary diffusion inside a variety of consumer electronic devices has been made possible, thanks to the continued evolution of both sensors' architectures and electronics' readout performances. A lot of circuital solutions, in fact, have been developed with the purpose to convert the capacitive displacement or resistance variation [2, 3] provided by such sensors into an analog or digital electrical signal [4–6] in an endless research of reduced power consumption and increased accuracy [7].

The prototype of a readout system typically provides the connection of the sensor to an analog front-end (AFE) stage followed by an analog-to-digital converter (ADC) with the eventual interposition of an anti-aliasing filter (AAF) depending on the nature of the ADC. Continuous-time delta-sigma ADCs save this AAF, thanks

R. Modaffari (✉) · P. Pesenti · G. Nicollini
STMicroelectronics, Cornaredo (MI), Italy
e-mail: roberto.modaffari@st.com

P. Harpe et al. (eds.), *Biomedical Electronics, Noise Shaping ADCs, and Frequency References*, https://doi.org/10.1007/978-3-031-28912-5_12

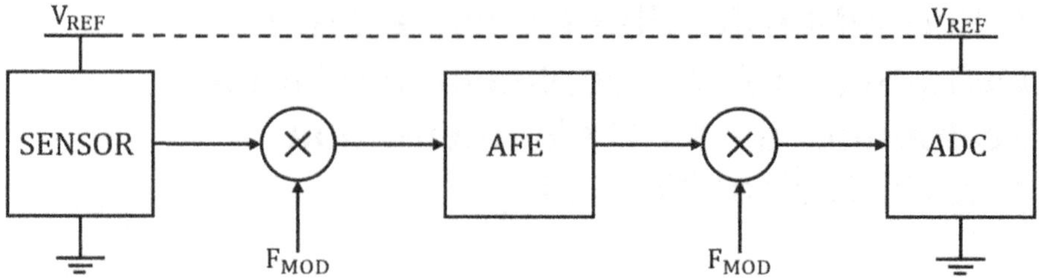

Fig. 1 Correlation of sensor and ADC reference voltage at system level

to the intrinsic low-pass filter that distinguishes their signal transfer function (STF), and this feature is one of the main reasons that led to the choice of this architecture for the converter object of this chapter. This is partially true since the final delta-sigma modulator uses a hybrid CT/DT loop filter, but the concept that the AAF is saved remains valid.

Before focusing on the ADC description, it is worth having a look at the block scheme of the sensing chain where such converter is employed, shown in Fig. 1. In this figure, two architectural solutions deserve to be highlighted:

1. Sensors are typically passive components that must be supplied with some electrical signal to generate an electrical output. In this design, such signal is the reference voltage V_{REF} which is also used in the sigma-delta ADC to generate its feedback current. This fact introduces a correlation that provides a first-order cancellation of the low-frequency noise that afflicts V_{REF}.
2. The signal that comes from the sensor is modulated at the frequency F_{MOD} and then amplified by the AFE. The demodulation removes the flicker noise of both AFE and ADC input resistances, since it is performed on the ADC input current as explained in Sect. 12.3.1.

Acting at system level, these two simple solutions mitigate the overall low-frequency noise and relax the area requirement of AFE. Other techniques to address these specifications have been implemented at ADC level and will be discussed in detail in the following sections.

This chapter is organized as follows. In Sect. 12.2, an overview of the ADC architecture is described, while the circuit blocks together with the detailed description of the resistive feedback DAC are presented in Sect. 12.3. In Sect. 12.4, the new excess loop delay (ELD) compensation technique, required from the architectural choice of using a SAR ADC as quantizer, is explained by exploiting the impulse invariance concepts. The self-calibration procedure of the ADC coefficients is reported in Sect. 12.5. Finally, the physical implementation of the ADC and the measurement results are reported in Sect. 12.6, followed by conclusions in Sect. 12.7.

2 ADC Architecture

The general architecture of the converter is presented in the block scheme of Fig. 2. The hybrid second-order loop filter is composed of the cascade of a continuous-time (CT) integrator and a switched-capacitor (SC) integrator, with a feedforward path that simplifies the hardware enabling the adoption of a single-feedback topology (CIFF structure). The 5bit quantizer is a classical SAR ADC with capacitive DAC and provides the numerical output of the modulator to the digital part of the ADC that embeds the SAR control logic, the dynamic element matching (DEM) hardware, and the phase generator that clocks the whole system. The resistive feedback DAC closes the loop of the modulator generating rectangular NRZ current pulses. This shape reduces the jitter sensitivity of the ADC (also mitigated from the high number of quantization levels) and relaxes the performance of the local oscillator [8].

Three main aspects of this design are relevant:

1. The usage of a SAR ADC saves area and current consumption with respect to the usage of a classical flash quantizer. As described in Sect. 12.3.2, the longer conversion time results in an ELD, whose compensation is presented in Sect. 12.4.
2. As described in Sect. 12.3.3, the resistive feedback DAC minimizes the low-frequency noise of the system. However, it will be shown that this solution requires the segmentation of such DAC to limit both area and parasitic capacitances.
3. Since the input stage of the modulator is a CT integrator, the tuning of its coefficients is needed to manage the process spreads that occur in mass production. A fast and effective way to perform this tuning will be discussed in Sect. 12.5.

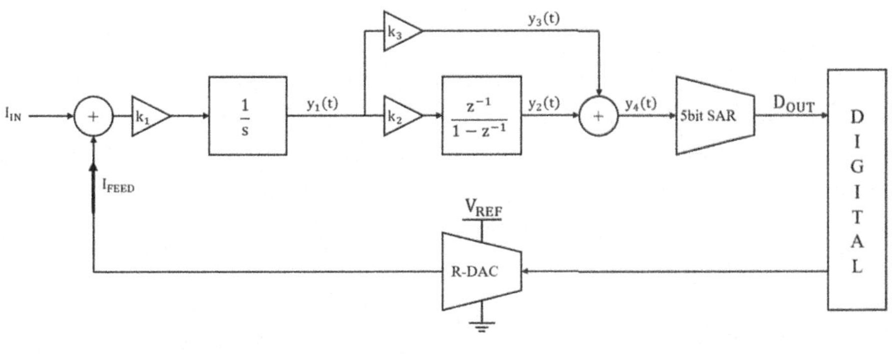

Fig. 2 Block scheme of the analog-to-digital converter

3 ADC Block Description

3.1 Loop Filter

The loop filter of the modulator is shown in Fig. 3. It uses a hybrid architecture where the first CT integrator, implemented using a classical active-RC topology, provides the anti-alias filtering. The input signal V_{IN}, which is modulated at the frequency F_{MOD} as presented in Sect. 12.1, is converted into a current through the ADC input resistances R_{IN} that can be tuned to adjust the sensitivity of the modulator. This input current is baseband demodulated and injected in the virtual ground of the first integrator cleaned up from the low-frequency noise of both AFE and ADC input resistances. The first integrator OTA is chopped as well, and its flicker noise sources are removed by modulation.

To prevent any possible variation of the modulator coefficients, the feedback capacitances of the first integrator can be trimmed. This operation is done changing the number of connected capacitive modules, allowing the correction of the process parameters spread that could take place in production. Notice that since they are the only tunable elements introduced in the modulator, their calibration range must be properly addressed considering the process variations of all the blocks of the ADC.

The second integrator uses switched-capacitor circuitry. This choice makes the coefficients of both integral and feedforward paths referred to as a stable capacitive ratio, avoiding the introduction of a second couple of tuned capacitances. The "ELD compensation" path is defined by a capacitive ratio as well, and its working principle will be discussed in Sect. 12.4.

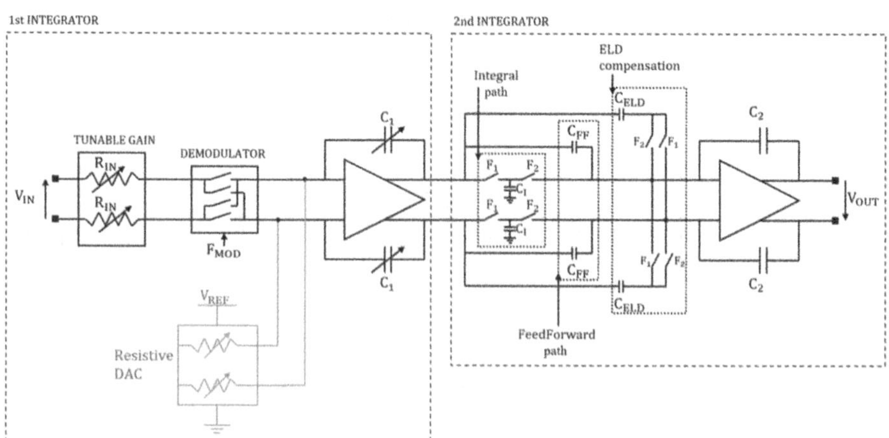

Fig. 3 Loop filter architecture

3.2 SAR Quantizer

The quantizer is a classical 5bit fully differential SAR ADC whose general structure is shown in Fig. 4. The positive and negative reference voltages of the converter are the buffered signal V_{REFQ} and ground, respectively. The signal is sampled on the "upper" part of the segmented capacitive DAC composed of the 16 + 16 = 32 capacitances C_0 to C_4 related to the 5 bits of the ADC. Notice that the number of such capacitances is halved with respect to the 31 + 31 = 62 elements ideally required for a 5bit binary DAC, thanks to single-ended driving of the LSB capacitance C_0. In the "lower" part of the DAC, the capacitance C_D is single-ended driven as well, and it is connected to the upper part through the bridge capacitance C_B. This capacitance is used to inject the dither signal of the sigma-delta modulator, whose weight of ¼ LSB is determined by the capacitive load C_L, directly on the quantizer.

As is known, the analog-to-digital conversion operated by SAR ADCs relies on a dichotomic search that needs a number of comparisons at least equal to the number of desired bits. These operations require time and are clocked in the proposed quantizer by the signal indicated in the timing diagram of Fig. 4 as COMP, while the signal SMP is used in the modulator to sample the output of the loop filter. As indicated in the figure, there is a certain amount of time between this sampling operated on the falling edge of SMP and the generation of the associated digital word which is presented to the input of the resistive feedback DAC on the rising

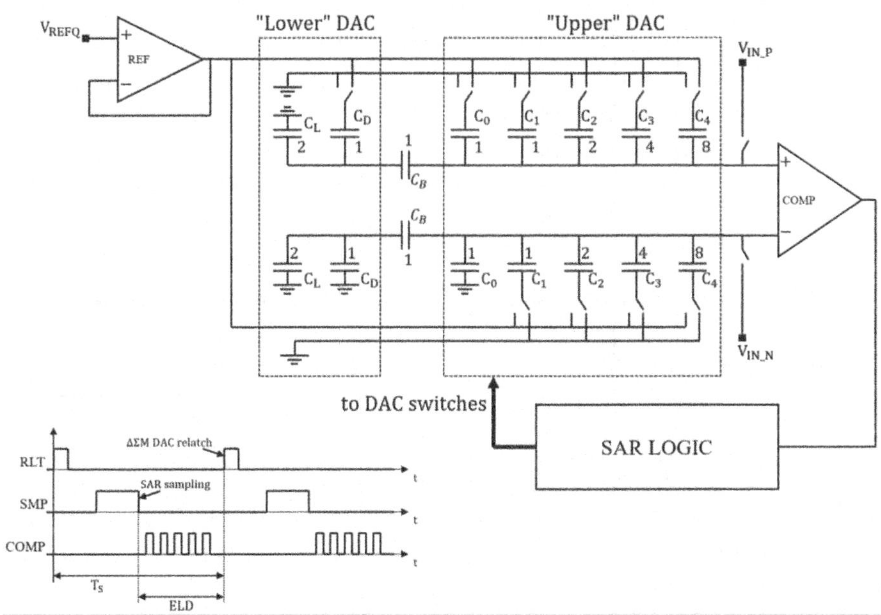

Fig. 4 Architecture of the SAR quantizer and related timing diagram

edge of the signal RLT. This time is equal to half of the modulator sampling period T_S and corresponds to an excess loop delay (ELD) that must be compensated, as discussed in Sect. 12.4.

3.3 Resistive Feedback DAC

The architectural choice of sharing the voltage reference between the ADC and the sensor introduces a correlation that enables a first-order cancelation of the flicker noise injected in the system from all those service blocks, such as bandgap reference and voltage buffers, that contribute to the generation of V_{REF}. The usage of a resistive DAC makes this cancelation even more effective since the ADC feedback current, directly generated from the voltage drop of V_{REF} across the DAC resistances, is injected in the first integrator saving both the current generator circuit and the current mirrors that would be required from a current-steering architecture. In Fig. 5 the conceptual scheme of a 5bit resistive DAC provides the connection of each resistance between the virtual ground of the first integrator and the three-state analog switch that connects the DAC element to V_{CM}, V_{REF}, or GND. Each switch

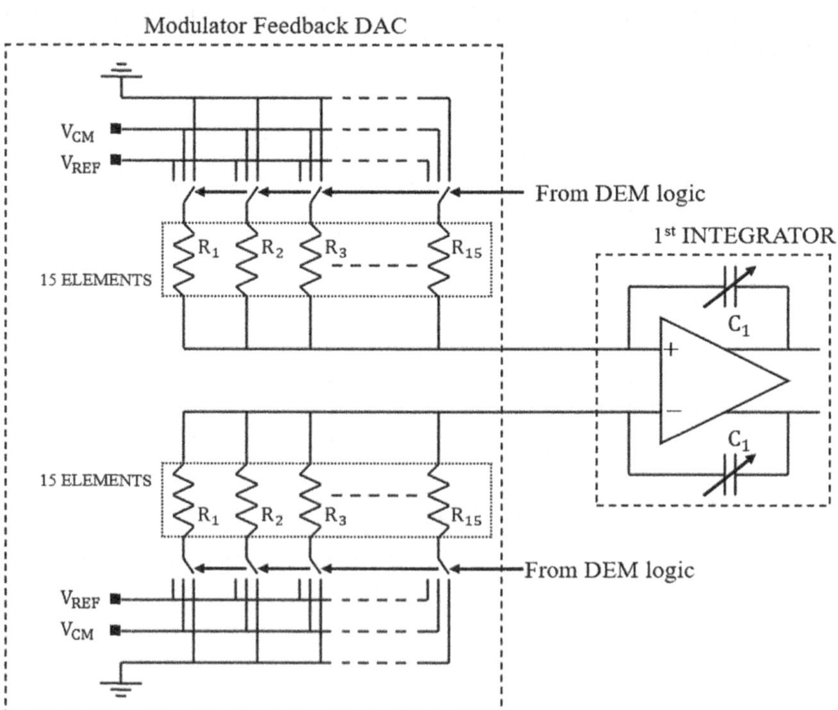

Fig. 5 Architecture of a 5bit fully differential resistive DAC

is driven by the output bus of the DEM logic with a complementary codification between positive and negative side. In total, 15 + 15 = 30 DAC elements are required to cover the 31 quantization levels, considering the code "0" where all the resistances are connected to V_{CM} and removing the code "-16" by a digital saturation.

Despite its simplicity, this architecture is not suitable for this design, presenting a practical limitation that can be pointed out with the following numerical considerations. The current budget of the whole ADC is set by the application to few tens of μA resulting in a maximum current of the feedback DAC in the order of some μA. It follows that the single DAC element has a current of few hundreds of nA, requiring a resistance that spaces from few MΩ to tens of MΩ depending on the voltage drop imposed from V_{REF}. With such voltage that usually is in the hundreds of mV to some volts range, the total DAC resistance can reach hundreds of MΩ. Considering now the techniques adopted for the cancelation of the low-frequency noise such as the usage of a common reference voltage and the chopping of both AFE and first integrator OTA, the DAC resistance turns out to be the ultimate flicker noise source of the whole system. This contributor can become a serious problem from those designs, as the one here discussed, where aggressive noise corner frequencies in the order of tens of mHz are addressed. Poly-silicon resistors with relative low density in the order of few hundreds of Ω/□ can be used to address this contributor, thanks to their good flicker noise coefficient [9], but this solution is clearly incompatible with the implementation of the big resistances required from the low current situation described before. Besides the unacceptable amount of area, in fact, the physical implementation of such big resistances would result in large parasitic time constants able to smooth the rectangular current pulse ideally provided by the DAC, altering the ADC coefficients and introducing an additive ELD. Moreover, big parasitic capacitances in combination with the finite resistance of the DAC switches can produce the degradation of the modulator THD with the appearance of inter-symbol interference effects (ISI).

This situation has been overcome with the DAC segmentation [10] shown in Fig. 6 where the resistive DAC is split into two parts where the unitary element of the "coarse" part has a weight that is four times the weight of the "fine" one. In order to shape the gain mismatch error among the two parts, the 9-level coarse codification (from +4 to −4) is generated by the first-order digital sigma-delta modulator of Fig. 7, and the fine codification (whose values spaces from −3 to +3) is simply calculated as the difference between the input code and the coarse DAC code. A DWA DEM algorithm is then provided to improve linearity of the two DACs. The discussed segmentation is very effective when combined with resistive DACs since in these circuits, increasing the weight of the elements essentially means using smaller resistances, reducing area and parasitic capacitances (the situation would be the opposite in the case of current-steering DACs). Considering this design, the upper resistances are 4 times smaller than those in the lower part, and then the resultant DAC counts 8 equivalent elements instead of 30 elements required from the original implementation, saving almost a factor 4 in area.

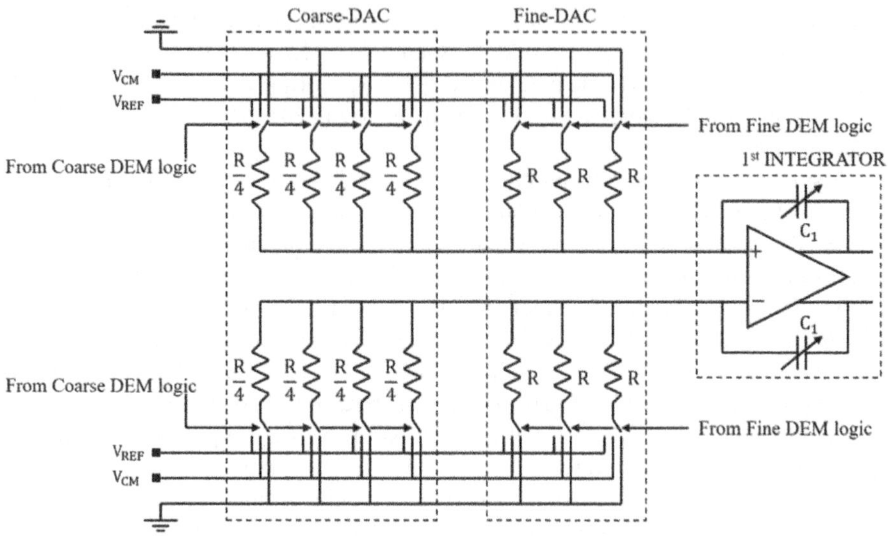

Fig. 6 Segmented DAC architecture

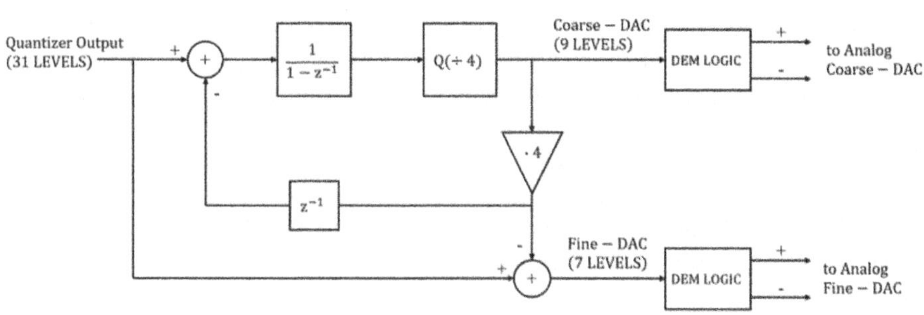

Fig. 7 Logic for coarse DAC and fine DAC data generation

4 ELD Compensation

The impulse invariance is a well-known method for the calculation of the coefficients of CTDSMs based on the equivalence that subsists between CT and DT systems that show the same impulse response (IR) [11]. With this approach, starting from the NTF of a given DT modulator assumed as reference, it is possible to calculate the coefficients of the equivalent CT modulator imposing the condition that the impulse responses (IR) of the two systems must coincide in every sampling instant. This general method, still valid in the case of hybrid loop filter such as the one used in this design, will be now used to evaluate the effect of the SAR-related ELD on the IR of the modulator. For this purpose, the conceptual analysis is done opening the loop and forcing a single digital pulse to the input of the feedback DAC as shown in Fig. 8. The propagation of the resultant NRZ current pulse, presented for

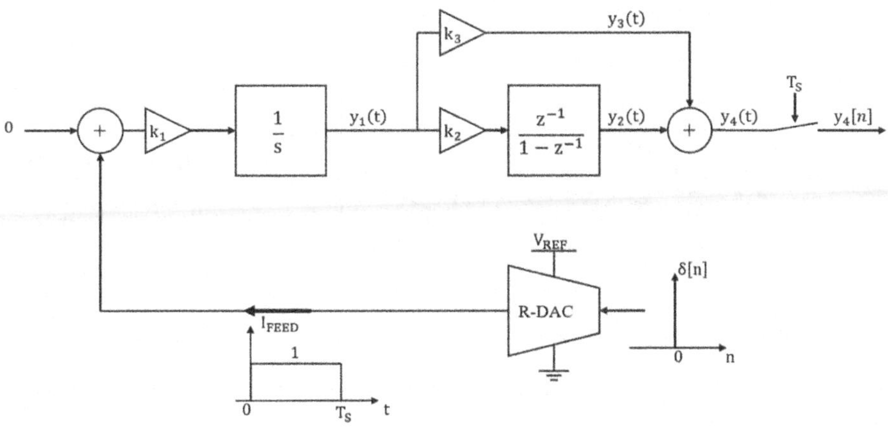

Fig. 8 Block scheme for the impulse response evaluation of the modulator

simplicity with normalized amplitude and without any ELD, generates in every node of the loop filter the waveforms shown in Fig. 10a. The voltage ramp on the output of the first integrator $y_1(t)$ is carried to the output of the feedforward path $y3(t)$ merely amplified by the coefficient k_3. The description of the integral path output $y_2(t)$, instead, requires at least the definition of the architecture used for the second SC integrator and its timing. The topology is shown in Fig. 9, and the phases F1 and F2 are indicated in Fig. 10a as SC PHASES. This circuit samples the output of the first integrator on the capacitances C_I during the phase F1 and shares the sampled charge on the feedback capacitances C_2 during the phase F2. It follows that $y_2(t)$ resembles a staircase signal with fixed step defined by the product of the coefficients $k_1k_2T_S$, where k_2 is the ratio C_I/C_2. The loop filter output signal $y_4(t) = y_2(t) + y_3(t)$ is the IR of the system, and a proper choice of the coefficients k_1, k_2, and k_3 makes this piecewise curve intercept the points of the reference DT IR that have been indicated in the chart by the crosses. The complete derivation of the loop filter coefficients is beyond the purpose of this text, and the detailed description of the reference DT modulator is omitted: the only fundamental information regards the points of the reference impulse response as reported in Fig. 10a.

Let's now consider the real case where an ELD of half the sampling period is introduced in the loop to accommodate the operations of the SAR quantizer, as discussed in Sect. 12.3.2. To take into account the new situation, the waveforms of the circuit have been modified assuming the shapes reported in Fig. 10b. Here, the feedback DAC pulse is delayed of $\frac{1}{2}T_s$ with respect to the first sampling event resulting in a delayed ramp on the output of the first integrator. The integral path that generates the signal $y_2(t)$ can easily recover from this shift with the inversion of the phases F1 and F2 in order to sample the signal $y_1(t)$ at the end of the voltage ramp, keeping in this way the same phase relationship of the ELD-free situation.

The compensation of the feedforward path is more laborious. The curves $y_3(t)$ and $y_4(t)$, in fact, show that due to the ELD, the voltage ramp has reached just half

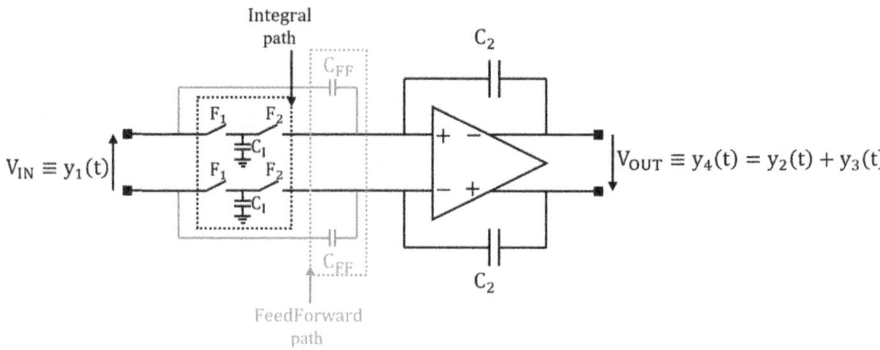

Fig. 9 Circuital implementation of the second SC integrator completed with the feedforward path

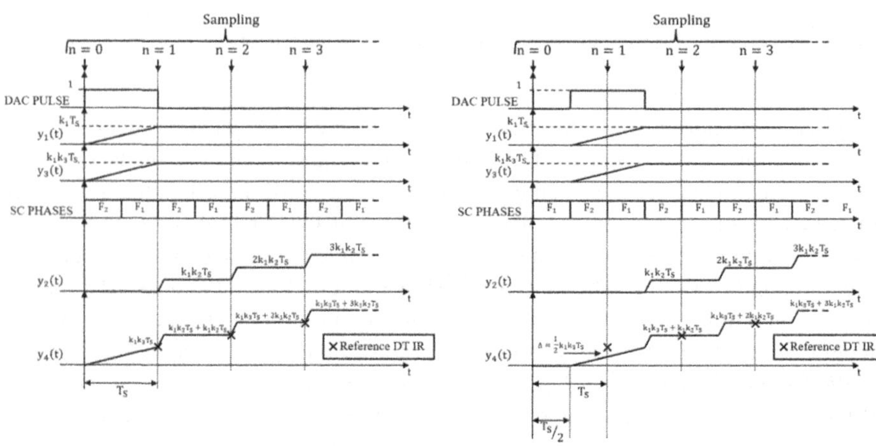

a) ELD-free modulator b) Modulator with ELD=1/2Ts

Fig. 10 IR of the modulator with ELD = 0 (**a**) and ELD = $1/2T_S$ (**b**)

of its final value when the sampling event $n = 1$ occurs. It follows an error on the first no-null sample of the IR equal to $½k_1k_3T_S$. Notice that this error cannot be compensated just doubling one of the two coefficients k_1 or k_3 since this operation would correct just the first point of the IR making wrong all the other samples.

The discussed problem is well known in literature [11, 12], and several techniques have been developed to address it. A comparative study with some examples is reported in [13]. In [14], the authors proposed a classical solution based on the introduction of a feedback path around the quantizer. This method allows a perfect compensation of the ELD, but it costs an additional feedback DAC and a differential summing amplifier. In [15], the differential summing amplifier is saved since the compensation signal is injected on the input of the second integrator, thanks to the introduction of a digital differentiation, but the additional DAC is still required. In [16], the authors present a proportional-plus-integral (PI) compensation based

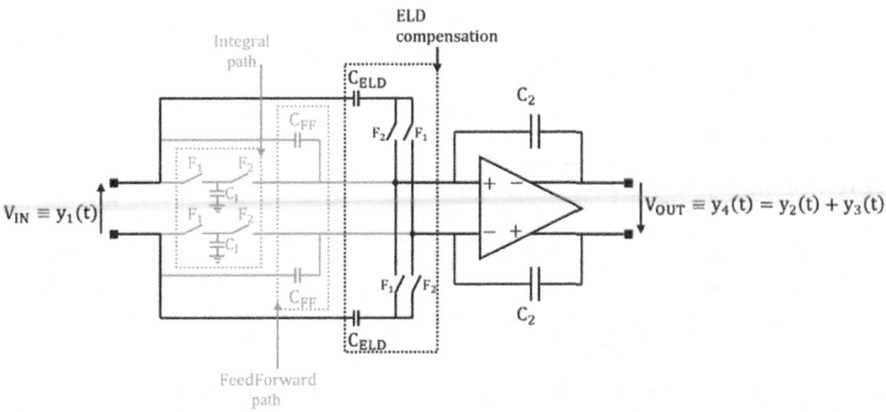

Fig. 11 ELD compensation network in the second integrator

on the usage of a resistance as proportional element in series to the feedback capacitance of the last integrator, but this solution turns out to be not suitable for single-loop architectures. In general, the compensation techniques, as the ones reported in the literature, either require complex extra hardware or present some topological limitations. The approach used in this work [17] overcomes these two drawbacks modifying the second integrator circuit as shown in Fig. 11. The additive network, composed of two capacitances and four switches, injects the output of the first integrator $y_1(t)$ in the second integrator with positive sign in the phase F2 and with negative sign in the phase F1, producing on the output of the loop filter the net contribution highlighted as ELD_COMP in Fig. 12. This signal has a peak equal to $\frac{1}{2}k_1k_ET_S$, where $\frac{1}{2}k_1T_S$ is due to the first integrator half-ramp and k_E is simply the coefficient C_{ELD}/C_2. Since the error to be corrected is $\frac{1}{2}k_1k_3T_S$, where $k_3 = C_{FF}/C_2$ is the feedforward coefficient, choosing $C_{ELD}=C_{FF}$ for the compensation capacitance returns a contributor that provides the perfect compensation of the ELD as expressed in Fig. 12 by the waveform $y_4(t)$.

5 Coefficients' Self-Trim Procedure

Let's consider the example of a given consumer electronics device whose mass production has a target capability of one million devices/day with a test-and-trim (T&T) time of 10 s for each sample. In the new generation of the same product, designers could decide to switch from a DTDSM to a CTDSM approach to take advantage of embedded AAF, reduced current consumption, and lower noise. However, the coefficients of DT modulators are typically referred to as capacitive ratios, whereas the ones of CT modulators usually depend on time constants which are afflicted from resistance and capacitance process variations. It follows that the

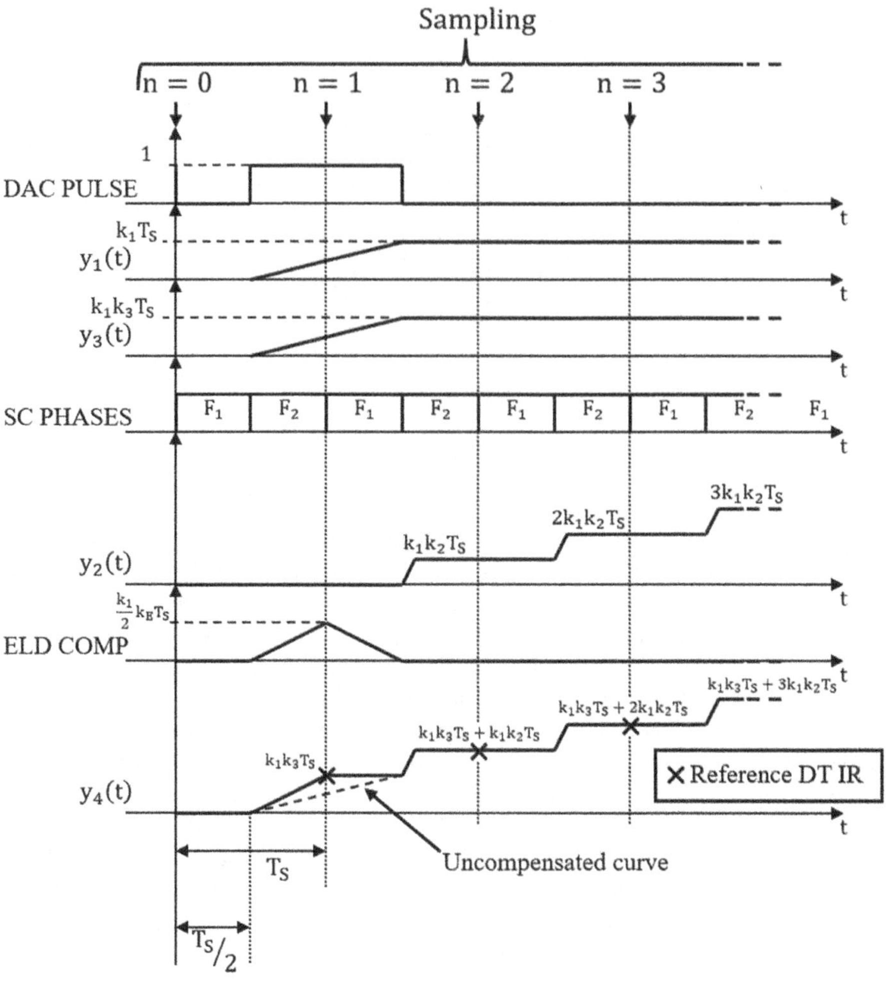

Fig. 12 Impulse response of the compensated modulator

new ADC requires a calibration to tune its coefficients that could take, for instance, 100 ms. As a result of the extra testing time, there could be a shortage of 10,000 devices/day. This simple example has been used to highlight that the usage of CT circuitry can require additive calibrations and the introduction of calibrations must be carefully evaluated considering the impact on the testing time.

Calibration routines implemented by fully-on-chip dedicated hardware are in general faster than procedures that outsource total or part of the elaboration to the automatic testing equipment (ATE), since they save the time required for the communication between the ATE and the device and they usually rely on deeply optimized hardware. For this reason, in the proposed ADC, a fully-on-chip trimming strategy based on a successive-approximation algorithm has been developed with

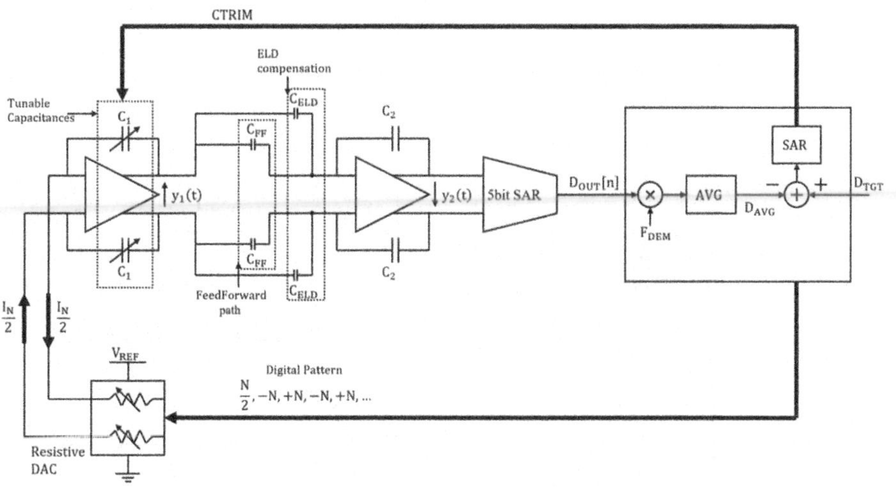

Fig. 13 ADC configured in calibration mode

the purpose to identify the optimum number of first integrator feedback capacitances that retrieves the proper modulator coefficients in the shortest time possible [18]. The working principle of the solution can be described considering the scheme in Fig. 13 where the modulator is configured in its "calibration mode." The digital logic provides to the input of the resistive feedback DAC a fixed pattern whose first samples are ½ N, −N, +N, −N, +N, −N, etc., where N is an integer that avoids the saturation of the analog stages. This pattern, shown in the first line of Fig. 14, corresponds to the fully differential output current of the DAC I_N whose amplitude can be expressed as:

$$I_N = \frac{V_{REF}}{R_{DAC}} \tag{1}$$

where R_{DAC} is simply the equivalent resistance expressed by DAC when the code N is applied at its input and V_{REF} is the DAC reference voltage. This current is injected in the virtual grounds of the first integrator, which has been properly disconnected from the input signal, and then integrated on the feedback capacitances C_1 generating the triangular voltage signal indicated in Fig. 14 as $y_1(t)$ whose peak amplitude is:

$$V_P = \frac{I_N T_S}{2C_1} = \frac{V_{REF} T_S}{2R_{DAC} C_1} \tag{2}$$

In the calibration mode, the phases of the second SC integrator are stopped in order to make it work as a simple analog-gain stage configured as in Fig. 13, with a gain defined by the ratio between the sum of the input capacitances $C_{FF} + C_{ELD}$ and

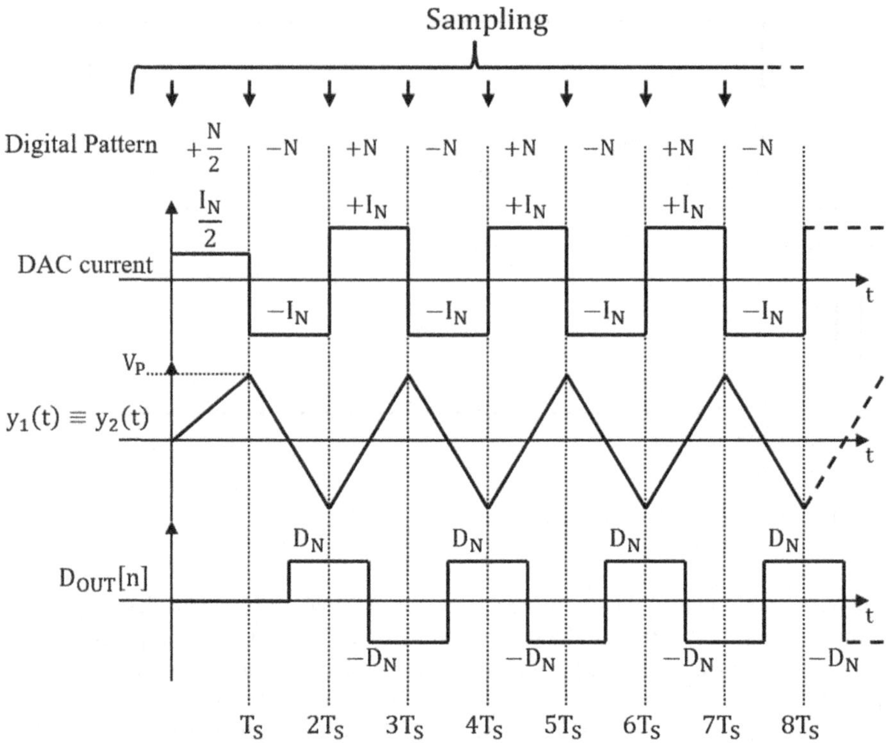

Fig. 14 Waveforms for the circuit of Fig. 13

the feedback capacitance C_2. Assuming this gain unitary, the triangular waveform on the output of the first integrator is carried unaltered to the input of the quantizer that, considering its timing, samples the triangular wave in its positive and negative peaks. The digital output signal $D_{OUT}[n]$ is then a square wave generated with a delay equal to the ELD introduced by the SAR quantizer and presents an amplitude that, neglecting the quantization, can be expressed as:

$$D_N \approx \frac{V_P}{V_{LSB}} = \frac{V_{REF} T_S}{2 R_{DAC} C_1 V_{LSB}} \tag{3}$$

where V_{LSB} is simply the value of the quantizer LSB. It is important to notice that this expression considers all the variables involved in the definition of the modulator coefficients and a tuning based on this information returns a perfect calibration of the ADC. Following the signal path of Fig. 13, the square wave $D_{OUT}[n]$ is demodulated in the digital domain and filtered by a moving-average filter that increases the resolution of the calibration generating the output word D_{AVG}. This signal is then subtracted from the reference digital word D_{TGT}, which represents the target value extrapolated from spice simulations that D_{AVG} would have in the case of ideal ADC parameters. Based on the resultant sign of this operation, a SAR logic drives the digital bus CTRIM that, acting on the feedback capacitances of the first integrator, modifies the amplitude of the triangular wave $y_1(t) = y_2(t)$ making D_{AVG} match D_{TGT}. Finally, at the end of the calibration procedure, the word CTRIM can be saved using a flash or OTP memory.

6 Measurement Results

The ADC has been realized in 130 nm CMOS process and covers an area of 130,000 μm^2, where 26,000 μm^2 are dedicated to the ADC gain trim through the input resistance calibration and 21,000 μm^2 are taken from the digital part that includes the SAR quantizer control logic, the digital delta-sigma modulator in the segmented DAC, the DEM logic, and the phase generator that clocks the whole system. The micrograph of the converter is shown in Fig. 15.

The total power consumption, including all the digital blocks, is only 57.6 μW. With a sampling frequency of 512 kHz, and an input signal frequency of 1 kHz, the ADC achieves a maximum SNDR of about 90.2 dB on a 4 kHz bandwidth, as depicted in Fig. 16. The ADC performances are summarized in Table 1.

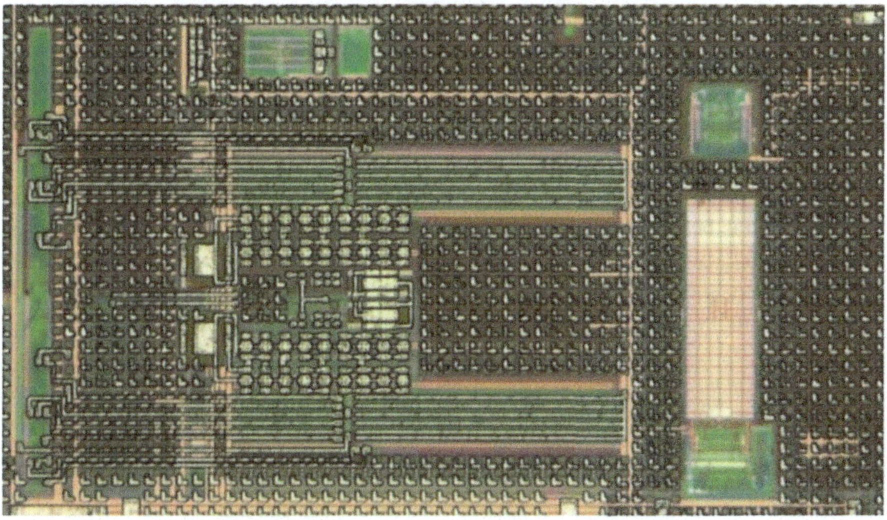

Fig. 15 Micrograph of the ADC

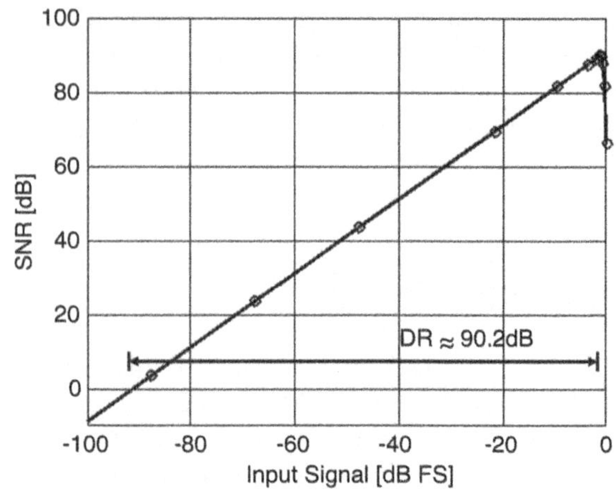

Fig. 16 Measured SNR

Table 1 ADC performance summary

Signal bandwidth/clock rate	4 kHz/512 kHz
Dynamic range	90.2 dB
Area (including gain trim and digital blocks)	0.13 mm^2
Process/supply voltage	130 nm CMOS/1.6 V
Power dissipation (including digital blocks)	57.6 μW

7 Conclusions

In this chapter, the design of a hybrid delta-sigma modulator has been presented, with particular focus on the solution adopted for the flicker noise reduction including the architecture of the segmented resistive DAC. A novel and simple compensation of the ELD has been developed to manage the usage of a 5bit SAR quantizer, and a fully-on-chip calibration of the modulator coefficients has been implemented to save time in the test-and-trim phase of the production. Finally, some physical information and measurement data that validate the design have been reported.

References

1. Status of the MEMS Industry 2021 Report, Yole Développement, 2021
2. D.K. Shaeffer, MEMS inertial sensors: A tutorial overview. IEEE Commun. Mag. **51**(4), 100–109 (2013)
3. S.C. Chen, V.P.J. Chung, D.J. Yao, W. Fang, Vertically integrated CMOS-MEMS capacitive humidity sensor and a resistive temperature detector for environment application. Transducers, 1453–1456 (2017)
4. E.M. Boujamaa, B. Alandry, S. Hacine, L. Latorre, F. Mailly, P. Nouet, *A low power interface circuit for resistive sensors with digital offset compensation* (Proceedings of 2010 IEEE International Symposium on Circuits and Systems, 2010), pp. 3092–3095

5. M. Lobur, A. Holovatyy, *Overview and analysis of readout circuits for capacitive sensing in MEMS gyroscopes (MEMS angular velocity sensors)* (International Conference on Perspective Technologies and Methods in MEMS Design, 2009)
6. M. Schipani, P. Bruschi, G.C. Tripoli, T. Ungaretti, *A low power CMOS interface circuit for three-axis integrated accelerometers* (Proc. Ph.D Res. Microelectron. Electron. Conf, Bordeaux, 2007), pp. 117–120
7. C. Valzasina et al., *A compact 6-axis IMU combining low-power operation with high stability and low-noise* (Proc. of IEEE International Symposium on Inertial Sensors and Systems (INERTIAL))
8. J.A. Cherry, W.M. Snelgrove, Clock jitter and quantizer metastability in continuous-time delta-sigma modulators. IEEE Trans. Circuits Syst. II Analog Digit. Signal Process **46**(6), 661–676 (1999)
9. R. Brederlow, W. Weber, C. Dahl, D. Schmitt-Landsiedel, R. Thewes, Low-frequency noise of integrated polysilicon resistors. IEEE Trans. Electron Devices **48**(6), 1180–1187 (2001)
10. R. Adams, K.Q. Nguyen, A 113-dB SNR oversampling DAC with segmented noise-shaped scrambling. IEEE J. Solid State Circuits **33**(12), 1871–1878 (1998)
11. J.A. Cherry, W.M. Snelgrove, Excess loop delay in continuous-time delta-sigma modulators. IEEE Trans. Circ. Syst. **II**, 376–389 (1999)
12. S. Pavan, Excess loop delay compensation in continuous-time delta-sigma modulators. IEEE Trans. Circuits Syst. II, Express Briefs **55**(11), 1119–1123 (2008)
13. M. Keller, A. Buhmann, J. Sauerbrey, M. Ortmanns, Y. Manoli, A comparative study on excess-loop-delay compensation techniques for continuous-time sigma-delta modulators. IEEE Trans. Circuits Syst. I Regul. Papers **55**(11), 3480–3487 (2008)
14. P. Benabes, M. Keramat, R. Kielbasa, *A methodology for designing continuous-time sigma-delta modulators* (European Design and Test Conference, 1997. ED TC 97. Proceedings, 1997), pp. 46–50
15. G. Mitteregger et al., A 20-mW 640-MHz CMOS continuous-time sigma-delta ADC with 20-MHz signal bandwidth, 80-dB dynamic range and 12-bit ENOB. IEEE J. Solid State Circuits **41**(12), 2641–2649 (2006)
16. M. Vadipour, *A 2.1mW/3.2mW Delay-Compensated GSM/WCDMA Analog-Digital Converter* (Proc. 2008 VLSI Circuits Symp, 2008), pp. 180–181
17. R. Modaffari, P. Pesenti, G. Nicollini, *Compensation circuit for delta-sigma modulators, corresponding device and method* (U.S. Patent Number 11290124)
18. R. Modaffari, P. Pesenti, *Self-Calibration technique for Continuous-Time Delta-Sigma modulators coefficients* (U.S. Patent Application Number 17721110)

Part III
Frequency References

The third section of this book discusses the design of integrated frequency references. Since they can be implemented in standard CMOS, frequency references based on RC time constants continue to be the subject of much research, and this is the subject of the first three chapters. However, their residual temperature dependency is typically limited to a few ppm/°C. For significantly better stability, extending down to the ppb/°C range, frequency references based on MEMS resonators and quartz crystals are required. In the last three chapters, recent advances in the design of such frequency references are presented.

In Chap. 13, Youngcheol Chae and Woojun Choi present a dual RC frequency reference. It consists of a frequency-locked loop FLL) in which the frequency of a digitally controlled oscillator is locked to a temperature-independent phase shift derived from two RC networks with different temperature dependencies. The design ach(ieves an inaccuracy of ±200 ppm from −40 °C to 85 °C after two-point trimming.

In Chap. 14, Taekwang Jang and his colleagues present two architectures for compensating for the higher-order temperature dependency of RC-based frequency references. Conventionally, their first-order temperature dependency is canceled by combining two resistors with complementary temperature dependencies. By using three different resistors in an R-RC oscillator, and combining a look-up table with two resistors in a nonlinearity-aware dual phase-locked loop, temperature dependencies of 7.9 ppm/°C and 8.7 ppm/°C, respectively, were achieved.

In Chap. 15, Kyusang Park and Pavan Hanumolu describe how the temperature dependency of RC-based frequency references can be compensated by using parallel combinations of switched resistors with different temperature dependencies, whose weights are digitally controlled by pulse-density modulated sequences. Prototype oscillators fabricated in a 65 nm CMOS process show that their inaccuracy can be reduced to better than ±140 ppm from −40 °C to 95 °C (2.1 ppm/°C).

In Chap. 16, Danielle Griffith describes the use of CMOS-compatible bulk acoustic wave (BAW) resonators in frequency references as an alternative to off-chip crystals, allowing a reduction in the footprint of electronic systems and enabling new applications. The chapter covers several aspects of designing with BAW resonators, including oscillator topologies, frequency tunability and stability, and temperature compensation. System-level advantages such as fast startup, higher frequency, and improved robustness will also be discussed.

In Chap. 17, Kamran Souri and Sassan Tabatabaei present the design of a MEMS-based temperature-compensated oscillator (TCXO) which achieves a frequency stability of ±100 ppb from −45 °C to 105 °C. A recently improved version even achieves ±10 ppb, which matches the performance of entry-level quartz-based oven-controlled oscillators (OCXOs). This level of performance is enabled by a highly stable, high-resolution temperature-to-digital converter (TDC) based on two MEMS resonators with different temperature dependencies.

In Chap. 18, Wim Kruiskamp presents an alternative to the de facto standard for crystal oscillators: the three-point Pierce oscillator. Despite its simplicity and good performance, it has some drawbacks. It requires large capacitors, suffers from slow and inefficient startup, and is a single-ended circuit with poor common-mode rejection. This chapter discusses these drawbacks and presents an alternative circuit. The proposed switched-capacitor oscillator requires about four times less capacitance, enables fast and efficient startup, and is fully differential.

RC Frequency References Based on Dual RC FLLs

Youngcheol Chae and Woojun Choi

Abstract This chapter introduces a dual RC frequency reference. It consists of a digital frequency-locked loop (FLL) in which the frequency of a digitally controlled oscillator is locked to a temperature-independent phase shift derived from two RC networks. It covers the operation principle of the dual RC architecture, key design techniques, and an implemented prototype. The measurement results show the dual RC frequency references achieve an inaccuracy of ±200 ppm over a temperature range from −40 °C to 85 °C after two-point trimming. It has great potential to be used in battery-powered wireless sensor nodes.

1 Introduction

The on-chip frequency references without external components are becoming more popular and starting to cover a wide range of applications. In this context, CMOS frequency references have been investigated extensively, and their performance has been greatly improved in recent years. Among the options available in a standard CMOS, RC frequency references offer several good properties, such as low-cost integration and high energy efficiency. Frequency reference using a frequency-locked loop (FLL) architecture is getting popular, because the output frequency is locked to the output of an accurate frequency-to-voltage converter (FVC), relaxing oscillator requirements. In addition, the inaccuracy of the temperature-compensated FLLs has been greatly improved from ±1000 to ±200 ppm over the temperature range from −45 °C to 85 °C by using a digital temperature compensation scheme. This chapter introduces the dual RC frequency reference, which is a temperature-compensated digital FLL. Its outlook is also given in the light of latest results.

Y. Chae (✉)
Department of Electrical and Electronic Engineering, Yonsei University, Seoul, South Korea
e-mail: ychae@yonsei.ac.kr

W. Choi
Information Technology and Electrical Engineering, ETH Zürich, Zürich, Switzerland
e-mail: wochoi@iis.ee.ethz.ch

P. Harpe et al. (eds.), *Biomedical Electronics, Noise Shaping ADCs, and Frequency References*, https://doi.org/10.1007/978-3-031-28912-5_13

2 Dual RC Frequency-Locked Loop

First, this chapter starts with an overview of the basic FLL architecture for on-chip clock generation. Thereafter, Sects. 2.2 and 2.3 describe two key techniques, a modified FLL architecture using phase information and a phase digitizer based on a phase-domain delta-sigma modulator. Section 2.4 explains the concept of digital temperature compensation schemes for FLLs. Finally, Sect. 2.5 provides the dual RC frequency references.

2.1 *Basic FLL Architecture*

Figure 1a shows the basic architecture of an FLL, which consists of two frequency-to-voltage converters (FVCs), a voltage detector, a loop filter, a voltage-controlled oscillator (VCO), and a frequency divider [1]. It takes an input frequency F_{IN}, and an output frequency F_{OUT} is divided by the frequency divider with a division ratio of N to provide the feedback signal. The voltage detector measures the error between the F_{IN} and the F_{OUT}/N via two FVCs and sets the loop filter that controls the output frequency of the VCO. At the steady state, the F_{OUT} of the FLL is locked to the $N \times F_{IN}$.

For on-chip clock generation, however, the F_{IN} does not exist anymore, and thus, the FLL can be modified by having a voltage source V_{IN} instead of F_{IN} and an FVC, as shown in Fig. 1b. Since the error between FVC output and the V_{IN} determines the

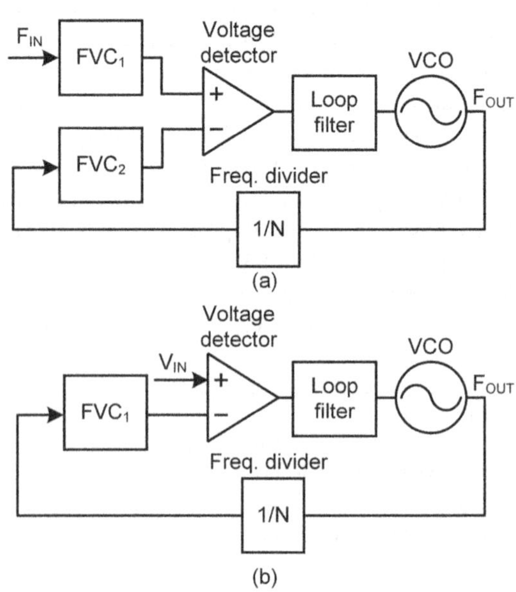

Fig. 1 Block diagram of (**a**) a basic FLL and (**b**) a modified FLL without F_{IN}

control voltage for the VCO's loop filter, the FLL is then locked to the V_{IN}. The FVC in combination to the V_{IN} mainly determines the frequency stability, and various topologies have been investigated to implement them accurately. A popular way to implement a FVC is to apply matched currents into a reference resistor R_{REF} and a switched-capacitor resistor C_S, resulting in a reference voltage and a frequency-dependent voltage [2, 3]. At the steady state, the F_{OUT} of the FLL is locked to $N/(R_{REF} \times C_S)$, resulting in a RC oscillator. Therefore, the large temperature coefficients (TCs) of the integrated resistors, typically more than 2000 ppm over the industrial temperature range, must be canceled (or corrected) for the FLL used as on-chip frequency references. The FVC can also be implemented with a Wheatstone bridge circuit [4, 5], which can eliminate the sources by using resistors. However, it should be noted that the role of FVC and V_{REF} can be performed in many other ways, for instance, by using phase information [6–12].

2.2 Modified FLL Using Phase Information

Alternatively, the FVC can be replaced with a filter and a synchronous phase detector as shown in Fig. 2 [6, 7]. The output of a VCO is divided by N and used to generate in-phase and quadrature-phase signals. The in-phase signal drives the filter, and the filter's phase shifted outputs are synchronously demodulated by the quadrature-phase signals. The demodulated output is integrated into the loop filter, and its output sets the VCO. The loop forces the average output of the demodulated signal to be zero, which corresponds to a *fixed* phase shift in the filter. Therefore, at the steady state, the F_{OUT} is locked to $N \times F_{IN}$ like the other FLLs. Since the F_{OUT} is insensitive to the amplitude variations of the filter outputs, a simple square wave can be directly used for the drive signal without using a dedicated reference voltage. The filter is often implemented with an RC filter, such as first-order RC filter, Wien bridge (WB) filter, and polyphase filter (PPF) [7–12]. By exploiting the temperature dependence of the RC filters, many FLL-based temperature sensors have been reported [6–9, 12–16], with results showing that they are small area, low power, and energy efficient and well suited for dense thermal monitoring.

In order to accurately measure the phase information, the synchronous phase detector requires an output filter, such as an analog low-pass filter (LPF), but this

Fig. 2 FLL with a synchronous phase detector [6]

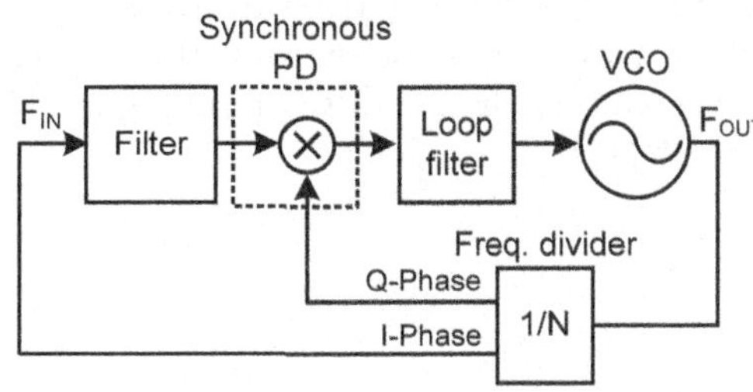

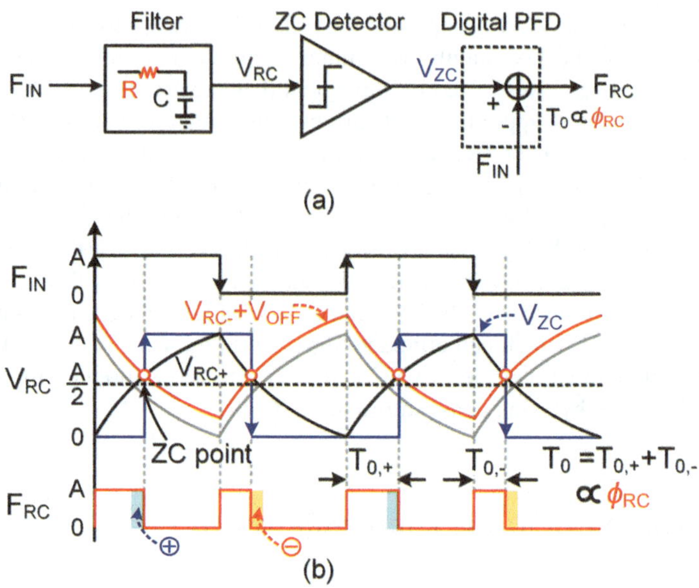

Fig. 3 (**a**) The ZC detection scheme with an RC filter; (**b**) timing diagram of the dual ZC detection scheme [9]

often requires large capacitors. However, after driving the RC filter, the resulting output signal has a different shape of waveform depending on the filter's time constant, but the output frequency with a duty cycle T_O always contains the desired phase information. Therefore, the synchronous phase detector can be implemented with a zero-crossing (ZC) detector and a digital phase/frequency detector (PFD) as shown in Fig. 3a [8]. This ZC detection scheme can be very effective, because it does not need a narrowband LPF, resulting in a compact and energy-efficient implementation [8, 9]. Another advantage is that the ZC scheme is well suited for a digital-intensive FLL, which can have a small area and low power consumption by minimizing the associated analog circuitry [10, 11]. One potential drawback of ZC scheme is that the offset error of ZC detector causes phase errors, which can be translated into output frequency errors. Since the ZC point either lags or leads depending on its polarity, the output phase exploits a constant phase error. To mitigate the offset error, a chopping technique can be applied to the input and output of the ZC detector [10]. However, by noting the symmetric behavior of ZC points at rising and falling edges, the *dual* ZC scheme, which averages the results from both edges, cancels the ZC offset, resulting in offset-canceled ZC outputs as shown in Fig. 3b [12].

2.3 Phase Digitizer Based on Phase-Domain $\Delta\Sigma$ Modulator

The phase shifts of RC filters are based on the TCs of the resistors in the standard CMOS process, highly depending on the type of resistors. Typically, the resistor's TC imposes high-order nonlinearities such as third- to fifth-order polynomials. After linear compensation by combining two resistors with opposite TCs, a large residual TC exists, limited to about 1% over the industrial temperature range, and the composite resistor is sensitive to process variations. To address this, digital temperature compensation schemes are proposed, resulting in significantly improved inaccuracy (±200 ppm) over a similar temperature range. To facilitate the digital temperature compensation, the phase shifts of the RC filter are digitized using a phase-to-digital converter and incorporated into a digital FLL. Since the RC filter's output is periodic, it can be digitized by a phase-domain delta-sigma modulator (PD$\Delta\Sigma$M). The PD$\Delta\Sigma$M was developed for the readout circuits of thermal diffusivity-based temperature sensors [13] and was used for resistor-based temperature sensors [14, 15].

The PD$\Delta\Sigma$M consists of a phase DAC, a loop filter, and a quantizer, as shown in Fig. 4a. The RC filter's output driven at the frequency F_{IN} is multiplied by the feedback signal. This down-converts the signal into DC by selecting that the frequency of the feedback signal is the same as the input but has a phase difference. The quantizer selects one of the phase references that span the maximum phase changes of the RC filter. The resulting output is proportional to the phase difference and applied to a loop filter. The $\Delta\Sigma$ loop forces the loop filter input to be minimized, and the bitstream output shows a digital representation of the RC filter's phase shift. In many applications [13–16], the first- or second-order loop filter is chosen for the PD$\Delta\Sigma$M to achieve a target resolution in a required conversion time. In [14], a temperature sensor using the PD$\Delta\Sigma$M readout achieves a 410 μK resolution in a 5 ms conversion time. The use of a high-resolution PD$\Delta\Sigma$M results in good results but requires a significant area (0.7 mm^2) and power consumption (0.16 mW) due to its analog-intensive approach [20]. A digital PD$\Delta\Sigma$M solves most of the issues in analog PD$\Delta\Sigma$Ms and can be implemented in a very compact area [15, 16, 21, 22], as shown in Fig. 4b.

2.4 Digital Temperature Compensation for FLLs

A digital temperature compensation scheme can be incorporated into a digital FLL using an explicit digital temperature sensor (TS), as shown in Fig. 5 [17, 18]. To compensate for the nonlinear TC of the RC filter in the digital domain, the filter's phase shift must be digitized firstly by a high-resolution PD$\Delta\Sigma$M, and the resulting phase shift is then compensated with an appropriate code from the digital TS. This leads to a temperature-independent phase shift output, so does the frequency output F_{OUT} of the FLL. The required polynomial at the TS output is fifth or sixth order

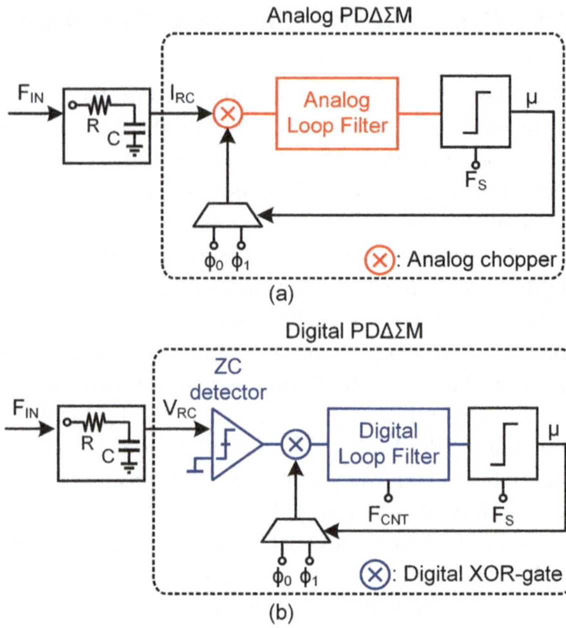

Fig. 4 (**a**) An analog PDΔΣM and (**b**) a digital PDΔΣM [22]

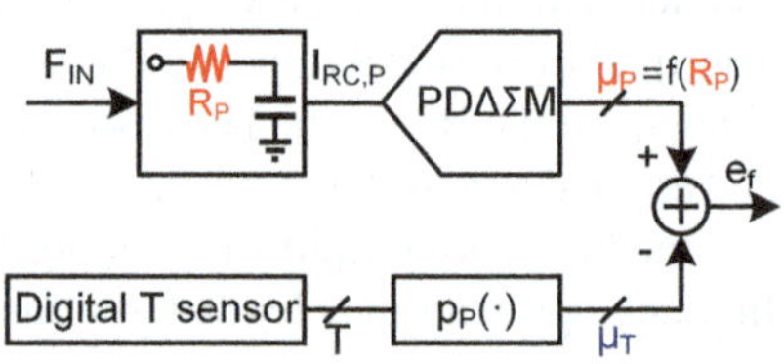

Fig. 5 Digital temperature compensation with a digital TS [14, 18]

to achieve an inaccuracy of ±200 ppm from −45 °C to 85 °C [19]. In general, the TS should be accurate enough to avoid any frequency errors of the frequency reference and can be implemented with BJTs, resistors, or RC filters [17–19]. When the TS is realized with a resistor-based TS with the same resistor type, the required polynomial can be simplified to first order, and the resulting frequency reference in [18] can reach an inaccuracy of ±400 ppm over a similar temperature range, but it seems to require higher-order polynomials to achieve better inaccuracy.

The need for an accurate TS can be eliminated by using a dual RC architecture as shown in Fig. 6 [20, 21]. This includes two RC filters with different TCs, and the filter's phase shifts are digitized and combined to accurately cancel the temperature dependencies. This should be possible to recall that the bandgap reference contains both PTAT and CTAT voltages and results in a constant voltage output and the BJT-based TS is also based on this fundamental principle [23]. The phase shift of each RC filter is digitized by a PDΔΣM, and the digital output is linearized by its polynomial function. The resulting linear TCs of two RC filters can be added to create a temperature-independent phase reference. The frequency references with

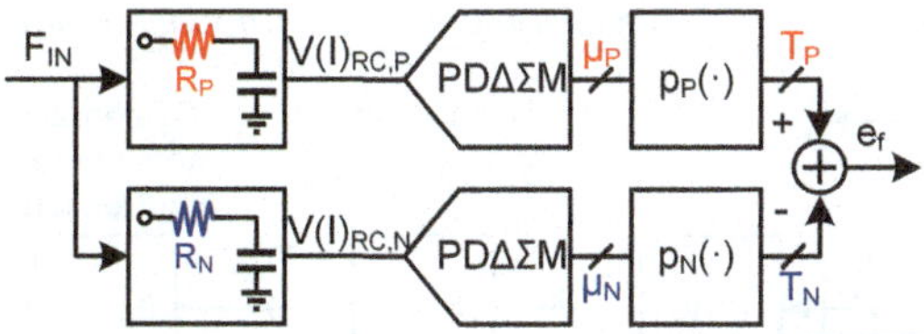

Fig. 6 Dual RC-based digital compensation [20, 21]

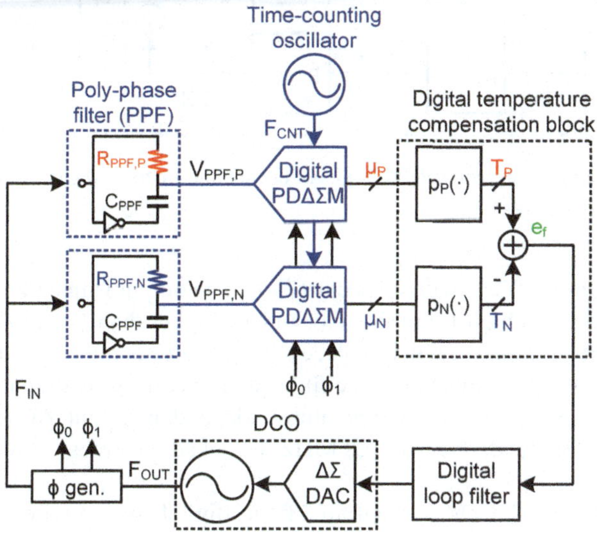

Fig. 7 Block diagram of the dual RC-based frequency reference [21]

dual RC architecture can achieve an inaccuracy of ±200 ppm from −40 °C to 85 °C [20, 21], which has a great potential to be used in many applications.

2.5 *Dual RC Frequency Reference*

Figure 7 shows the dual RC frequency reference, essentially a digital FLL, with the output frequency F_{OUT} of a digitally controlled oscillator (DCO) precisely locked to a temperature-independent phase shift [20–22]. The phase shift is derived from two RC filters, and the outputs of the PDΔΣMs are then combined in the digital domain. Each RC filter is configured as a polyphase filter (PPF), and their complementary TCs are generated by using different resistors such as silicide p-poly and p-poly resistors. The choice of the PPF over the WB filter is simply because the PPF requires fewer components and has a larger voltage swing. The PPF's phase shifts are digitized by digital PDΔΣMs. In a digital PDΔΣM, the chopper phase detector of an analog PDΔΣM can be replaced by a ZC detector and a digital XOR gate, and the XOR gate detects the phase difference between the PPF's phase shift and the

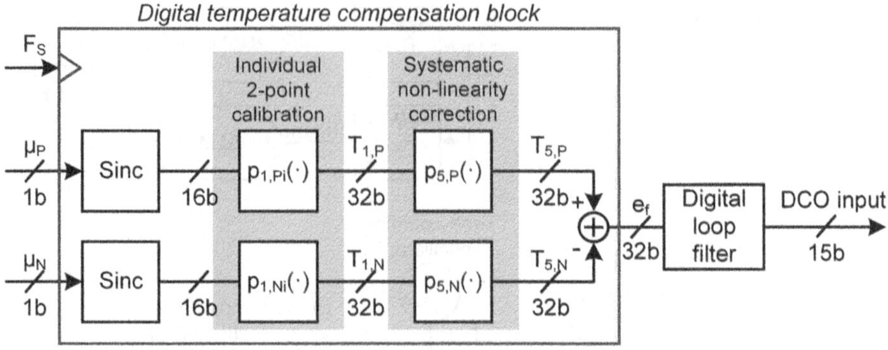

Fig. 8 Digital temperature compensation block: decimation filters, a digital polynomial engine, and a digital loop filter [22]

reference phases ($\pm$ 22.5° with respect to F_{IN}). The output is integrated by a digital counter running on a high-frequency clock (10$\times$ higher than F_{OUT}) with negligible quantization noise penalty, which replaces an analog integrator. The digital outputs of the PD$\Delta\Sigma$M are decimated by sinc filters and linearized by digital polynomials. The combined result provides a temperature-independent output. At steady state, the DCO's output frequency F_{OUT} (= 28 MHz) is locked regardless of the temperature drift.

Figure 8 shows the block diagram of the digital temperature compensation, and its details are well described in [21]. The output bitstream of the PD$\Delta\Sigma$M is decimated by sinc filter implemented as a 16-bit counter, which aims to achieve a conversion rate of 3 Hz. Therefore, if the target thermal transient of the frequency reference is faster, the conversion time of the PD$\Delta\Sigma$M must be adjusted accordingly. To compensate for the process spread on PPFs, first-order polynomials are employed from a two-point calibration. The systematic polynomials are removed by the following fixed fifth-order polynomials. The subtracted frequency error signal is then integrated into the digital loop filter. The loop filter output is converted into a 15-bit fixed point word to drive the DCO. The expected truncation error from 1 LSB of the digital word is about 360 Hz, which corresponds to 13 ppm of the output frequency.

The inaccuracy analysis of dual RC frequency reference has been described in [20], which is estimated from a linear model as shown in Fig. 9. The errors in each RC filter are mainly due to the spread in resistors (σ) and in their TCs (TC$_P$ and TC$_N$). The relative frequency error is given by

$$\frac{\sigma_f}{f_0} = \frac{\sqrt{\sigma_P^2 + \sigma_N^2}}{f_0} \cdot \left| \frac{1}{\frac{\delta\phi_P}{\delta f}\frac{\delta T}{\delta\psi_P} - \frac{\delta\phi_N}{\delta f}\frac{\delta T}{\delta\psi_N}} \right| = \frac{\sqrt{\sigma_P^2 + \sigma_N^2}}{\left| \frac{1}{\mathrm{TC}_P} - \frac{1}{\mathrm{TC}_N} \right|}, \tag{1}$$

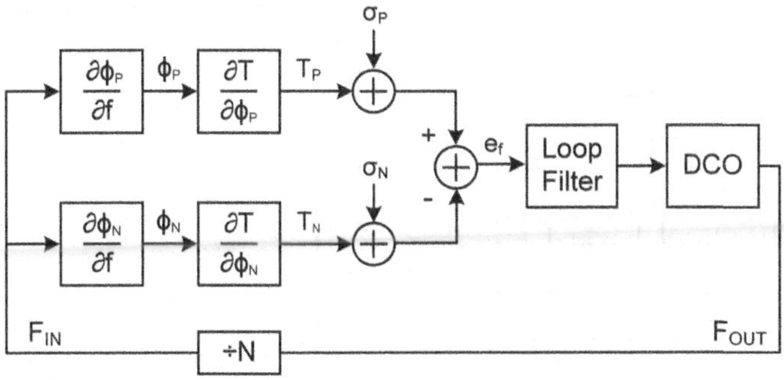

Fig. 9 Linearized model of the dual RC-based FLL [20]

where σ_f denotes the output frequency error and the $\delta T/\delta\psi$ is the phase-to-temperature gain of each PPF at f_0. It shows that the opposite TCs with different signs reduce the frequency error, which is a clear advantage compared to other temperature-compensated FLL architecture. Ideally, assuming two resistors with exactly matched TCs of opposite sign, the expected inaccuracy of the dual RC FLL can be $1/\sqrt{2}$ that of the other TC-compensated FLL. Of course, if the TCs of two resistors have a large asymmetry, i.e., $TC_P >> TC_N$, this advantage is compromised.

3 Implementation and Measurement Results

This chapter provides implementation details for key building blocks of the dual RC FLL. Measurement results from a prototype chip, implemented in a 65 nm CMOS process, are also discussed.

3.1 Digital Phase-Domain Delta-Sigma Modulator

The schematic of the digital PD$\Delta\Sigma$M is shown in Fig. 10 [22]. The RC PPF is implemented with $R = 120$ kΩ and $C = 8.4$ pF. The DCO output frequency F_{OUT} is divided by 128, resulting in a driving frequency F_{IN} (= 218.75 kHz). The phase references (ϕ_0 and ϕ_1) for the feedback signal and the sampling clock F_S for the quantizer are also generated by the F_{OUT}. Considering two PPF's phase shifts over the target temperature range, the phase references are set to ±22.5°. After driving the PPF with F_{IN}, the PPF's phase shift is detected with the ZC detector, and the phase difference F_{XOR} between the ZC output and the feedback signal F_{DAC} is integrated by a 13-bit up/down counter. The counter's MSB is samples at F_S, resulting in the output bitstream *BS*. The resulting *BS* selects the phase references in a $\Delta\Sigma$ loop and accurately measures the PPF's phase shift.

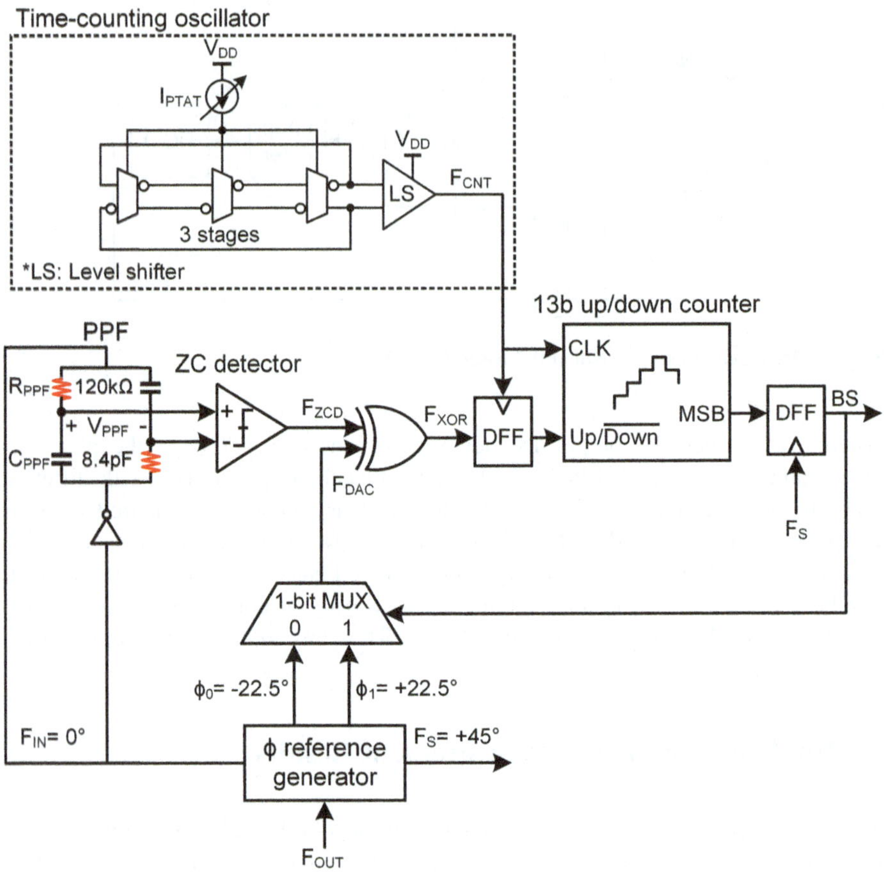

Fig. 10 Schematic diagram of the digital PDΔΣM [21]

Since the digital PDΔΣM uses an up/down counter instead of an analog integrator, the quantization error appears at each up/down operation. Therefore, the clock frequency of the counter must be increased to meet the required phase resolution. In this work, the counter clock F_{CNT} is set to be around 280 MHz, which is generated by a free-running current-controlled oscillator (CCO), as shown in Fig. 11. This oscillator is asynchronous to F_{XOR}, and therefore, the resulting errors are relatively random and considered as additive white noise. As a result, the digital PDΔΣM achieves a temperature resolution of <10 mK with a conversion time of 3 Hz. Since the ϕ_{PPF} varies by ~21° over the temperature range, an 11-bit counter is required to cover the maximum output code. However, considering the variations of F_{CNT}, a 13-bit counter is chosen, which is synthesized using the standard cell library in a 65 nm CMOS, resulting in a power consumption of 15.6 μW from a 0.9 V supply.

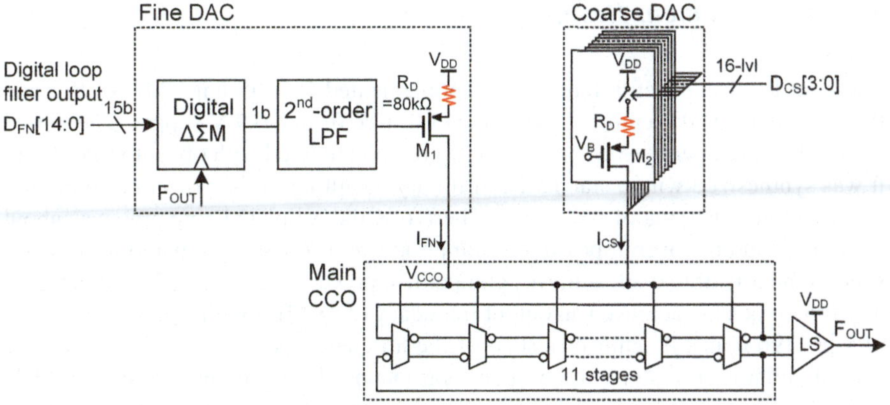

Fig. 11 Schematic diagram of the DCO [22]

The ZC detector is implemented with a two-stage comparator and an output buffer that has a bandwidth of 127 MHz, consuming 5.3 μA from a 0.9 V supply. The associated temperature error of the ZC detector is around 0.6 °C, which means that the target inaccuracy of ±200 ppm can be achieved. It turns out that the supply sensitivity of ZC detector determines the overall supply sensitivity of the frequency reference. In the prototype in [21], the designed ZC detector achieves an output frequency error of <480 ppm with the supply voltage variations of 0.85–1.05 V. This can be improved with additional circuit technique.

3.2 *Digitally Controlled Oscillators*

The schematic diagram of the DCO is shown in Fig. 11 [22], which consists of a CCO controlled by a 4-bit current-steering DAC and a 15-bit fine ΔΣ DAC. The coarse DAC tunes the output frequency shift of ±50% mainly over process corners, and the fine DAC covers ±12% over temperature variations. The DAC offers high resolution to meet target inaccuracy and sufficient tuning range to account for process and temperature variations. The ΔΣ DAC is implemented in a very compact area while achieving high resolution. It is driven by a digital ΔΣM implemented in a second-order error-feedback architecture, and the output second-order LPF removes the shaped quantization noise properly. After the LPF, the control signal is applied via a source-degenerated g_m, and this generates the bias current of the CCO. The jitter level of the frequency reference is mainly determined by the jitter of the free-running CCO and should be well matched to the uncertainty level. This aspect is often ignored in many previous works. The target period jitter of 7 ps corresponds to a frequency error of 200 ppm considering the output period. The resulting period jitter of the DCO is designed to be 6.3 ps.

3.3 Measurement Results

The dual RC frequency reference was implemented in a 65 nm CMOS [21], and the core area was 0.06 mm^2. It consumes 142 μW from a 0.9 V supply. The digital backends for temperature compensation were implemented with off-chip FPGA, but it was synthesized with a standard cell library, resulting in the power consumption of 34 μW and the area of 0.07 mm^2. Therefore, the estimated core area was about 0.13 mm^2, and the overall power consumption was 176 μW. Compared to previous state of the arts, this work achieves competitive power efficiency and smallest area. The prototype chip achieved an output frequency of 28 MHz with a period jitter (1 σ) of 7 ps (196 ppm). It achieves a state-of-the-art energy per cycle of 5pJ/cycle. The overall calibration is done in an open-loop manner by using an external 28 MHz reference clock. After two-point trimming, the measured output frequency errors of 16 samples were less than ±200 ppm (+190/−130 ppm) over the temperature range from −40 °C to 85 °C. This corresponds to the TC of 2.56 ppm/°C. Due to the open-loop nature of the calibration, there were some residual errors. Therefore, better results can be achieved by using a closed-loop calibration. It also achieved the line sensitivity of 0.29%/V. The measured Allan deviation reaches a floor of <2 ppm over 40 s average time.

4 Conclusions

A dual RC frequency reference achieves excellent frequency stability through the use of digital temperature compensation and is extremely energy efficient due to its highly digital architecture. The prototype chip operates in a 0.9 V supply and achieves a low energy consumption of 5pJ/cycle. It also achieves an inaccuracy of ±200 ppm from −40 to 85 °C after two-point trimming. The dual RC frequency reference has great potential for use in many wireless sensor nodes.

References

1. A. Djemouai, M.A. Sawan, M. Slamani, New frequency-locked loop based on CMOS frequency-to-voltage converter: Design and implementation. IEEE Trans. Circuits Syst. II **48**(5), 441–449 (2001)
2. M. Choi, T. Jang, S. Bang, Y. Shi, D. Blaauw, D. Sylvester, A 110nW resistive frequency locked on-chip oscillator with 34.3 ppm/°C temperature stability for system-on-chip designs. IEEE J. Solid State Circuits **51**(9), 2106–2118 (2016)
3. D.S. Truesdell, A. Dissanayake, B.H. Calhoun, A 0.6-V 44.6-fJ/cycle energy-optimized frequency-locked loop in 65-nm CMOS with 20.3-ppm/°C stability. IEEE Solid-State Circuits Lett. **2**(10), 223–226 (2019)
4. M.H. Perrott et al., A temperature-to-digital converter for a MEMS-based programmable oscillator with <0.5-ppm frequency stability and 1-ps integrated jitter. IEEE J. Solid State Circuits **48**(1), 276–291 (2013)

5. K. Pappu, G.P. Reitsma, S. Bapat, *Frequency-locked-loop ring oscillator with 3 ns peak-to-peak accumulated jitter in 1 ms time window for high-resolution frequency counting* (IEEE ISSCC Dig. Tech. Papers, Feb, 2017), pp. 92–93
6. K.A.A. Makinwa, M.F. Snoeij, A CMOS temperature-to-frequency converter with an inaccuracy of less than 0.5°C from 40°C to 105°C. IEEE J. Solid State Circuits **41**(12), 2992–2997 (2006)
7. P. Park, D. Ruffieux, K.A.A. Makinwa, A thermistor-based temperature sensor for a real-time clock with ±2 ppm frequency stability. IEEE J. Solid State Circuits **50**(7), 1571–1580 (2015)
8. W. Choi, Y. Lee, S. Kim, S. Lee, J. Jang, J. Chun, K. Makinwa, Y. Chae, *A 0.53pJK^2 7000μm^2 resistor-based temperature sensor with an inaccuracy of ±0.35°C (3σ) in 65nm CMOS* (IEEE ISSCC Dig. Tech. Papers, 2018), pp. 322–324
9. W. Choi, Y. Lee, S. Kim, S. Lee, J. Jang, J. Chun, K. Makinwa, Y. Chae, A compact resistor-based CMOS temperature sensor with an inaccuracy of 0.12 °C (3σ) and a resolution FoM of 0.43pJ·K^2 in 65-nm CMOS. IEEE J. Solid State Circuits **53**(12), 3356–3367 (2018)
10. M. Ding, Z. Zhou, S. Traferro, Y.-H. Liu, C. Bachmann, K. Philips, F. Sebastiano, A 0.7-V 0.43-pJ/cycle wakeup timer based on a bang-bang digital-intensive frequency-locked-loop for IoT applications. IEEE Solid-State Circuits Letters **1**(2), 30–33 (2018)
11. M. Ding, Z. Zhou, S. Traferro, Y.-H. Liu, C. Bachmann, F. Sebastiano, A 33-ppm/°C 240-nW 40-nm CMOS wakeup timer based on a bang-bang digital-intensive frequency-locked-loop for IoT applications. IEEE Transactions on Circuits and Systems I: Regular Papers **67**(7), 2263–2273 (2020)
12. Y. Lee, W. Choi, T. Kim, S. Song, K.A.A. Makinwa, Y. Chae, A 5800-μm^2 resistor-based temperature sensor with a one-point trimmed inaccuracy of ±1.2 °C (3σ) from −50 °C to 105 °C in 65-nm CMOS. IEEE Solid-State Circuits Letters **2**(9), 67–70 (2019)
13. C.P.L. van Vroonhoven, K.A.A. Makinwa, *A CMOS Temperature-to-Digital Converter with an Inaccuracy of ± 0.5° C (3σ) from −55 to 125°C* (IEEE ISSCC Dig. Tech. Papers, 2008), pp. 576–637
14. S. Pan, Y. Luo, S. Heidary Shalmany, K.A.A. Makinwa, A resistor-based temperature sensor with a 0.13 pJK^2 resolution FoM. IEEE J. Solid State Circuits **53**(1), 164–173 (2018)
15. J.A. Angevare, K.A.A. Makinwa, A 6800-μm^2 resistor-based temperature sensor with ±0.35 °C (3σ) inaccuracy in 180-nm CMOS. IEEE JSSC **54**(10), 2649–2657 (2019)
16. J.A. Angevare, Y. Chae, K.A.A. Makinwa, *A highly digital 2210μm2 resistor-based temperature sensor with a 1-point trimmed inaccuracy of ± 1.3°C (3 σ) from −55 °C to 125 ° C in 65nm CMOS* (ISSCC Dig. 2021, Tech. Papers), pp. 76–78
17. S.M. Kashmiri et al., A scaled thermal-diffusivity-based 16 MHz frequency reference in 0.16μm CMOS. IEEE J. Solid State Circuits **47**(7), 1535–1545 (2012)
18. Ç. Gürleyük et al., *A 16 MHz CMOS RC frequency reference with ±400ppm inaccuracy from −45°C to 85°C after digital linear temperature compensation* (ISSCC Dig, 2020), pp. 64–65
19. H. Jiang et al., *A 0.14mm^2 16 MHz CMOS RC frequency reference with a 1-point trimmed inaccuracy of ±400 ppm from −45°C to 85°C* (ISSCC Dig, 2021), pp. 436–437
20. Ç. Gürleyük, L. Pedala, S. Pan, F. Sebastiano, K.A.A. Makinwa, A CMOS dual-RC frequency reference with ±200-ppm inaccuracy from −45°C to 85°C. IEEE J. Solid State Circuits **53**(12), 3386–3395 (2018)
21. W. Choi, J. Angevare, I. Park, K.A.A. Makinwa, Y. Chae, *A 0.9 V 28 MHz dual-RC frequency reference with 5 pJ/cycle and ±200 ppm inaccuracy from −40°C to 85°C* (ISSCC Dig, 2021), pp. 434–435
22. W. Choi, J. Angevare, I. Park, K.A.A. Makinwa, Y. Chae, A 0.9-V 28-MHz highly digital CMOS dual-RC frequency reference with ±200 ppm inaccuracy from −40 °C to 85 °C. IEEE J. Solid State Circuits **57**(8), 2418–2428 (2022)
23. K. Souri, Y. Chae, K.A.A. Makinwa, A CMOS temperature sensor with a voltage-calibrated inaccuracy of ±0.15°C (3σ) from −55°C to 125°C. IEEE J. Solid State Circuits **48**(1), 292–301 (2013)

RC Oscillators with Non-linear Temperature Compensation

Giorgio Cristiano, Can Livanelioglu, Youngwoo Ji, Jiawei Liao, and Taekwang Jang

Abstract Various IoT applications require on-chip time references to achieve a small form factor. An on-chip RC oscillator is one of the most promising candidates, thanks to its full compatibility with standard CMOS technology. Conventional on-chip RC oscillators combine resistors with complementary temperature coefficients to achieve a low-temperature coefficient (TC). While this can successfully eliminate the first-order temperature dependency after a two-point trim, it leaves a residual non-linear dependency, which ranges from 20 to 50 ppm/°C. Consequently, several techniques have been proposed to eliminate this residual temperature dependency to achieve less than 10 ppm/°C temperature stability. This chapter reviews two recent designs, an R-RC oscillator (Ji et al., Second-order temperature-compensated on-chip R-RC oscillator achieving 7.93ppm/°C and 3.3pJ/Hz in −40°C to 125°C temperature range. In: 2022 IEEE international solid-state circuits conference (ISSCC), vol 65, pp 1–3. https://doi.org/10.1109/ISSCC42614.2022.9731730, 2022) and a non-linearity-aware dual phase-locked loop (Cristiano et al., A 8.7ppm/°C, 694nW, one-point calibrated RC oscillator using a nonlinearity-aware dual phase-locked loop and DSM-controlled frequency-locked loops. In: 2020 IEEE symposium on VLSI circuits, pp. 1–2. https://doi.org/10.1109/VLSICircuits18222.2020.9162838, 2020), which achieve 7.9 ppm/°C and 8.7 ppm/°C temperature stability using analog and digital techniques, respectively.

1 Introduction

Process, supply voltage, and temperature (PVT) variations degrade the frequency accuracy of on-chip RC oscillators. This is typically overcome by using calibration

G. Cristiano · C. Livanelioglu · J. Liao · T. Jang (✉)
ETH Zurich, Zurich, Switzerland
e-mail: tkjang@iis.ee.ethz.ch

Y. Ji
Hanbat National University, Daejeon, South Korea

P. Harpe et al. (eds.), *Biomedical Electronics, Noise Shaping ADCs, and Frequency References*, https://doi.org/10.1007/978-3-031-28912-5_14

circuits or trimming techniques. For example, the effect of process variation can be eliminated by adopting a one-point trimming after fabrication. Supply voltage dependency can also be improved by combining supply-insensitive oscillator topologies with a low-drop-out regulator. This leaves their temperature dependency as the primary source of variation, which needs to be compensated to achieve an accurate output frequency.

To achieve temperature compensation, a series combination of proportional-to-absolute temperature (PTAT) and complementary-to-absolute temperature (CTAT) resistors is commonly employed to cancel out their first-order dependencies. However, the remaining high-order or non-linear temperature dependency will still limit the accuracy of the output frequency. Recently, several designs have been introduced to improve the non-linear temperature dependency. RC oscillators using delta-sigma modulator-based trimming techniques demonstrate single-digit temperature coefficient (TC) by employing look-up tables [2, 3] or polynomial correction [4–6]. Also, an analog second-order cancellation technique using an R-RC oscillator was introduced in [1]. In this chapter, we will describe the operating principles of [1, 2].

2 R-RC Oscillator for Second Order

Figure 1a shows the conventional RC branch of a relaxation oscillator. It consists of a first-order temperature-compensated resistor (R_1) and a capacitor (C). The oscillation cycle is defined by the time it takes for V_{RC} to charge from GND to a reference voltage, V_{REF}. Therefore, the oscillation period varies according to the second-order temperature dependency of R_1. For example, when R_1 becomes smaller at a middle temperature (T_M), the oscillation period becomes shorter. In order to remove this second-order temperature dependency, the R-RC configuration introduced in [1] employs an additional first-order compensated resistor (R_2) that has a stronger second-order temperature dependency than that of R_1. In this R-RC

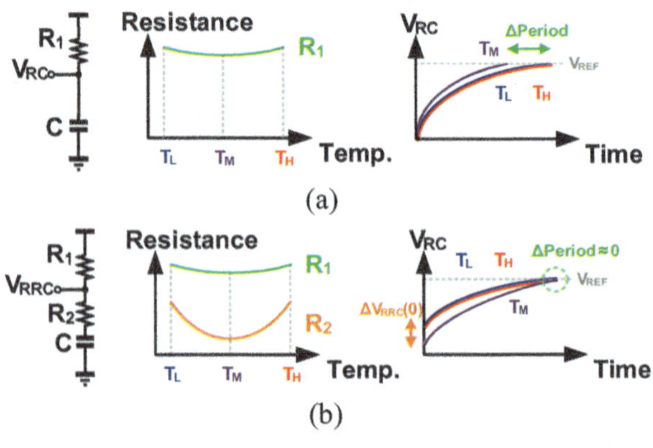

Fig. 1 Settling behaviors of (**a**) conventional RC circuit and (**b**) R-RC circuit

configuration, as shown in Fig. 1b, $V_{\text{R-RC}}$ then starts increasing from:

$$V_{\text{R-RC}}(0) = V_{\text{DD}}\left(\frac{R_2}{R_1 + R_2}\right) \cong V_{\text{DD}}\left(\frac{R_2}{R_1}\right) \quad (1)$$

rather than from GND.

This change in starting point can be designed to cancel the effect of the second-order reduction in R_1 so that the time it takes $V_{\text{R-RC}}$ to reach V_{REF} (T_{Period}) is independent of the temperature.

$$V_{\text{R-RC}}\left(T_{\text{period}}\right) = V_{\text{DD}}\left(1 - \frac{R_1}{R_1 + R_2} e^{-\frac{T_{\text{period}}}{(R_1+R_2)C}}\right) \quad (2)$$

For example, although R_1 becomes smaller at T_{M}, the time it takes to reach V_{REF} becomes less sensitive to temperature, thanks to the lower start point as R_2 is further decreased. By properly choosing the R_1 and R_2 ratio, the second-order coefficient of the two resistors can be eliminated, as shown in Fig. 1b.

The correct ratio is thus found by equating the impacts that the temperature variation has on the starting point and on the transition speed. The former is dominated by R_2, thanks to its larger second-order temperature dependency, while the latter is affected mostly by R_1 due to its larger nominal value.

Then, by defining each resistance as:

$$R = R_0\left(1 + \alpha\Delta T + \beta\Delta T^2\right) \quad (3)$$

where R_0 is the nominal value at room temperature and α and β are the first- and second-order temperature coefficients, respectively, we can find the R_2 nominal value that eliminates the second order as:

$$R_{0,2} \cong \sqrt{\frac{\beta_1}{\beta_2} R_{0,1}\left(R_{0,1} - \frac{T_{\text{period}}}{C}\right)} \quad (4)$$

It is important to note how R_2 is proportional to β_1. In fact, the faster the transition becomes due to a more non-linear R_1 (larger $\beta_1 R_1$), the lower the starting point must be, thanks to a smaller $\beta_2 R_2/\beta_1 R_1$.

Additionally, it is worth noticing that R_1 and R_2 do not need to be implemented with four resistors (i.e., two first-order compensated resistors). In fact, an implementation consisting of three resistors is sufficient to have a period not affected by either first- or second-order variation. In such implementation, R_2 is chosen as per (4) to minimize the second-order variation, while the resistor R_1 is split into a PTAT and a CTAT component ($R_{1,\text{PTAT}}$ and $R_{1,\text{CTAT}}$, respectively). In this work, $R_{1,\text{PTAT}}$ and $R_{1,\text{CTAT}}$ are implemented with p-poly resistors with and without silicide, and R_2 is implemented with an n-well resistor. Then, by adjusting the ratio of these two,

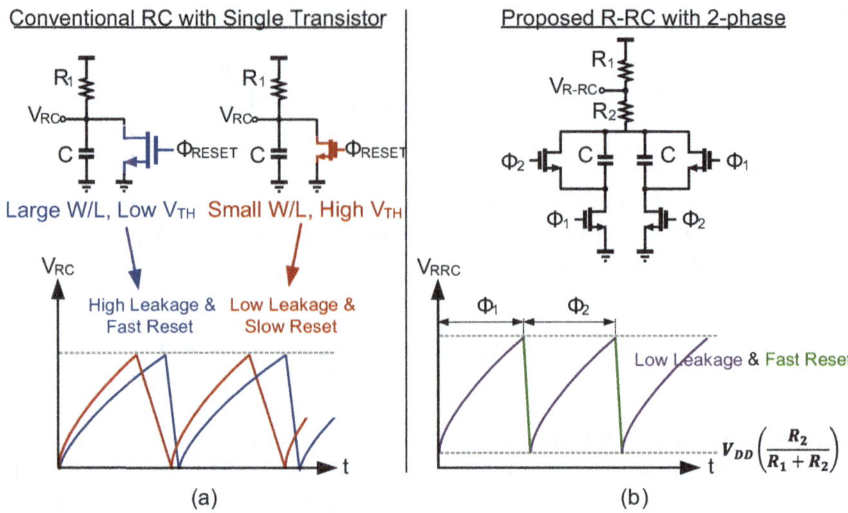

Fig. 2 (**a**) Conventional RC circuit reset switching with transistor design trade-off. (**b**) R-RC circuit with two-phase operation

we can eliminate the first-order effect resulting from all the three resistors in the system.

Figure 2 compares the conventional RC circuit with a single transistor reset and the R-RC circuit using two-phase operation. The oscillation period of the conventional RC circuit is affected by the large reset time for a weak switching transistor, whereas a strong pull-down transistor alters the period due to its large off leakage. To circumvent this design trade-off, the R-RC circuit adopts a dual-phase operation by alternating two capacitors between reset and RC transition modes. Here, the idle capacitor is in full reset state before it is switched into operation; therefore, the output voltage can attain a sharp transition to the initial point of the R-RC charging phase. This offers a relaxed reset time requirement, which can be as high as the R-RC transition time, with no adverse impact on the output frequency. Hence, the switching transistors can be reduced in size to achieve ultra-low leakage. Despite their small size, the two bottom transistors can still construct a low resistance connection to ground owing to their large V_{GS}.

Figure 3 illustrates the conventional RC and proposed R-RC oscillator circuits with their timing control circuitry and relevant waveforms. For the conventional RC oscillator shown in Fig. 3a, the reset pulses are generated via a continuous time comparator whose delay adds to the overall oscillation period, thereby causing non-linear temperature dependency. To make the temperature-dependent comparator delay negligible compared to the RC time constant, a power-hungry comparator is required.

The comparator power versus accuracy trade-off is addressed in the R-RC oscillator via a bang-bang delay-locked loop (DLL)-assisted reset pulse generation

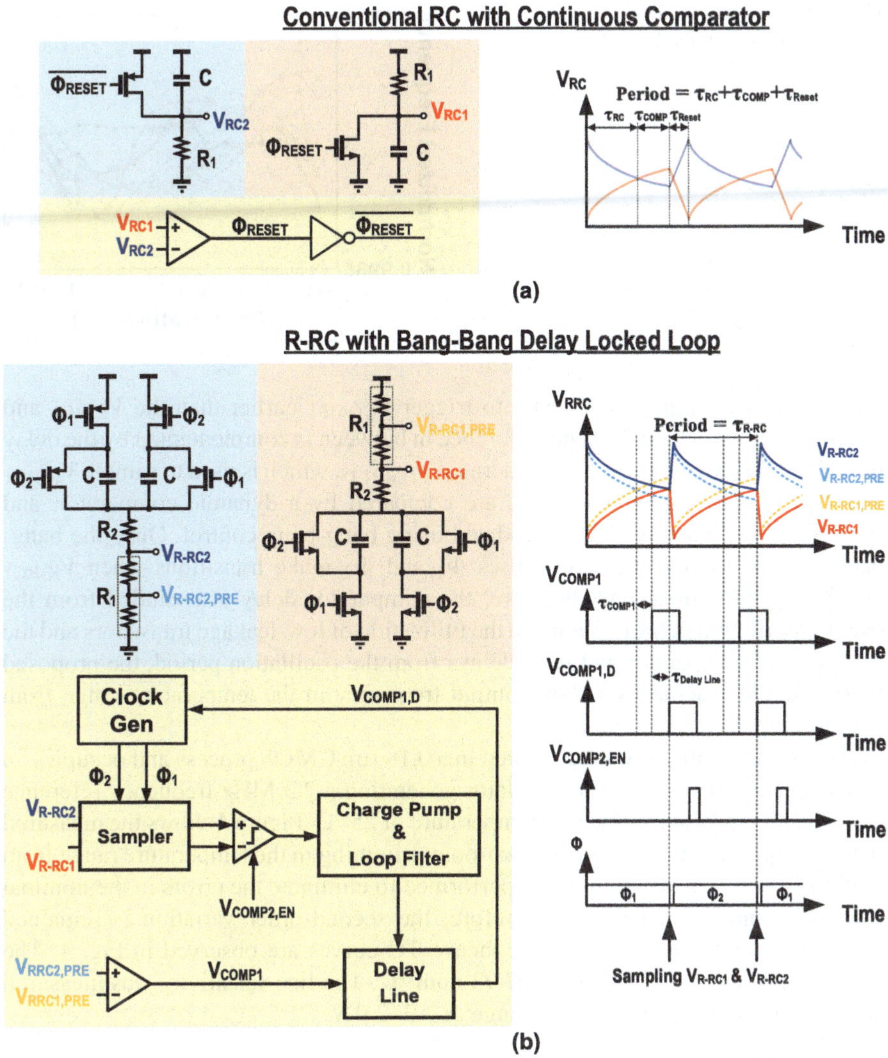

Fig. 3 (**a**) Circuit diagram of the conventional RC circuit and continuous comparator for reset pulse generation and RC oscillation waveforms. (**b**) R-RC oscillator with bang-bang delay-locked loop for reset pulse generation and corresponding waveforms

circuit, depicted in Fig. 3b. As shown in the figure, the R_1 resistors in R-RC cores are split into two resistors in series, with their middle nodes yielding two additional voltages $V_{\text{R-RC1,PRE}}$ and $V_{\text{R-RC2,PRE}}$. These voltages have shorter time constants than $V_{\text{R-RC1}}$ and $V_{\text{R-RC2}}$, and hence they trigger the output of the continuous comparator, V_{COMP1}, several tens of ns earlier than the crossing point of $V_{\text{R-RC1}}$ and $V_{\text{R-RC2}}$. Note that the division of R_1 does not need to be precise because the purpose of

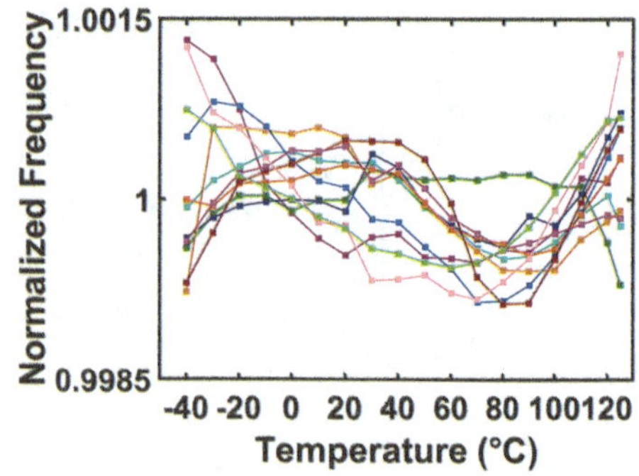

Fig. 4 Measured normalized frequency after two-point trimming

$V_{\text{R-RC1,PRE}}$ and $V_{\text{R-RC2,PRE}}$ is only to trigger V_{COMP1} earlier than the $V_{\text{R-RC1}}$ and $V_{\text{R-RC2}}$ crossing, where the time difference in between is complemented by the delay line. Then, V_{COMP1} is delayed to generate $V_{\text{COMP1,D}}$, which is used to sample $V_{\text{R-RC1}}$ and $V_{\text{R-RC2}}$. The sampled voltages are compared by a dynamic comparator, and the delay line control voltage is updated using bang-bang control. Once the bang-bang DLL is locked, the reset pulses Φ_1 and Φ_2 make transitions when $V_{\text{R-RC1}}$ and $V_{\text{R-RC2}}$ are equal, and therefore, the comparator delay is excluded from the overall oscillation period. Thanks to the utilization of low-leakage transistors and the exclusion of comparator and reset delays from the oscillation period, the proposed R-RC oscillator achieves a stable output frequency in the temperature range from −40 to 125 °C.

The R-RC oscillator was fabricated in a 0.18 μm CMOS process and occupies an active area of 0.07 mm^2. The oscillator generating a 2.3 MHz frequency reference consumes 7.6 μW at the nominal temperature of 25 °C. Figure 4 shows the measured output frequency of 11 chips after two-point trimming in the temperature range from −40 to 125 °C. The trimming was performed to eliminate the errors in the nominal frequency and its first order. Therefore, the second-order variation is remained untrimmed, and both convex and concave TC curves are observed in Fig. 4. The average temperature coefficient is 7.93 ppm/°C. The line sensitivity was measured at 0.51%/V in the supply voltage range of 1.3–2.0 V.

3 Non-linearity-Aware Dual Phase-Locked Loop

The block diagram of a non-linearity-aware dual-phase locked loop is shown in Fig. 5. It consists of two FLL-based RC oscillators, OSC_A and OSC_B, which generate temperature-dependent periods of P_{OSCA} and P_{OSCB}, respectively. The oscillation frequencies are controlled by the frequency control words, FCW_A and FCW_B, through delta-sigma modulators, DSM_A and DSM_B, respectively. The two oscillator frequencies are equalized by using a digital dual PLL labeled as PLL_A and PLL_B, whose integral path gains are made proportional to the oscillators' TCs,

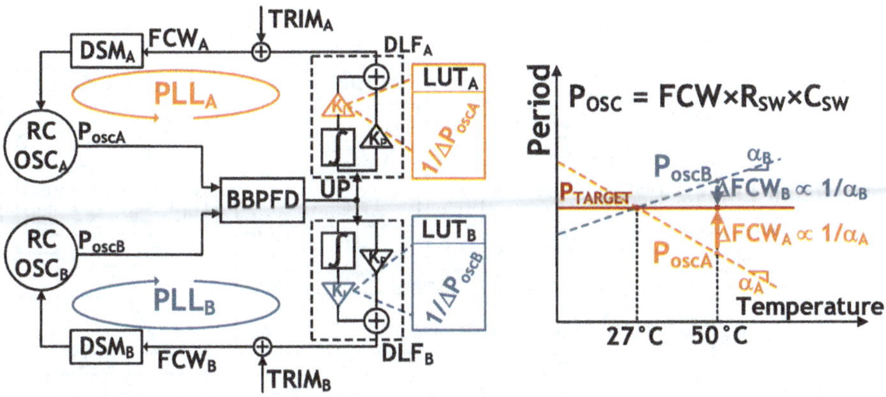

Fig. 5 Block diagram and operation principle of a non-linearity-aware dual phase-locked loop

labeled as α_A and α_B in Fig. 5, so that P_{OSCA} and P_{OSCB} always meet at the target period, P_{Target}, at any given temperature point. To further improve the temperature compensation, the integral paths are implemented by means of two LUTs (LUT_A and LUT_B). Then, the UP signal coming from the BBPFD goes to an integrator, and the accumulated value is used to address the LUTs and update the two FCWs. Assuming the variation of TCs for the resistors in the given technology is low enough, only the nominal value of the resistance needs to be trimmed. Hence, a 1-point calibration would be sufficient. In this work, the calibration of each oscillator is performed by means of the inputs $TRIM_A$ and $TRIM_B$, which are adjusted so that the oscillating period of each individual FLL matches with the target period at room temperature.

The schematic of each RC oscillator is depicted in Fig. 6. The system is based on an FLL that equates the currents generated by the duty-cycled resistor (I_R) and the switched capacitor (I_C). To achieve this, the voltages on both the duty-cycled resistor and switched capacitor are set to an on-chip voltage reference V_{ref} by the fully differential amplifier A1. Hence, the currents are $I_R = V_{ref}/(\mathrm{FCW} \times R_{SW})$ and $I_C = V_{ref} C_{SW} F_{OSC}$. Then, since the FLL will lock when $I_R = I_C$, the locked frequency will be:

$$F_{osc} = \frac{1}{\mathrm{FCW} \times R_{SW} \times C_{SW}}$$

Any temperature dependency coming from the FLL and not from the resistor cannot be cancelled by the PLL. For this reason, the charge pump contains a chopping scheme to minimize the variation induced by device mismatch in the mirroring process (N1–N2, P1–P2) and amplifier offset (A1). To make the control voltage constant within a divider cycle and mitigate the non-idealities due to capacitor and resistor switching, a sampler is placed between charge pump and VCO. Finally, to ensure better stability under process, voltage, and temperature

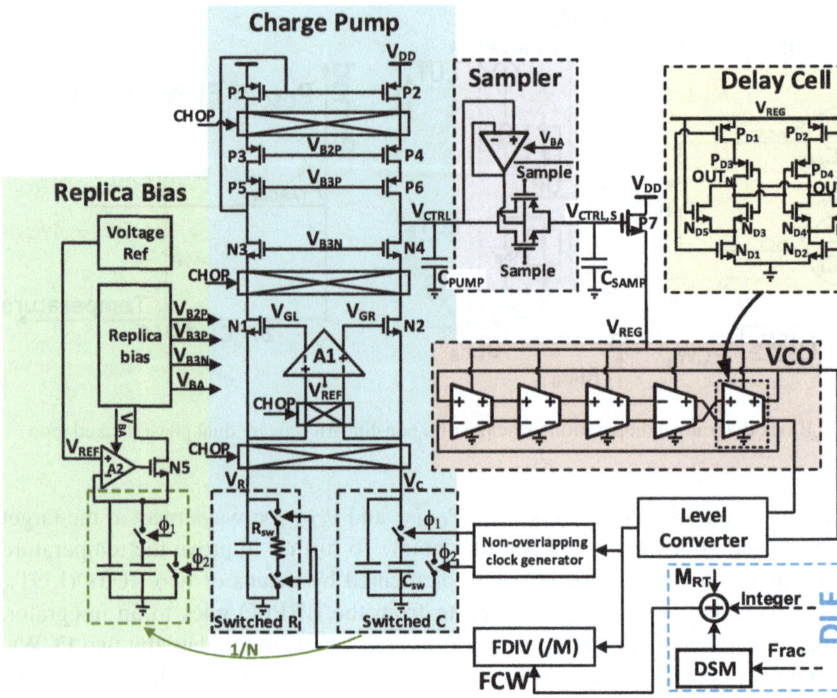

Fig. 6 Schematic of the FLL-based RC oscillator

variation, the biasing circuit is implemented with a switched capacitor replica circuit.

In Fig. 7, the locking behavior of the dual PLL is described. When the temperature is changed from 13 to 25 °C, F_{OSCA} is lower than F_{OSCB} due to the TC difference. Note that the dual PLL can correctly lock to the target frequency as far as the TCs of the two oscillators are different from each other. In this example, the two resistors are implemented as high-resistance p-poly and p-poly, so the TC of both oscillators is positive, with OSC_B having a stronger dependency on the temperature. Then both FCW_A and FCW_B increase due to the integral paths of the PLLs, but at the rates proportional to the two oscillators' TC, respectively. The non-linear TC is piece-wise modeled in the look-up tables; thus, when the accumulator output exceeds the boundary, the slope values, SS_A and SS_B, are updated accordingly. Eventually, FCW_A and FCW_B cancel the resistance change as a function of temperature, and both oscillators are locked at the target frequency in the steady state.

The proposed design was fabricated in 0.18 μm CMOS and occupied 1.2 mm^2. The proposed design only uses a single pair of supply and ground voltages and does not require any additional reference. The measured output frequency is shown in Fig. 8. A total of ten samples from two different wafers are measured between −15

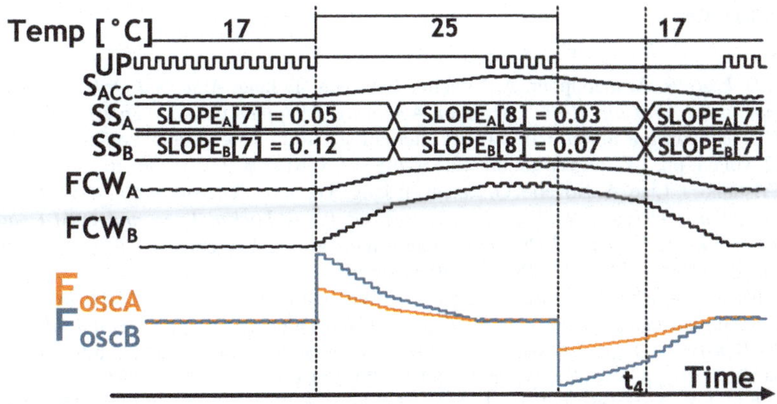

Fig. 7 Conceptual waveforms describing the locking behavior of the dual phase-locked loop

Fig. 8 Measured output frequency while varying the ambient temperature from −15 to 85 °C

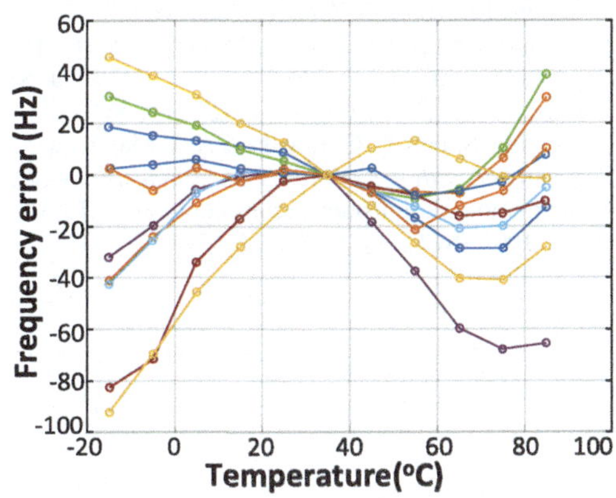

and 85 °C. While generating a 116 kHz output frequency, the worst TC with one-point calibration was measured to be 8.7 ppm/°C. The design consumes less than 0.7 µW, and the measured line sensitivity is less than 0.4%/V.

4 Conclusion

In this chapter, we introduced two recent designs proposing the calibration of high-order temperature dependency. While the analog approach using R-RC calibration is area and energy efficient, its calibration has limitations in achieving precise cancellation as it relies on the intrinsic matching of second-order coefficients of two resistor types. On the other hand, the digital approach using dual PLL and LUT can achieve higher accuracy but occupies a larger area. Still, both techniques significantly improve the temperature stability compared to the conventional RC oscillators and can serve as a precise frequency reference for miniaturized IoT applications.

References

1. Y. Ji, J. Liao, S. Arjmandpour, A. Novello, J.-Y. Sim, T. Jang, *A Second-Order Temperature-Compensated On-Chip R-RC Oscillator Achieving 7.93ppm/°C and 3.3pJ/Hz in −40°C to 125°C Temperature Range*, 2022 IEEE international solid-state circuits conference (ISSCC), 2022, vol 65, pp. 1–3. https://doi.org/10.1109/ISSCC42614.2022.9731730
2. G. Cristiano, J. Liao, A. Novello, G. Atzeni, T. Jang, *A 8.7ppm/°C, 694nW, One-Point Calibrated RC Oscillator Using a Nonlinearity-Aware Dual Phase-Locked Loop and DSM-Controlled Frequency-Locked Loops*. 2020 IEEE symposium on VLSI circuits, 2020, pp. 1–2. https://doi.org/10.1109/VLSICircuits18222.2020.9162838
3. A. Khashaba, J. Zhu, M. Ahmed, N. Pal, P.K. Hanumolu, *3.5 A 34μW 32MHz RC Oscillator with ±530ppm Inaccuracy from −40°C to 85°C and 80ppm/V Supply Sensitivity Enabled by Pulse-Density Modulated Resistors*. 2020 IEEE international solid- state circuits conference – (ISSCC), 2020, pp. 66–68. https://doi.org/10.1109/ISSCC19947.2020.9062942
4. W. Choi, J.A. Angevare, I. Park, K.A.A. Makinwa, Y. Chae, *31.2 A 0.9V 28MHz Dual-RC Frequency Reference with 5pJ/Cycle and ±200 ppm Inaccuracy from −40°C to 85°C*. 2021 IEEE international solid-state circuits conference (ISSCC), 2021, vol 64, pp. 434–436. https://doi.org/10.1109/ISSCC42613.2021.9366021
5. Ç. Gürleyük, S. Pan, K.A.A. Makinwa, A 16 MHz CMOS RC frequency reference with ±90 ppm inaccuracy from −45 °C to 85 °C. IEEE J. Solid State Circuits **57**(8), 2429–2437 (2022). https://doi.org/10.1109/JSSC.2022.3142662
6. H. Jiang, S. Pan, Ç. Gürleyük, K.A.A. Makinwa, *31.3 A 0.14mm2 16MHz CMOS RC Frequency Reference with a 1-Point Trimmed Inaccuracy of ±400ppm from −45°C to 85°C*. 2021 IEEE international solid-state circuits conference (ISSCC), 2021, vol 64, pp. 436–438. https://doi.org/10.1109/ISSCC42613.2021.9365795

RC Frequency References Based on Pulse-Density Trimmed Resistors

Kyu-Sang Park and Pavan Kumar Hanumolu

Abstract Stable frequency references serve many essential purposes, such as time-keeping for calendar functions and reference clocks for radios and microcontrollers. While quartz crystal-based references can achieve the desired performance, they are bulky and expensive and suffer from a long start-up time. Hence, techniques for implementing power-efficient frequency references that can be fully integrated in standard CMOS technology and achieve excellent frequency stability across a wide temperature and voltage range are needed. Highly stable on-chip RC frequency references offer the possibility of replacing crystal oscillators in many cost- and form factor-constrained applications. However, achieving good frequency stability in a power-efficient manner across process, voltage, and temperature variations poses many design challenges. In this paper, we describe these challenges and present methods for improving the frequency accuracy of RC oscillators by overcoming them. We show that the impact of the resistor temperature coefficient on the output frequency accuracy can be mitigated by using a parallel combination of switched resistors digitally controlled by pulse-density modulated sequences. Finally, we present experimental results obtained from prototype oscillators fabricated in a 65 nm CMOS process and use them to illustrate that reducing the inaccuracy to better than ±140 ppm over −40 °C to 95 °C (2.1 ppm/°C) is feasible without compromising long-term stability (1.3 ppm Allan deviation) and power efficiency (1 μW/MHz).

1 Introduction

Frequency references are needed in almost all electronic systems with varying degrees of accuracy and stability requirements. For example, microcontrollers need a clock source with ±1 % frequency inaccuracy, while local interconnect network

K.-S. Park (✉) · P. K. Hanumolu
University of Illinois at Urbana-Champaign, Champaign, IL, USA
e-mail: kyusang2@illinois.edu

P. Harpe et al. (eds.), *Biomedical Electronics, Noise Shaping ADCs, and Frequency References*, https://doi.org/10.1007/978-3-031-28912-5_15

protocol used in automotive requires better than ±5000 ppm inaccuracy [1]. Real-time clocks [2] need better than ±250 ppm inaccuracy, and wireless applications mandate even more stringent requirements with even the most relaxed specification requiring inaccuracies smaller than ±40 ppm [3]. Crystal oscillators are used as de facto frequency references in most of these systems today. However, crystals are bulky, hard to integrate, and not cost-efficient. In view of these disadvantages, there have been significant efforts to develop alternate frequency references that are better suited for monolithic integration. Among them, frequency references built using on-chip time constants are gaining popularity as possible replacements to crystal oscillators in low-cost applications. Some of the earlier work has shown that LC time constant-based references can achieve very stable output frequencies with less than ±10 ppm inaccuracy [4]. However, their output frequency is typically in the GHz range, and they consume milliwatts of power [4, 5]. Scaling the LC oscillator frequency to the MHz range needs a huge LC tank, which significantly increases the chip area. Alternatively, RC time constant-based references [1, 6–10] occupy smaller area, consume lower power, and are well suited for integration in any standard CMOS process. But their performance is very susceptible to variations in temperature and supply voltage, which has precluded them from being used widely. Frequency-locked loop (FLL)-based closed-loop RC oscillators [1, 9–12], compared to open-loop relaxation oscillators, have higher immunity to supply voltage variations. However, their performance is also limited by the temperature dependence of the resistor. This limitation was addressed in the prior art by using a composite resistor made from resistors with opposing temperature coefficients (TCs) [9]. Unfortunately, TC of the composite resistor is highly dependent on the process-sensitive sheet resistance of resistors comprising it. Consequently, composite resistors are implemented using a vast array of resistors and then trimmed on a sample-by-sample basis. While this approach helps alleviate process sensitivity, two critical factors limit the achievable frequency inaccuracy to about 5000 ppm: First, the accuracy with which TC is canceled across process corners is fundamentally limited by the finite number of resistor combinations that can be practically implemented on a chip. Second, leakage currents in the switches used to select the resistors and the mixing of TCs of individual resistors in the array generate large higher-order TCs that are difficult to compensate. This chapter presents a method for precise and robust cancellation of the resistor TC across process corners using a parallel combination of switched resistors that are digitally controlled by pulse-density modulated sequences [13, 14]. Because the number of switches used to perform temperature compensation is reduced to just two or three switches, the compensation method does not introduce additional higher-order TCs or degrade supply sensitivity. Consequently, it makes it possible to achieve excellent accuracy with only first-order or second-order temperature compensation. Using the switched resistors, a prototype FLL-based 100 MHz oscillator fabricated in a 65 nm CMOS process achieves an inaccuracy of ±140 ppm (2.1 ppm/°C), 83 ppm/V voltage sensitivity, and 1.3 ppm Allan deviation floor with an power efficiency of 1 μW/MHz.

2 First-Order Temperature-Compensated Oscillator

The first-order temperature-compensated oscillator (TCO) using two switched resistors is shown in Fig. 1. It is composed of a frequency-locked loop (FLL) that locks the period of a voltage-controlled oscillator (VCO) to a scaled reference RC time constant, $R_{REF}C_R$, where R_{REF} is the reference resistor and C_R is the capacitance of switched-capacitor resistor (SCR). The TCO architecture has two voltage dividers. The first voltage divider is formed by a SCR, and a temperature-compensated reference resistor, R_{REF}, generates a frequency-dependent voltage, V_{FB}. The second voltage divider is formed by two resistors $3R_D$ and R_D of the same type and generates a fixed reference voltage $V_{REF} = V_{DD}/4$. The integrator integrates the difference between V_{FB} and V_{REF} and produces a control voltage V_C to the VCO. The divided output clock of the VCO is fed to the SCR via a nonoverlapping clock generator. In steady state, thanks to the significant loop gain of the FLL, the ratio of average SCR resistance, R_{SCR}, and R_{REF} is kept constant such that:

$$R_{SCR} : R_{REF} = 3R_D : R_D \tag{1}$$

where $R_{SCR} = N/F_{OUT}C_R$ and N is the division ratio of feedback clock divider in the FLL loop. TCO's output frequency, F_{OUT}, can be calculated from (1) as:

$$F_{OUT} = \frac{N}{3R_{REF}C_R}. \tag{2}$$

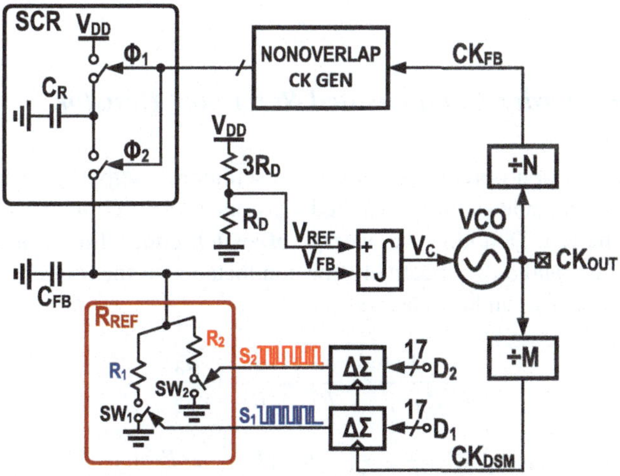

Fig. 1 First-order temperature-compensated oscillator

The above equation reveals the accuracy of F_{OUT} is primarily dictated by the temperature and supply sensitivity of $R_{REF}C_R$. R_{REF} is implemented using a parallel combination of two switched resistors R_1 and R_2 [13]. R_{REF} is tuned by controlling SW_1/SW_2 with pulse-density modulated signals (S_1/S_2) such that the opposing TCs of R_1 and R_2 are canceled. Denoting the average of S_1 and S_2 by α_1 and α_2:

$$\frac{1}{R_{REF}} = \frac{\alpha_1}{R_1} + \frac{\alpha_2}{R_2}. \tag{3}$$

From (2) and (3), the output frequency can be expressed as:

$$F_{OUT} = \alpha_1 F_1 + \alpha_2 F_2 \tag{4}$$

where:

$$F_i = \frac{N}{3R_i C_R}, i = 1, 2. \tag{5}$$

For some optimum value of $\alpha_1 = \alpha_{1opt}$ and $\alpha_2 = \alpha_{2opt}$, the output frequency, F_{OUT}, can be made nearly independent of temperature. The terms α_{1opt} and α_{2opt} are calculated using a two-point trimming procedure, quantized to 17-bit digital words D_1 and D_2 and fed to the $\Delta\Sigma$ modulators. The $\Delta\Sigma$ modulators convert D_1/D_2 into pulse-density modulated signals (S_1/S_2) and controls SW_1/SW_2, thus making F_{OUT} nearly independent of temperature. The compensation scheme can significantly reduce the impact of first-order resistor TC by eliminating most of the switches.

2.1 *Second-Order Compensated Reference Resistor*

The first- and second-order TCs of the oscillator's output frequency can be compensated using at least three switched resistors with different TCs [14] as shown in Fig. 2 on the left. Denoting the average of switch-controlling sequences S_1, S_2, and S_3 by α_1, α_2, and α_3, respectively, the conductance of the reference resistor and the output frequency can be written as:

$$\frac{1}{R_{REF}} = \frac{\alpha_1}{R_1} + \frac{\alpha_2}{R_2} + \frac{\alpha_3}{R_3} \tag{6}$$

$$F_{OUT} = \frac{N}{3R_{REF}C_R} = \alpha_1 F_1 + \alpha_2 F_2 + \alpha_3 F_3 \tag{7}$$

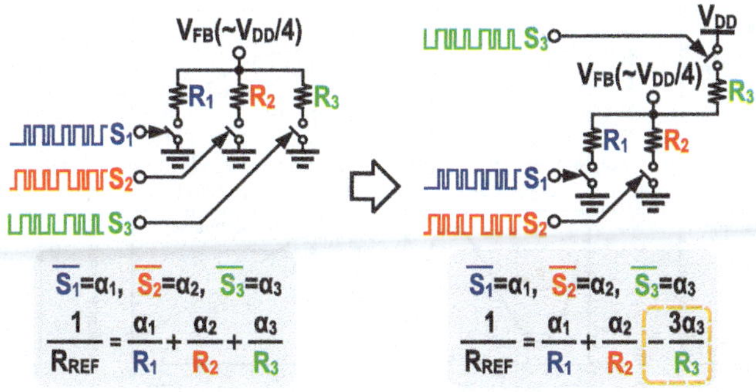

Fig. 2 Example illustration of the method to realize a negative scaling factor for R_3

where:

$$F_i = \frac{N}{3R_iC_R},\ i = 1, 2, 3. \tag{8}$$

There exists some optimum α_1, α_2, and α_3 at which the TCO output frequency, F_{OUT}, is insensitive to the temperature at least to the first- and second-order. These optimum alphas can be determined uniquely by using a three-point trimming method. To this end, F_{OUT} is measured at three temperatures, T_1, T_2, and T_3, and the desired frequency, F_{DES}, is expressed in terms of the unknown α_{1OPT}, α_{2OPT}, and α_{3OPT} as follows:

$$F_{DES} = \alpha_{1OPT}F_1(T_1) + \alpha_{2OPT}F_2(T_1) + \alpha_{3OPT}F_3(T_1) \tag{9}$$

$$F_{DES} = \alpha_{1OPT}F_1(T_2) + \alpha_{2OPT}F_2(T_2) + \alpha_{3OPT}F_3(T_2) \tag{10}$$

$$F_{DES} = \alpha_{1OPT}F_1(T_3) + \alpha_{2OPT}F_2(T_3) + \alpha_{3OPT}F_3(T_3). \tag{11}$$

Solving the above three equations yields α_{1OPT}, α_{2OPT}, and α_{3OPT}. However, when these calculations were performed with p-poly, n-diffusion, and p-diffusion resistors for R_1, R_2, and R_3, respectively, α_{3OPT} turned out to be negative, as shown in Fig. 3a. Unfortunately, such a negative α_{3OPT} cannot be realized because the average of sequence S_3 can only be between [0 and 1). Note that even though the p-poly resistor has an opposite sign first- and second-order TCs compared to other two resistors, it cannot guarantee that all α_{1OPT}, α_{2OPT}, and α_{3OPT} are positive. Another combination consisting of only positive TC resistors (n-poly, n-diffusion, and p-diffusion) also results in a negative α_{3OPT} (see Fig. 3b). Therefore, a method to realize a negative α_{OPT} is needed for performing second-order compensation. Figure 2 shows the method to address the issue of negative α_{OPT}. Switched resistor R_3, which was previously connected to the ground terminal, is now connected to V_{DD}. As a result, it pulls up node V_{FB} instead of pulling it down, behaving as a negative resistor as indicated by the negative sign in the last term of the R_{REF}

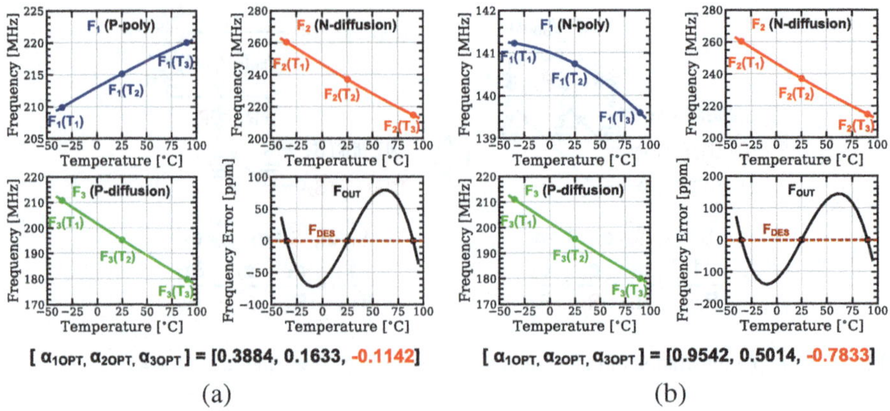

Fig. 3 Simulated F_1, F_2, F_3, three-point trimmed F_{OUT}, and α_{OPT}s (**a**) with p-poly resistor and (**b**) with n-poly resistor

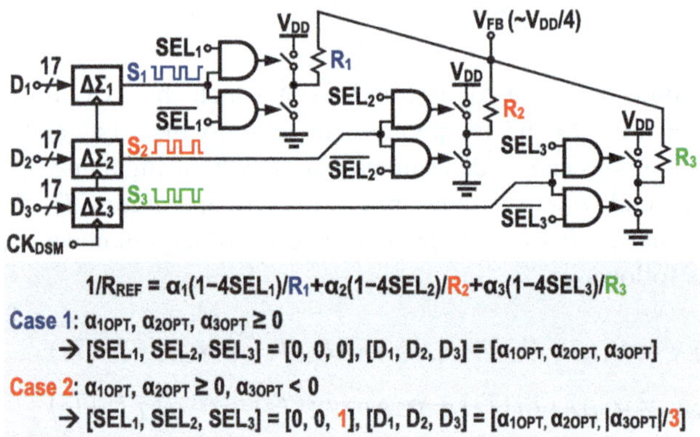

Fig. 4 Reference resistor schematic

equation shown at the bottom of Fig. 2. The 3× factor is a result of V_{FB} being forced to be equal to $V_{REF} = V_{DD}/4$ by the FLL in steady state. The details of the complete reference resistor are shown in Fig. 4. Resistors R_1, R_2, and R_3 are controlled by sequences, S_1, S_2, and S_3 generated by three first-order $\Delta\Sigma$ modulators. Each resistor can be either connected to ground or V_{DD} using low-leakage switches controlled by signals SEL_1, SEL_2, and SEL_3. If the calculated α_{1OPT}, α_{2OPT}, and α_{3OPT} are between zero and one (see case 1 in Fig. 4), all the SEL signals are set to zero, and the α_{OPT}s are quantized to 17-bit digital words D_1, D_2, and D_3 and fed to the $\Delta\Sigma$ modulators. If some α_{OPT}s are negative, the corresponding SEL signal is set to one, and the corresponding digital word is set to $|\alpha_{OPT}|/3$ to compensate for the 3× factor described earlier. For example, if α_{OPT3} is negative as shown in case 2 of Fig. 4, SEL_3 is set to one, and D_3 is set to $|\alpha_{OPT3}|/3$, resulting in R_{REF} and F_{OUT} of:

$$\frac{1}{R_{REF}} = \frac{\alpha_{1OPT}}{R_1} + \frac{\alpha_{2OPT}}{R_2} - \frac{|\alpha_{3OPT}|}{R_3} \tag{12}$$

$$F_{OUT} = \alpha_{1OPT}F_1 + \alpha_{2OPT}F_2 - |\alpha_{3OPT}|F_3. \tag{13}$$

This programmability eliminates the need for resistors with opposite TCs.

2.2 *Design Considerations*

Having presented the architecture, we now describe circuit-level imperfections that degrade the TCO performance and elucidate design trade-offs involved in overcoming them. Specifically, SCR nonidealities, leakage currents in the switches, and integrator offset can introduce errors that degrade temperature stability, long-term stability, and supply sensitivity. These impairments and the methods to mitigate them will be described in details.

2.2.1 Finite Capacitance at SCR's Switching Node

We have so far assumed SCR's effective resistance to be $N/(F_{OUT}C_R)$; however, this expression is not accurate especially if the capacitance at the switching node, C_{FB}, is not much larger than C_R [13, 15]. To elucidate this further, consider the SCR branch that operates in two nonoverlapping phases, ϕ_1 and ϕ_2, as shown in Fig. 5a. Capacitor C_R is connected to V_{DD} and V_{FB}, during ϕ_1 and ϕ_2 phases, respectively. In steady state, $V_{FB}(t)$ voltage is periodic with period of $T = N/F_{OUT}$ and can be written as:

$$V_{FB}(t) = \begin{cases} V_P\, e^{\frac{-t}{R_{REF}(C_{FB}+C_R)}} & 0 < t < T/2 \\ V_P\, e^{\frac{-T/2}{R_{REF}(C_{FB}+C_R)} + \frac{-t+T/2}{R_{REF}C_{FB}}} & T/2 < t < T \end{cases} \tag{14}$$

where V_P is the peak value of $V_{FB}(t)$ and can be calculated from the charge-sharing equation at $t = T$ as follows:

$$V_P = V_{FB}(T)\frac{C_{FB}}{C_R + C_{FB}} + V_{DD}\frac{C_R}{C_R + C_{FB}} \tag{15}$$

Substituting (14) into (15) and rearranging the equation gives:

$$V_P = V_{DD}\frac{C_R}{C_R + C_{FB}\left(1 - e^{\frac{-T/2}{R_{REF}(C_{FB}+C_R)} + \frac{-T/2}{R_{REF}C_{FB}}}\right)} \tag{16}$$

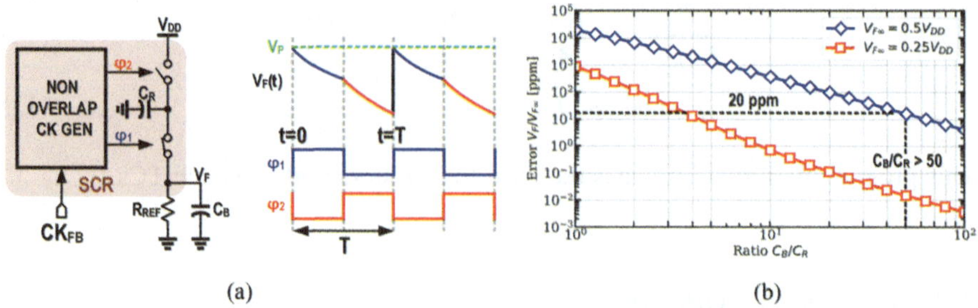

(a) (b)

Fig. 5 Effect of finite capacitance at switching node: (**a**) Model. (**b**) Simulation results

The above equation reveals that the average value of $V_{FB}(t)$, $\overline{V_{FB}}$, converges to its ideal steady-state value of $V_{FB} = V_{DD}/(1+N/(R_{REF}C_RF_{OUT})) = V_{DD}/4$ when $C_{FB} \gg C_R$. However, for smaller values of C_{FB}, $\overline{V_{FB}}$ deviates from the ideal value, which causes SCR conductance a nonlinear function of the switching frequency, and therefore F_{OUT}. To quantify this effect further, the normalized error of $\overline{V_{FB}}$ is plotted for two different output voltages in Fig. 5b, where $V_{FB\infty}$ represents $\overline{V_{FB}}$ when $C_{FB} = \infty$. To keep the error below 20 ppm, C_{FB} must be at least 50 times C_R.

2.2.2 Incomplete Settling of SCR Internal Nodes

Effective resistance of the SCR is equal to T/C_R only if capacitor C_R is fully charged to V_{DD} and discharged to V_{FB} in ϕ_1 and ϕ_2, respectively. However, if the voltage across C_R does not settle completely within half the switching clock period (=T/2), SCR's resistance deviates from T/C_R. Consider the SCR branch connected between voltages V_1 and V_2 as shown in Fig. 6a, whose effective resistance can be expressed as:

$$R_{SCR} = \frac{V_2 - V_1}{I_{av}} \tag{17}$$

where I_{av} is the average current flowing between the two nodes V_2 and V_1. I_{av} can be defined in terms of the average charge transferred between V_2 and V_1 as follows:

$$I_{av} = \frac{Q_{av}}{T} = \frac{C_R(V_2' - V_1')}{T} \tag{18}$$

where V_2' and V_1' are the maximum and minimum voltages across C_R, respectively. Substituting (18) into (17) results in:

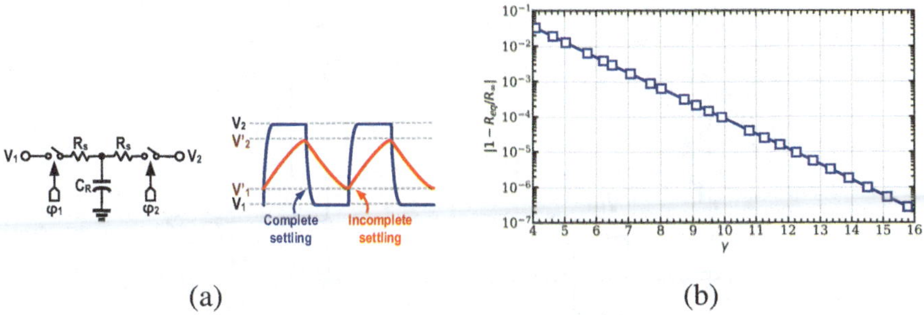

Fig. 6 Effect of incomplete settling of the SCR: (**a**) Model. (**b**) Simulation results

$$R_{SCR} = \frac{T}{C_R}\frac{V_2 - V_1}{V_2' - V_1'} \tag{19}$$

$V_2' - V_1'$ can be calculated from the charging and discharging equations resulting in:

$$V_2' - V_1' = (V_2 - V_1)\frac{1 - e^{-\gamma}}{1 + e^{-\gamma}} \tag{20}$$

where $\gamma = T/(2R_sC_R)$. Substituting (20) into (19) gives:

$$R_{SCR} = \frac{T}{C_R}\frac{1 + e^{-\gamma}}{1 - e^{-\gamma}} = \frac{T}{C_R} coth\left(\frac{\gamma}{2}\right) \tag{21}$$

The above equation shows that effective resistance of the SCR increases due to incomplete settling by a factor of $coth(\gamma/2)$. The same result was obtained in [16] using power loss analysis in switched-capacitor DC-to-DC converters. Since $coth(\gamma/2)$ depends on switch resistance R_S, which is a strong function of the supply, the supply sensitivity of the TCO is degraded in the presence of incomplete settling of C_R voltage. Plotting error due to incomplete settling as a function of γ, as shown in Fig. 6b, reveals that $\gamma > 12$ is needed to keep the error less than 20 ppm.

2.2.3 Finite Off Resistance of SCR Switches

The impact of leakage on the SCR can be modeled by a resistor R_P connected in parallel with the SCR, as shown in Fig. 7. In the presence of R_P, Eqs. (1) and (2) get modified as follows:

$$R_{SCR} \parallel R_P : R_{REF} = 3R_D : R_D \tag{22}$$

$$F_{OUT} = \frac{N}{3R_{REF}C_R} - \frac{N}{R_PC_R}. \tag{23}$$

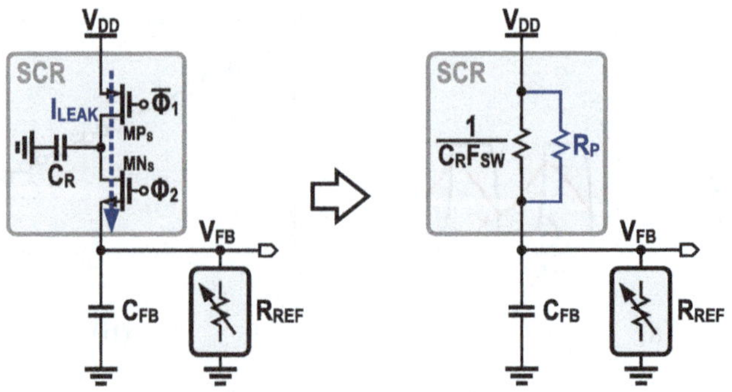

Fig. 7 Effect of switch leakage currents in a conventional SCR

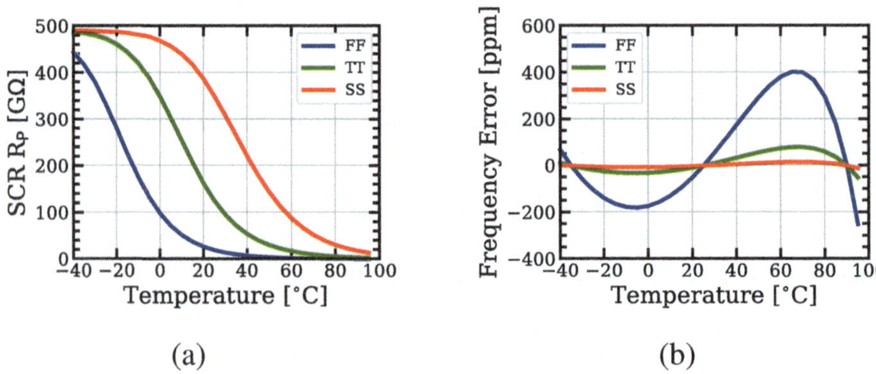

Fig. 8 Simulated SCR imperfection: (**a**) equivalent leakage resistor, R_P, versus temperature. (**b**) Frequency error caused by R_P after three-point trim

The above equation reveals that leakage-resistor R_P introduces an offset in the output frequency of $\Delta F_{OUT} = -N/R_P C_R$. Because R_P is a highly nonlinear function of temperature as illustrated by Fig. 8a, high-order TC compensation of F_{OUT} is required to overcome its impact. However, even with three-point trimming, an error caused by ΔF_{OUT} as shown in Fig. 8b significantly limits the frequency accuracy.To mitigate this effect, a low-leakage SCR shown in Fig. 9 is proposed. When Φ_1 is low and Φ_2 is high, the off-state leakage current of the PMOS switch (MP_C) is steered away from C_{FB} to the buffer output using a transmission gate (TG_P) [17, 18]. PMOS transistor (MP_S) is added to prevent C_R from being connected to the buffer output. Because MP_S source is connected to $V_{BUF} = V_{DD}/4$ and its drain is discharged to $V_{DD}/4$, V_{DS} of MP_S equals zero, thus significantly reducing its leakage. Similarly, NMOS transistor (MN_S) and transmission gate (TG_N) prevent leakage current of NMOS switch (MN_C) from flowing into C_{FB}. With the proposed low-leakage SCR, the magnitude of R_P is significantly increased, as shown in Fig. 10a, making its effect on frequency error negligible after three-point trimming, as illustrated in Fig. 10b.

Fig. 9 Proposed low-leakage SCR

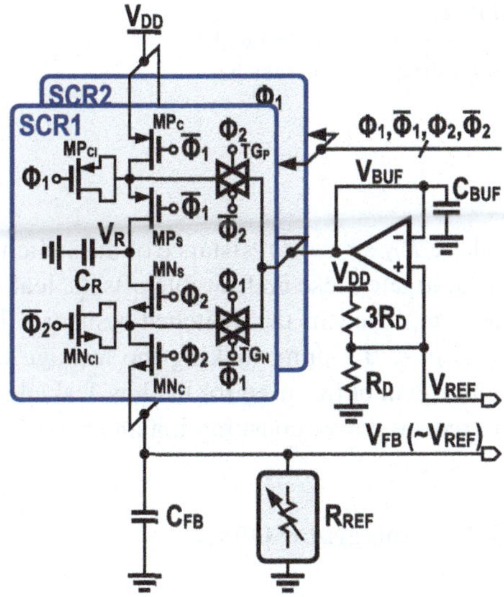

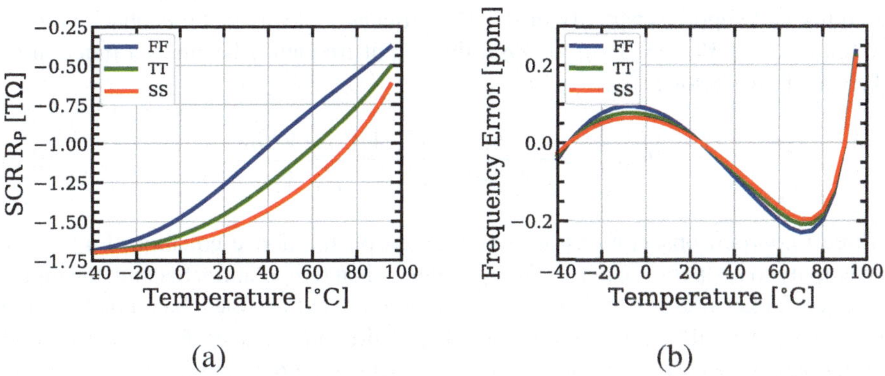

Fig. 10 Effect of leakage in the proposed SCR: (**a**) R_P versus temperature. (**b**) Frequency error caused by R_P after three-point trim

2.2.4 Finite Off Resistance of Switched Resistors

Finite off resistance of the switched resistors introduces temperature- and α-dependent offsets in the reference resistor and the output frequency. Denoting off resistance of the switches by R_{off} as depicted in Fig. 11, we can express the reference resistance as shown below:

$$\frac{1}{R_{REF}} = \frac{\alpha_1}{R_1} + \frac{1-\alpha_1}{R_{OFF}} + \frac{\alpha_2}{R_2} + \frac{1-\alpha_2}{R_{OFF}} + \frac{\alpha_3}{R_3} + \frac{1-\alpha_3}{R_{OFF}} \tag{24}$$

$$F_{OUT} = \alpha_1 F_1 + \alpha_2 F_2 + \alpha_3 F_3 + (3 - \alpha_1 - \alpha_2 - \alpha_3) F_{OFF} \tag{25}$$

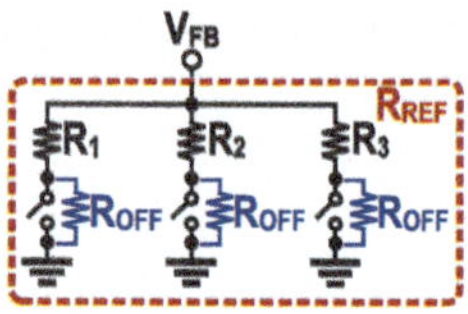

Fig. 11 Equivalent switched-resistor model with off resistance of the switches

where R_{OFF} is off resistance of the switched resistor and $F_{OFF} = N/(3R_{OFF}C_R)$. To mitigate these undesirable offsets, leakage reduction techniques used in the SCR are employed in the switched resistors. They help achieve $R_{OFF} > 0.9$ TΩ and $F_{OFF} < 0.5$ ppm, making the leakage effect on frequency error negligible. The unity-gain buffer used for the low-leakage SCR is shared with the switched resistors to reduce power consumption and save area.

2.2.5 Integrator Offset

Integrator input-referred offset voltage, V_{OS}, detrimentally impacts temperature stability and supply sensitivity of the TCO. Because steady-state feedback voltage is forced to be $V_{FB} = V_{REF} + V_{OS}$, the output frequency becomes a function of V_{OS} as shown below:

$$F_{OUT} = \frac{N}{3R_{REF}C_R}\frac{1+4(V_{OS}/V_{DD})}{1-\frac{4}{3}(V_{OS}/V_{DD})}. \tag{26}$$

Three important observations are made based on the above equation: First, V_{OS} sensitivity to temperature makes F_{OUT} to depend on temperature. Second, the input-referred flicker noise of the integrator can be modeled as slow-changing V_{OS} and degrades the oscillator's long-term stability. Third, V_{OS} makes F_{OUT} sensitive to variations in V_{DD}. To overcome these drawbacks, chopping at the integrator input is used to suppress V_{OS}, which improves oscillator's temperature and long-term stability and supply sensitivity.

2.2.6 Frequency-Locked Loop Design

The linearized FLL model shown in Fig. 12 is utilized to determine the loop parameters and ensure the TCO's stable closed-loop operation. The transfer function of SCR can be derived as a low-pass filter with a gain of K_{F2V} and a time constant of τ_{F2V} [13] where:

$$K_{F2V} = \frac{V_{DD}}{F_{FB}}\frac{R_{REF}C_RF_{FB}}{(1+R_{REF}C_RF_{FB})^2} \tag{27}$$

$$\tau_{F2V} = \frac{R_{REF}(C_{FB}+C_R/2)}{1+R_{REF}F_{FB}C_R}. \tag{28}$$

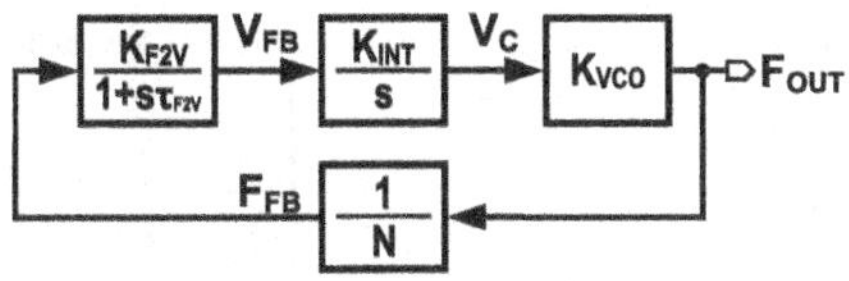

Fig. 12 Linearized FLL model

In steady state, thanks to FLL's large loop gain, the ratio of R_{SCR} to R_{REF} equals three to one, which allows K_{F2V} and τ_{F2V} to be rewritten as:

$$K_{F2V} = \frac{3V_{DD}}{16F_{FB}} = \frac{3V_{DD}}{16}\frac{N}{F_{OUT}} \tag{29}$$

$$\tau_{F2V} = \frac{1}{8}\left(1 + 2\frac{C_{FB}}{C_R}\right)\frac{N}{F_{OUT}}. \tag{30}$$

From Fig. 12, the loop gain, $LG(s)$, is given by:

$$LG(s) = \frac{K_{F2V}}{1 + s\tau_{F2V}}\frac{K_{INT}}{s}\frac{K_{VCO}}{N} \tag{31}$$

where K_{INT} is the integrator gain and K_{VCO} is the voltage-to-frequency gain of the VCO. In the TCO prototype, the unity gain frequency, ω_{UGF}, is designed to be much smaller than $1/\tau_{F2V}$ and is equal to:

$$\omega_{UGF} \approx \frac{3}{16}\frac{V_{DD}K_{INT}K_{VCO}}{F_{OUT}}. \tag{32}$$

To ensure FLL stability and suppress $\Delta\Sigma$ quantization error, ω_{UGF} is set to be about 1 kHz.

2.3 *Trimming Procedure*

The three-point trimming process is illustrated in Fig. 13. First, all switched resistors are connected to ground by setting SEL_1, SEL_2, and SEL_3 signals to zero. Then, the temperature is set to T_1, and output frequencies $F_1(T_1)$, $F_2(T_1)$, and $F_3(T_1)$ are measured by setting [S_1, S_2, S_3] = [1, 0, 0] (for measuring F_1), [S_1, S_2, S_3] = [0, 1, 0] (for measuring F_2), or [S_1, S_2, S_3] = [0, 0, 1] (for measuring F_3), respectively. This process is repeated at two other temperatures, T_2 and T_3, and the resulting nine measured frequencies are used to set up the system of equations shown in Fig. 13. Solving these equations yields an estimate of alpha values, which are then fine-tuned to obtain α_{1OPT}, α_{2OPT}, α_{3OPT}. They are quantized to 17-bit digital words D_1, D_2, and D_3. $\Delta\Sigma$ modulators are used to truncate the 17-bit digital words to 1-bit pulse-density modulated signals used to control the switched resistors, providing better

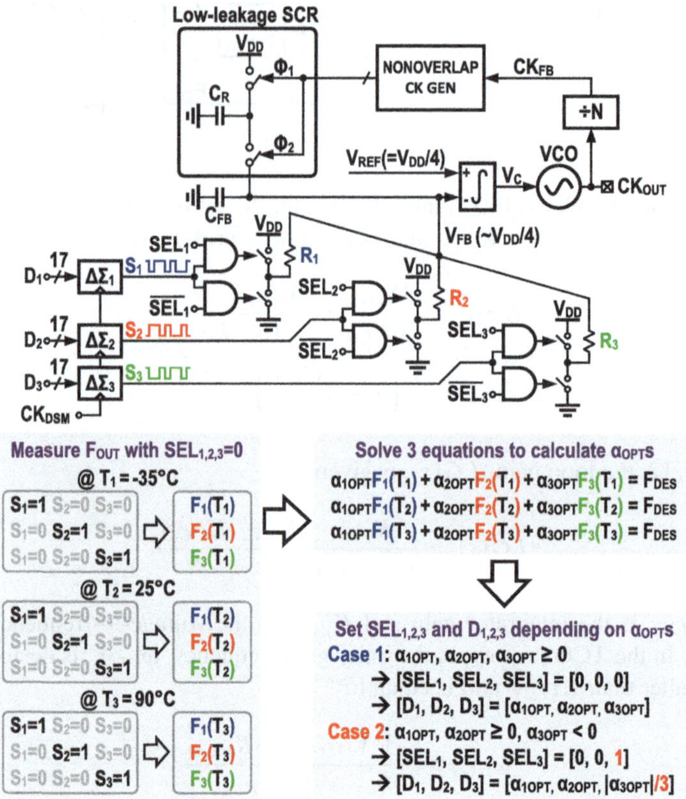

Fig. 13 Three-point trimming logic

than 8-ppm frequency tuning resolution. If an α_{iOPT} turns out to be negative, SEL$_i$ and D$_i$ are set to 1 and $|\alpha_{iOPT}|/3$, respectively (see Fig. 13).

2.4 Circuit Implementation

The complete architecture of the proposed TCO is shown in Fig. 14. A VCO generates a 100-MHz output clock and feeds to a frequency divider (division ratio of 126) that produces about 0.8 MHz feedback clock, CK_{FB}. A nonoverlapping clock generator produces phases Φ_1 and Φ_2 used in two low-leakage SCRs (SCR$_1$ and SCR$_2$) from CK_{FB}. Each SCR has a capacitance C_R implemented using a 0.5 pF MIM capacitor with a TC of -15.7 ppm/°C, whose effect is also compensated after trimming. To linearize the SCR's conductance versus switching frequency characteristic, a capacitor, $C_{FB} = 50$ pF, is connected at the V_{FB} node. The three resistors in the reference resistor are switched using three pulse-density modulated

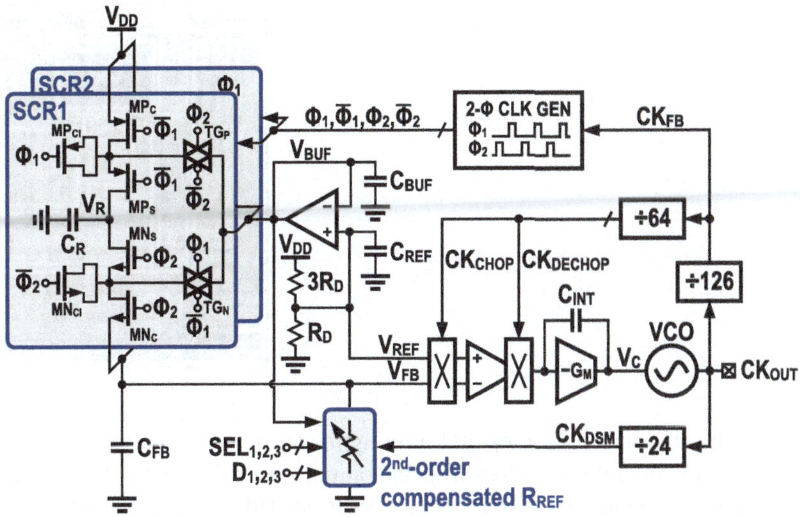

Fig. 14 Complete TCO architecture

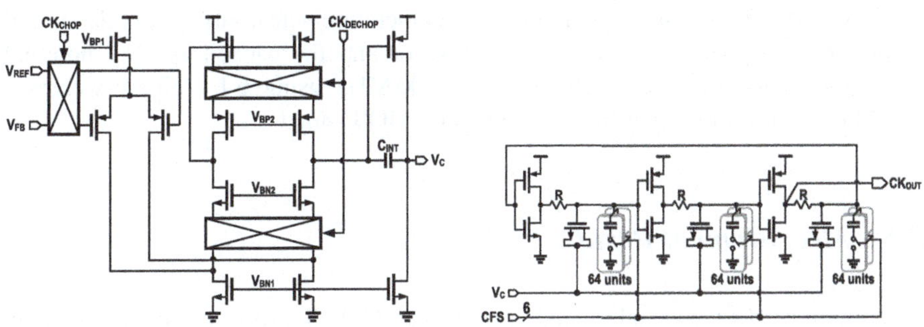

(a) (b)

Fig. 15 Schematic of (**a**) the two-stage integrator and (**b**) the voltage-controlled ring oscillator

sequences S_1, S_2, and S_3, each of which is generated by a first-order $\Delta\Sigma$ modulator. The $\Delta\Sigma$ modulator implemented using an error-feedback topology clocked at 4.2 MHz quantizes a 17-bit digital word, D_i, to a 1-bit pulse-density modulated sequence, S_i. The optimum average values for the three sequences are determined using the three-point trimming process described earlier in Sect. 2.3. The loop is closed by a high-gain two-stage integrator implemented using $G_M - C$ topology depicted in Fig. 15a. The integrator's first stage is chopped to suppress flicker noise and temperature-dependent offset, improving Allan deviation and temperature stability. The 12 kHz chopping/de-chopping clocks are generated by dividing the feedback clock. The second stage of the integrator is used for increasing the gain and implementing the large integrator time constant using a modest capacitor $C_{INT} = 30\,\text{pF}$. The Miller effect causes a dominant pole at the output of the first

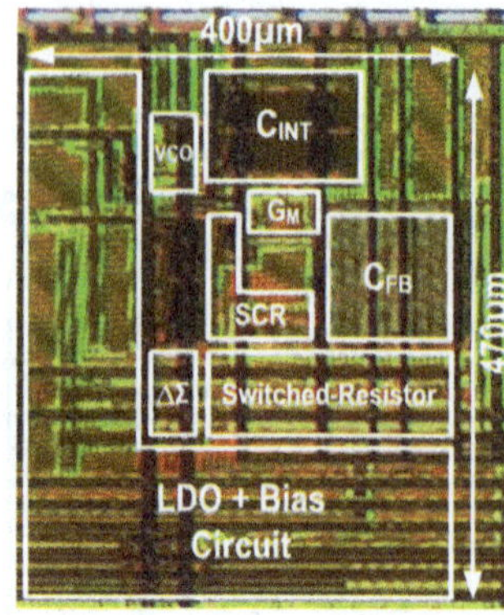

Fig. 16 Die micrograph

stage, thereby stabilizing the loop and limiting the loop bandwidth to 1 kHz needed to suppress ΔΣ modulators' shaped quantization error. Thanks to the integrator's high gain, the DC loop gain is more than 100 dB needed to suppress VCO's supply and temperature sensitivity adequately. The schematic of the VCO is shown in Fig. 15b. It comprises three delay cells implemented using an inverter, a fixed poly resistor, and controllable capacitors. The temperature sensitivity of the delay cell is dominated by the sensitivity of the resistor and the capacitors. The nominal oscillator frequency is coarsely tuned to 100 MHz using 6-bit binary-weighted MOM capacitor arrays and finely tuned using MOS varactors.

2.5 *Measurement Results*

A prototype TCO was fabricated in a 65 nm CMOS process and packaged in a plastic QFN. The die micrograph is shown in Fig. 16. To fully characterize the impact of resistor TC on frequency inaccuracy, eight different resistors (two poly resistors, two diffusion resistors, and four composite silicide resistors) were implemented in the prototype with the option of choosing any three out of them to be used in the TCO. The prototype occupies an area of 0.19 mm^2 and draws 84.5 μA from a 1.2 V supply with on-chip LDOs generating 0.9 V for the VCO and digital blocks and 1.0 V for analog blocks. The VCO consumes more than 50% of the total power, while the digital blocks such as the dividers, clock generators, and ΔΣ modulators (DSMs) consume 36.6 μW (∼37%) from the 1.2 V supply. SCR, the switched resistors, and the integrator consume only 5% of the total power.

2.5.1 Performance with Two-Point Trim

Two-point trim is performed to determine D_1 and D_2 for each sample by trimming the TCO at two temperatures (−40 °C and 95 °C) for $F_{OUT} = F_{DES} = 100$ MHz.

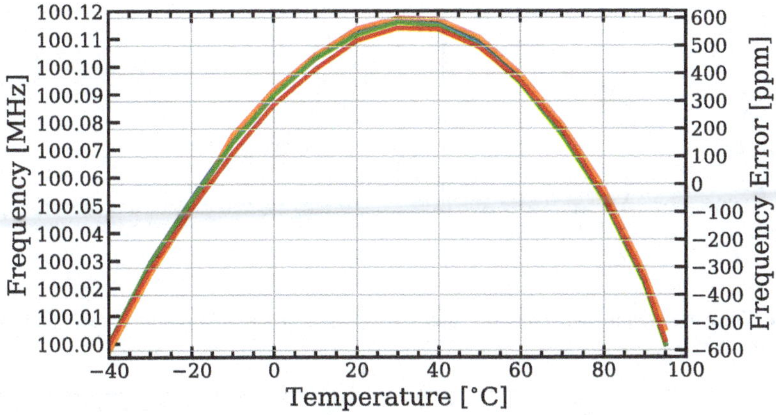

Fig. 17 Measured temperature stability after two-point trim

The measured F_{OUT} and residual frequency error for four samples using R_1 (189-kΩ negative TC p-poly resistor) and R_2 (positive TC 171-kΩ n-diffusion resistor) are shown in Fig. 17. The frequency inaccuracy is less than ±600 ppm over −40 °C to 95 °C, which corresponds to a residual TC of 8.9 ppm/°C in the box method.

2.5.2 Performance with Three-Point Trim

D_1, D_2, and D_3 are determined for each sample by trimming the TCO at three temperatures (−35 °C, 25 °C, and 90 °C) for $F_{OUT} = F_{DES} = 100$ MHz. Figure 18a shows the measured F_{OUT} and residual frequency error for twenty samples using R_1 (189-kΩ negative TC p-poly resistor), R_2 (positive TC 171-kΩ n-diffusion resistor), and R_3 (208-kΩ positive TC p-diffusion resistor). The frequency inaccuracy is less than ±140 ppm over −40 °C to 95 °C, which corresponds to a residual TC of 2.1 ppm/°C in the box method. It represents about 4× improvement over the first-order compensated TCO. The inaccuracy at three trim temperature points is not zero, which we believe is limited by ±0.2 °C short-term temperature variation of our temperature chamber (TestEquity Model 107) during trimming. Frequency inaccuracy is also measured with all positive TC resistors by replacing R_1 with a positive TC 289-kΩ n-poly resistor. The inaccuracy for 20 samples shown in Fig. 18b is less than ±316 ppm over −40 °C to 95 °C (4.7 ppm/°C in box method), which is larger than the inaccuracy with the p-poly resistor mainly due to the higher third-order TC of the n-poly resistor.

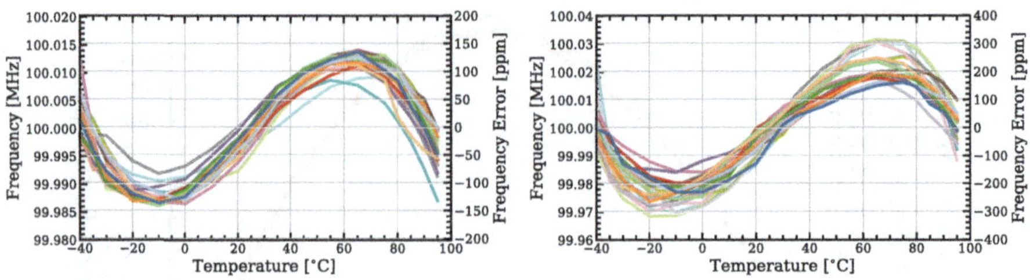

(a) (b)

Fig. 18 Measured temperature stability after three-point trim (**a**) with p-poly resistor and (**b**) with n-poly resistor

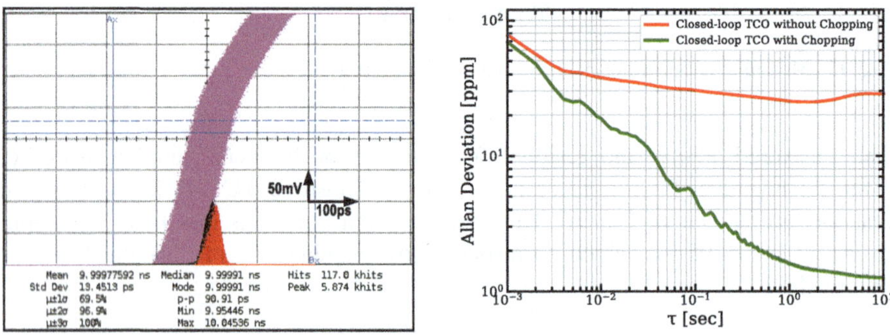

(a) (b)

Fig. 19 Output clock performance of the TCO: (**a**) period jitter and (**b**) Allan deviation

2.5.3 Output Clock Performance

The measured output period jitter is 13.4 ps_{rms} in Fig. 19a. The Allan deviation for a 1 s stride is 25.5 ppm without chopping, reduces to 1.6 ppm with chopping, and reaches a floor of 1.3 ppm over a 10 s stride in Fig. 19b.

3 Conclusions

A closed-loop ring oscillator-based temperature-compensated oscillator (TCO) is presented. Temperature compensation was performed by implementing the reference resistor using a parallel combination of switched resistors that are controlled by pulse-density modulated sequences. Because only one switch was used in series with the resistors, precise temperature compensation is achieved without introducing higher-temperature dependencies and without increasing supply sensitivity. A switched-capacitor resistor was used as a frequency-to-voltage converter, and the period of a ring oscillator was locked to a temperature-compensated RC time constant by embedding it in a high loop-gain feedback loop. The integrator was chopped to remove the flicker noise and its temperature-dependent offset, which helped achieve 1.3 ppm Allan floor. Using trimming at three temperature-insertion points, the FLL-based temperature-compensated oscillator achieves an inaccuracy of ±140 ppm from −40 °C to 95 °C (2.1 ppm/°C), 83 ppm/V voltage sensitivity, 1.3 ppm Allan deviation floor, and a power efficiency of 1 μW/MHz.

References

1. G. Zhang, K. Yayama, A. Katsushima, T. Miki, A 3.2 ppm/°C second-order temperature compensated CMOS on-chip oscillator using voltage ratio adjusting technique. IEEE J. Solid-State Circuits **53**(4), 1184–1191 (2018)
2. *Bluetooth Core Specification Version 5.1*, Bluetooth SIG, Kirkland, WA, USA, 2020
3. *IEEE 802.15 WPAN Task Group 4*. http://www.ieee802.org/15/pub/TG4.html. Accessed Mar 2004
4. E.O. Ates, A. Ergul, D.Y. Aksin, Fully integrated frequency reference with 1.7 ppm temperature accuracy within 0-80°C. IEEE J. Solid State Circuits **48**(11), 2850–2859 (2013)
5. M.S. McCorquodale et al., A 25-MHz self-referenced solid-state frequency source suitable for XO-replacement. IEEE Trans. Circuits Syst. Regul. Pap. **56**(5), 943–956 (2009)
6. A. Paidimarri, D. Griffith, A. Wang, G. Burra, A.P. Chandrakasan, An RC oscillator With comparator offset cancellation. IEEE J. Solid State Circuits **51**(8), 1866–1877 (2016)
7. A. Savanth, J. Myers, A. Weddell, D. Flynn, B. Al-Hashimi, A 0.68nW/kHz supply-independent relaxation oscillator with ±0.49%/V and 96ppm/°C stability, in *2017 IEEE International Solid-State Circuits Conference (ISSCC)* (2017), pp. 96–97
8. D. Griffith, P.T. Røine, J. Murdock, R. Smith, A 190nW 33kHz RC oscillator with ±0.21% temperature stability and 4ppm long-term stability, in *2014 IEEE International Solid-State Circuits Conference (ISSCC)* (2014), pp. 300–301
9. M. Choi, S. Bang, T.-K. Jang, D. Blaauw, D. Sylvester, A 99nW 70.4kHz resistive frequency locking on-chip oscillator with 27.4ppm/°C temperature stability, in *2015 Symposium on VLSI Circuits (VLSI Circuits)* (2015), pp. C238–C239
10. C. Gurleyuk, L. Pedala, S. Pan, F. Sebastiano, K.A.A. Makinwa, A CMOS dual-RC frequency reference with ±200-ppm inaccuracy from −45 °C to 85 °C. IEEE J. Solid-State Circuits **53**(12), 3386–3395 (2018)
11. T. Jang, M. Choi, S. Jeong, S. Bang, D. Sylvester, D. Blaauw, A 4.7nW 13.8ppm/°C self-biased wakeup timer using a switched-resistor scheme, in *2016 IEEE International Solid-State Circuits Conference (ISSCC)* (2016), pp. 102–103
12. P. Chen, D. Li, Z. Yu, Q. Jin, K. Yang, A 0.84pJ/cycle Wheatstone bridge based CMOS RC oscillator with reconfigurable frequencies, in *2019 IEEE Custom Integrated Circuits Conference (CICC)* (2019), pp. 1–4
13. A. Khashaba, J. Zhu, N. Pal, M.G. Ahmed, P.K. Hanumolu, A 32-MHz, 34-μW temperature-compensated RC oscillator using pulse density modulated resistors. IEEE J. Solid State Circuits **57**(5), 1470–1479 (2021)
14. K.-S. Park et al., A second-order temperature compensated 1μW/MHz 100MHz RC oscillator with ±140ppm inaccuracy from −40°C to 95°C, in *2021 IEEE Custom Integrated Circuits Conference (CICC)* (2021), pp. 1–2
15. H. Jiang, C.-C. Huang, M. R. Chan, D.A. Hall, A 2-in-1 Temperature and humidity sensor with a single FLL wheatstone-bridge front-end. IEEE J Solid State Circuits **55**(8), 2174–2185 (2020)
16. S. Ben-Yaakov, M. Evzelman, Generic and unified model of switched capacitor converters, in *2009 IEEE Energy Conversion Congress and Exposition* (2009), pp. 3501–3508

17. M. O'Halloran, R. Sarpeshkar, An analog storage cell with $5e^{-}/sec$ leakage, in *2006 IEEE International Symposium on Circuits and Systems* (2006), pp. 557–560
18. L. Xu, T. Jang, J. Lim, K. Choo, D. Blaauw, D. Sylvester, A 0.51nW 32kHz crystal oscillator achieving 2ppb Allan deviation floor using high-energy-to-noise-ratio pulse injection, in *2020 IEEE International Solid-State Circuits Conference (ISSCC)* (2020), pp. 62–64

Integrated BAW-Based Frequency References

Danielle Griffith

Abstract Nearly all electronic systems require a clock. Crystal oscillators have been the most common way to generate an accurate reference frequency for the last 100 years. Despite their ubiquity, crystal oscillators have several drawbacks, including cost, large size, degraded frequency stability at temperature extremes, and sensitivity to shock and vibration. Bulk acoustic wave resonators (BAW) have emerged as an alternative to crystals, allowing a reduction in the footprint of electronic systems and enabling exciting new applications. This chapter covers several aspects of designing with BAW resonators, including oscillator topologies, frequency tunability and stability, and passive and active temperature compensation. System-level advantages such as fast startup, higher frequency, and improved robustness will also be discussed. The chapter concludes with an introduction to applications for BAW that extend beyond frequency references.

1 Introduction

Electronic systems are an integral part of the modern world with wide-ranging use in infrastructure, transportation, industrial, and consumer applications. From 5G devices to wireless sensors powered by energy harvesting, and from electrical grid monitoring to toys, there is one thing in common with nearly all of them: the need for a reference clock. These systems demand ever-more integrated functions and higher data rates, requiring reference oscillators with ever better stability and lower phase noise. Moreover, electronic devices have rapidly evolved from yesterday's brick phone to today's Bluetooth earbud in just three short decades, leading to an ever-growing need for a smaller form factor, reduced power consumption, and low cost.

D. Griffith (✉)
Texas Instruments, Dallas, TX, USA
e-mail: danielle.griffith@ti.com

P. Harpe et al. (eds.), *Biomedical Electronics, Noise Shaping ADCs, and Frequency References*, https://doi.org/10.1007/978-3-031-28912-5_16

For the past 100 years, quartz crystal oscillators have been used when a stable reference clock is needed [1]. Quartz is a crystalline form of silicon dioxide, which, when cut at precisely the correct angles, creates a resonant frequency that is stable to 10–200 parts per million (ppm) over a wide temperature range. Crystal oscillators exhibit several key advantages. First, even though it is a mature technology, significantly predating integrated circuits, new research is published every year with advances in crystal oscillator design, technology, and applications [2, 3]. Second, crystals have a high quality factor, usually greater than 30,000, enabling very low power consumption from crystal oscillator designs. There are, however, several disadvantages as well. First, crystals are relatively expensive, equivalent to the cost associated with 1–2 mm^2 of silicon. This means one passive component costs as much as 20 million transistors in a 16 nm technology. In 2022, the most common crystal size is $2 \times 1.6mm^2$ which limits integration. Second, although frequency accuracy is approximately <20 ppm at room temperature, it degrades quickly at the temperature extremes, leading to violations in some wireless standards in automotive and industrial applications operating at very high or low temperatures. Third, although the high quality factor is an advantage for frequency stability, it leads to very long oscillator startup times which can be a disadvantage in duty-cycled systems. Finally, although less well known, crystals can be sensitive to shock and vibration and can fail in harsh environments. Given the disadvantages of crystal-based oscillators, there has long been a desire for a low-power, ultra-stable, and ultra-compact reference clock that can be fully integrated in CMOS technologies.

Microelectromechanical systems (MEMS) can be used to fabricate resonators with high quality factor, typically >1000, enabling low power oscillators with high frequency stability and low phase noise. Examples of MEMS in the marketplace are surface acoustic wave (SAW) resonators, FBAR, as shown in Fig. 1a, and bulk acoustic wave (BAW) resonators as shown in Fig. 1b. These resonators are used extensively in filter and duplexer products. The first BAW oscillator was published in 1976 [4] but was not released in commercial products until 2019 [5]. Today, BAW resonators are one of the best solutions available to address the limitations of crystal oscillators. A description of the fundamentals of BAW resonators is contained in Sect. 2. Section 3 will show examples of BAW oscillator topologies, while Sect. 4 addresses frequency compensation. Section 5 discusses limits to frequency accuracy. The advantages of BAW oscillators are covered in Sect. 6, and finally, Sect. 7 will discuss new applications of this technology.

2 Bulk Acoustic Wave Resonators

BAW resonators are passive devices consisting of a piezoelectric material, sandwiched between two metal electrodes as shown in Fig. 1. The electric excitation of the electrodes causes acoustic vibration in the piezoelectric layer, converting between electrical and mechanical energy through interaction with the high quality

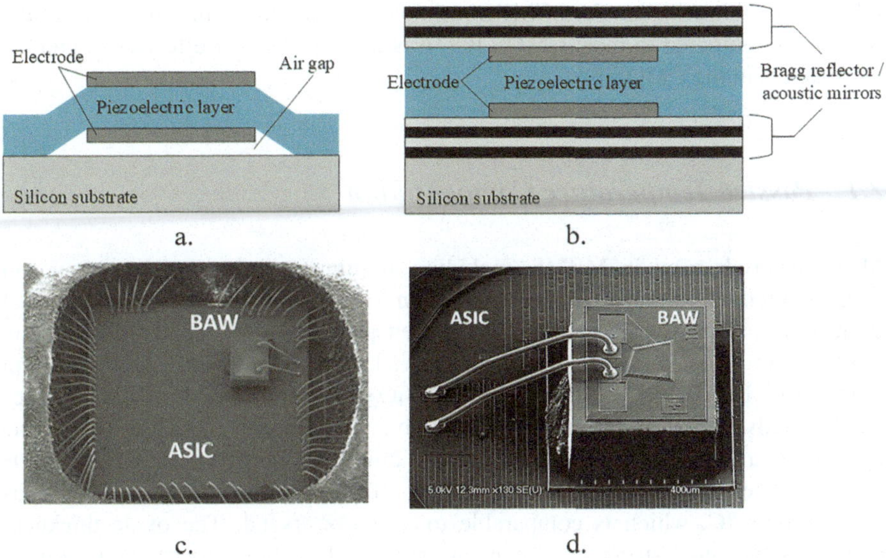

Fig. 1 (**a**) FBAR resonator cross section. (**b**) BAW resonator cross section. (**c**) and (**d**) BAW resonator co-packaged with ASIC

factor mechanical resonance of the BAW structure. Aluminum nitride is the most commonly used thin film piezoelectric material for BAW resonators. Bragg reflectors [6], or acoustic mirrors, are structures formed with alternating high and low acoustic impedance layers that reflect energy back into the aluminum nitride piezoelectric film. The better the mirrors are, the less energy leaks into the substrate below or the package material above, and the higher the quality factor that can be achieved. The dual Bragg structure in [7] protects the BAW resonator by encapsulating the device, thereby making it resilient to moisture, contamination, and He permeation. This structure eliminates the need for vacuum or hermetic packaging, significantly reducing the cost of fabrication. The mechanical resonant behavior for BAW resonators occurs in the bulk of the material in contrast to a SAW resonator, where the resonant behavior is on the surface. The resonant frequency, f_R, is dependent to the first order on the thickness of the aluminum nitride, t, and the acoustic velocity, v_L.

$$f_R \approx \frac{v_L}{2t}$$

To increase the resonance frequency, the film has to become thinner. The possible BAW frequencies are therefore set by the thin film manufacturing capabilities and are usually in the range of 1–5 GHz. The BAW resonator packaged together with an ASIC is shown in Fig. 1c, d. The irregular shape visible in Fig. 1d is optimized to reduce spurious resonance modes. The BAW resonator technology is fully compatible with standard CMOS technologies, but due to the small size of the resonators compared to most CMOS ICs, it is usually not cost-effective to include these layers in the CMOS flow.

2.1 Passive Temperature Compensation

Many materials used in MEMS, including aluminum nitride (AlN), soften and expand as temperature increases. Aluminum nitride has a temperature coefficient of frequency or TCF of −25 ppm/°C. Uncompensated, this corresponds to more than a 3000 ppm frequency shift over the industrial temperature range, corresponding to the black line in Fig. 2. Silicon dioxide (SiO_2) has a positive TCF and becomes mechanically stiff at high temperature. Passive temperature compensation is done by adding a SiO_2 layer to counteract the aluminum nitride TCF. If the SiO_2 layer thickness is perfectly controlled, the effective TCF reduces to as low as +/−0.5 ppm/°C, which is comparable to a quartz crystal. The oxide thickness control is the most stringent step in the entire fabrication process and requires extremely tight control over thin film thickness. An example of the different frequency variations that can be seen for SiO_2 layer thickness variation is shown in Fig. 2. Manufacturing tolerances are usually not precise enough to add exactly the right thickness of SiO_2, so frequency variations of up to 300 ppm over the entire temperature range can be observed. It is also important to note that a second-order temperature coefficient can remain even with perfect compensation. Passive temperature compensation, while very effective, is still not sufficient for all

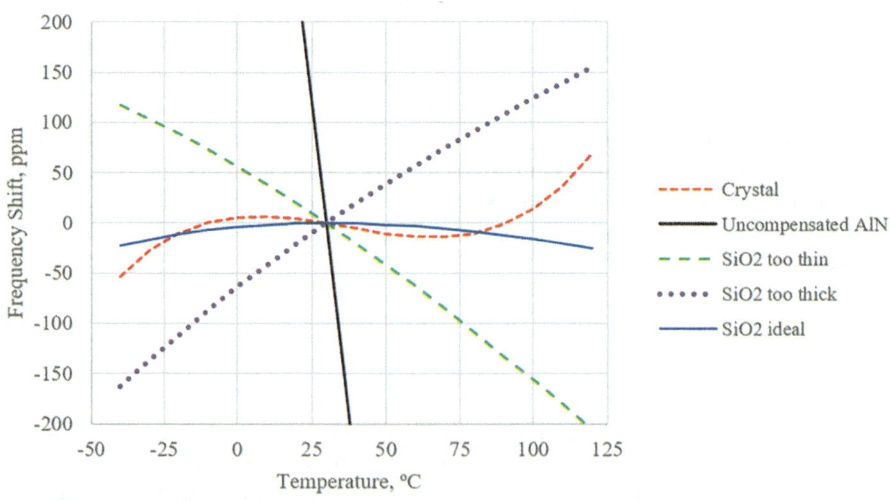

Fig. 2 BAW passive temperature compensation

Table 1 Typical BAW resonator comparison to 48 MHz crystal

	Frequency	Q	Size	Frequency variation −40 to 85 °C
BAW resonator	2.5 GHz	>1000	0.5 × 0.5 mm^2	<±150 ppm
Crystal	48 MHz	~30,000	2.0 × 1.6 mm^2 (packaged)	<±25 ppm

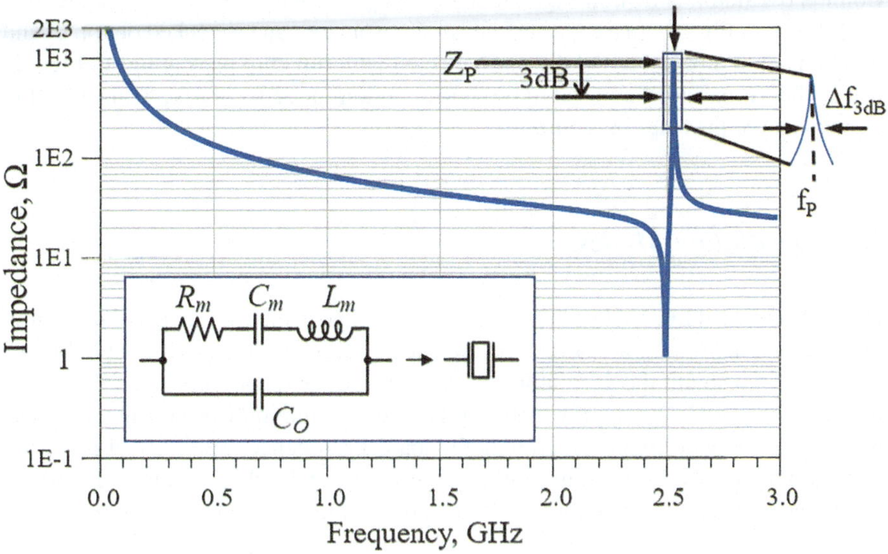

Fig. 3 Modified Butterworth-Van Dyke model and impedance plot for a BAW resonator

communication protocols. Active temperature compensation may also be needed and will be discussed in Sect. 4 (Table 1).

2.2 *Resonator Model*

Just like a crystal, the BAW resonator can be modeled by the four-component modified Butterworth-Van Dyke model topology shown in Fig. 3. The resonator appears inductive between the series and parallel resonance frequencies. The parallel quality factor Q_P is the resonance frequency divided by the 3 dB bandwidth. Also important to note is that the BAW resonator, unlike some other MEMS structures, requires only two pins. Ground and bias voltage pins are not required. Since the resonator exhibits a high impedance at DC, oscillator topologies used with inductive loads or LC tanks must be modified to work with BAW resonators.

3 BAW Oscillator Topologies

Although BAW oscillators were not used in commercial products until 2019, many papers on the technology have been published since the first one in 1976. Examples include [8–10]. Two common BAW oscillator topologies include the Colpitts oscillator, shown in Fig. 4a, and a complementary differential architecture [11] shown in Fig. 4b. In each case, the BAW resonator can be stacked on top of and wirebonded to a base CMOS die or connected with flip chip packaging. In Fig. 4b, two NMOS transistors are used to provide common-mode feedback, while a PMOS current source provides bias current.

3.1 Oscillation Modes

Unlike traditional LC-VCOs, there are four modes possible for this BAW oscillator. First, the capacitors C_d are added to capacitively decouple the cross-coupled pairs. If they are too small, the oscillator will latch. If, on the other hand, the value of C_d is greater than $nC_L/2$ [11], a parasitic relaxation oscillation can be triggered. The third mode is the desired oscillation at the loaded resonance frequency of the BAW resonator, typically in the multi-GHz range for most BAW resonators. Finally, a fourth mode can exist when there is a parasitic oscillation at a frequency f_B,

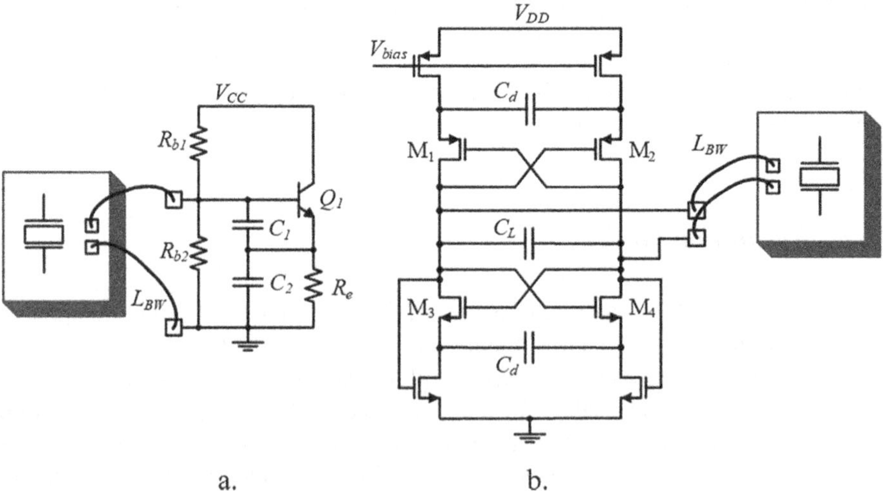

Fig. 4 Two commonly used BAW oscillator architectures: (**a**) Colpitts and (**b**) complementary differential

$$f_B = \frac{1}{2\pi\sqrt{L_{\text{BW}}\frac{C_L C_O}{C_L + C_O}}}$$

where the bondwire inductance L_{BW} and the equivalent load capacitance from the oscillator, C_L, and the shunt capacitance from the BAW resonator, C_O, form a resonance. Due to multiple possible modes, the design is more constrained than the typical GHz oscillator. Several design and manufacturing techniques can be implemented to avoid the parasitic oscillations. First, the bondwires can be kept short and spaced as close together as the packaging rules allow. This will increase the mutual inductance and, because the signals in each bondwire are out of phase, decrease the effective inductance, thereby increasing the parasitic resonance frequency. The higher this resonance frequency, the easier it is to ensure there is no gain at that frequency. If flip chip packaging or transformer coupling [13] is an option, the bondwires are removed, greatly simplifying the problem. The transistors M1–M4 in Fig. 4b are sized and biased so that there is no loop gain at any of the possible parasitic oscillation frequencies. Alternatively, if these techniques are not possible, an LC trap can be used instead of the decoupling capacitor C_d. This will cause the impedance between the sources of M3 and M4 to be low only at the desired oscillation frequency, resulting in very low loop gain at any other frequency.

3.2 Phase Noise

Phase noise of reference oscillators is most important at the low-frequency offsets, such as 1–10 kHz. Beyond this, other noise sources in the PLL will usually dominate. There is a tradeoff between phase noise and power consumption, with the oscillator figure of merit (FOM) that relates phase noise to power consumption given by

$$\text{FOM} = \left(\frac{f_o}{\Delta f}\right)^2 \frac{1}{L(\Delta f) \times P_{\text{DC}}}$$

The high quality factor of a BAW resonator enables a higher FOM than would be achievable with an LC tank consisting of a planar inductor and capacitor array. A summary of achievable FOMs from state-of-the-art GHz frequency MEMS-based oscillators is given in [12].

4 Active Temperature Compensation

Section 2 showed that a completely uncompensated BAW resonator had over 3000 ppm or 0.3% frequency variation with temperature. This is far too much vari-

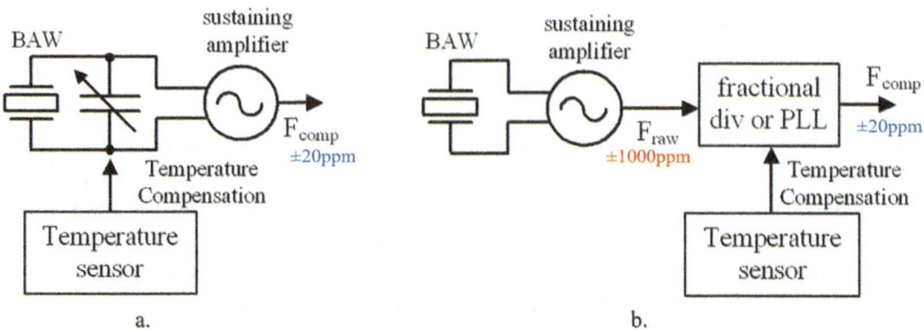

Fig. 5 (**a**) Temperature compensation implemented by capacitively pulling the BAW oscillator. (**b**) Temperature compensation implemented in a fractional divider or PLL

ation for nearly all communication standards. The addition of passive temperature compensation reduced the frequency variation to <300 ppm variation over the entire temperature range. Despite a 10× improvement over the uncompensated oscillator, this is still insufficient for many wireless communication protocols. To reduce the frequency variation further, both passive and active temperature compensation can be combined as shown in Fig. 5. Active temperature compensation can achieve an additional ~10× reduction in frequency variation [14], thereby improving the oscillator stability sufficiently to allow its use as a reference clock for a large number of wireless and wired communication protocols. Active compensation employs both a temperature sensor to measure the current temperature and a lookup table to apply a frequency correction code to the BAW oscillator based on that temperature. The limit to the accuracy achieved is impacted by the temperature sensor accuracy and how frequently the compensation is performed, both of which can be adapted based on the needs of a particular application.

4.1 Frequency Adjustment

Due to manufacturing tolerances, the BAW oscillator frequency will not be exactly at the desired frequency. A typical offset could be +/−1000 ppm or 0.1%. Also, the frequency must be adjusted as a part of the active temperature compensation. There are two methods for tuning the oscillator frequency. As shown in Fig. 5a, the first method involves applying the temperature compensation and frequency offset tuning to a capacitor array across the BAW resonator. The advantage of pulling the BAW frequency directly is that the complexity of the solution is low. This technique has two disadvantages, however. First, adding the capacitor tuning array across the BAW resonator lowers the quality factor of the tank, which degrades the phase noise and increases the power consumption. Second, if bondwires are used, adding a tuning capacitance makes it harder to avoid having gain at an undesired oscillation frequency. An alternative approach is to do the frequency tuning in a fractional divider which follows the oscillator as shown in Fig. 5b. This leads to the lowest power consumption and best phase noise in the oscillator. Also, it is easy to avoid any parasitic resonance modes while simultaneously having the tuning range as wide as needed. An additional benefit of this approach occurs when the BAW oscillator is used as a reference clock to a fractional PLL, as is often the case. The system can then be simplified by performing the frequency adjustment in the PLL feedback divider, leading to no circuit overhead for frequency offset and temperature compensation. Regardless of which method is used to implement the frequency adjustment, most wireless communication standards will require this active temperature compensation to remove the remaining second-order frequency variation that is left over after the passive compensation is implemented.

Implementing active compensation requires measuring the temperature sensor output at three temperatures during production test. Note that only the resolution of the temperature sensor is important, not the absolute accuracy. Three temperatures are used because a parabolic curve is needed to fit the residual error of a resonator with the passive temperature compensation. A parabolic curve fit is done using an equation of the form

$$\frac{\Delta f}{f_0}(T) = \mathrm{TC}_2(T - T_0)^2 + \mathrm{TC}_1\,(T - T_0) + \mathrm{TC}_0$$

where TC_2 is the curvature of the parabola, TC is the slope of the curve, and TC_0 is the offset due to process variation. During operation, temperature compensation can be done by interpolating at any point on the curve as needed. Figure 6 shows measured results on 80 devices with the frequency error plotted on the y-axis versus temperature on the x-axis. The red data points are with the active temperature compensation disabled. The passive temperature compensation from the BAW resonator

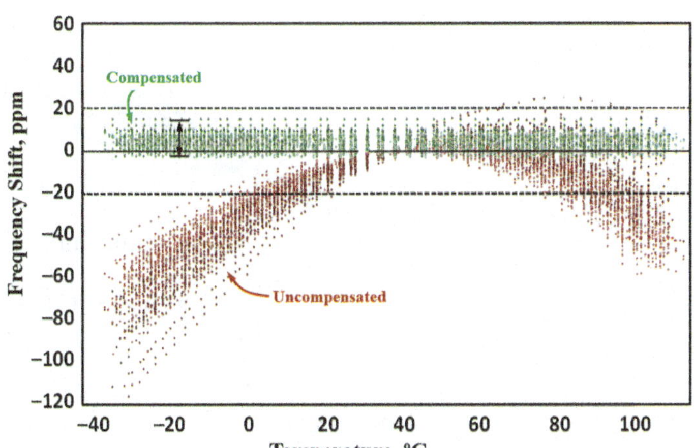

Fig. 6 Frequency vs. temperature, before and after active temperature compensation

clearly shows the expected residual second-order error. When active compensation is enabled, the total error becomes less than 10 ppm. This combination of active and passive temperature compensation takes an uncompensated resonator with 3000 ppm frequency variation and lowers it to 10 ppm, making it useful for wireless communication systems.

5 Frequency Error

Frequency references experience frequency drift over time as a result of aging. Aging can come from several sources. First, contamination can occur when particles land on the resonator, increasing the mass and decreasing the frequency. Conversely, outgassing can occur, decreasing the mass and increasing the frequency. Second, stress relaxation or change due to packaging or fabrication can shift the frequency. Finally, the active and passive components in the CMOS oscillator IC can age, shifting the oscillation amplitude and therefore the frequency due to nonlinear parasitic capacitance. An example of oscillator aging is shown in Fig. 7. The oscillator was operated continuously, and the resultant frequency is plotted vs. time in hours on a log scale. In this example, when the aging is extrapolated to 10 years of continuous operation, about 10 ppm of frequency shift on average is expected. This is similar to the drift that low-cost crystals experience. Combining this with other sources of error, the frequency accuracy of a BAW oscillator is still sufficient for many wireless communication standards. A comparison of frequency error sources between BAW resonators and crystals is shown in Table 2.

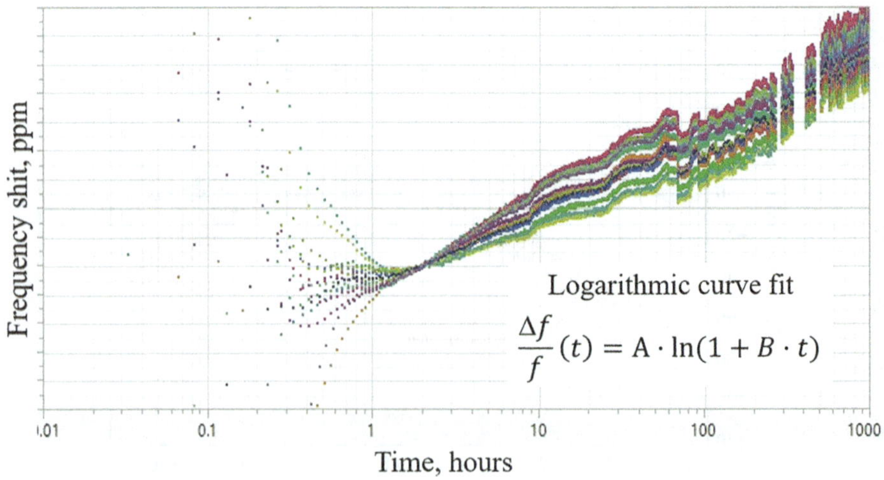

Fig. 7 Resonator frequency drift is logarithmic with time

Table 2 Comparison of frequency error sources between typical BAW resonator and crystal

	Manufacturing tolerance, $T = 25\ ^\circ C$	Frequency variation, −40 to 85 °C	Frequency drift (aging), 10 years at 25 °C
BAW resonator	<±1000 ppm	<±150 ppm	<10 ppm
Crystal	±20 ppm	<±25 ppm	<10 ppm

6 Advantages of BAW Oscillators

BAW oscillators have several system-level advantages over crystals. The most obvious is the smaller size which allows higher integration and potentially new applications when the space saved by removing the crystals opens up more room for sensors or other electronics. However, there are several other advantages.

6.1 High PLL Reference Frequency

BAW oscillator frequencies are usually between 1 and 3 GHz due to limitations in thin film manufacturing capabilities. Until recently, manufacturing capabilities have also limited crystal frequencies to under 100 MHz. Traditional Wi-Fi and Bluetooth products have used crystal oscillators as reference clocks for the RF synthesizer. However, their relatively low oscillation frequency can lead to poor phase noise performance. Figure 8a shows a basic PLL block diagram. To ensure PLL stability, the loop bandwidth is usually limited to about one tenth of the reference clock frequency, F_{REF}. In Fig. 8b, the pink curve is the phase noise at the output. In this example, the loop bandwidth is ~100 kHz, and the in-band phase noise is ~−100 dBc/Hz. However, if F_{REF} could be increased, the loop bandwidth could also be increased, lowering the in-band noise due to a lower feedback divider ratio N and lower noise from the PFD. This corresponds to the blue curve, where the loop bandwidth is approximately 10× higher and the in-band phase noise is more than 10 dB lower. Using this principle, the higher available BAW reference clock frequency allows the designer to choose the input divider ratio M to optimize F_{REF} based on the desired system performance rather than crystal availability.

6.2 Fast Startup

In many Internet of Things applications, the data throughput requirements are low compared to the channel bandwidth. This means the radio can duty-cycle aggressively to reduce average power. To minimize power between radio events, the reference clock, RF synthesizer, and entire radio are disabled except when needed. If a crystal oscillator is used as a reference clock, however, its startup time (100 s of microseconds) can be longer than the time required to exchange the data, greatly

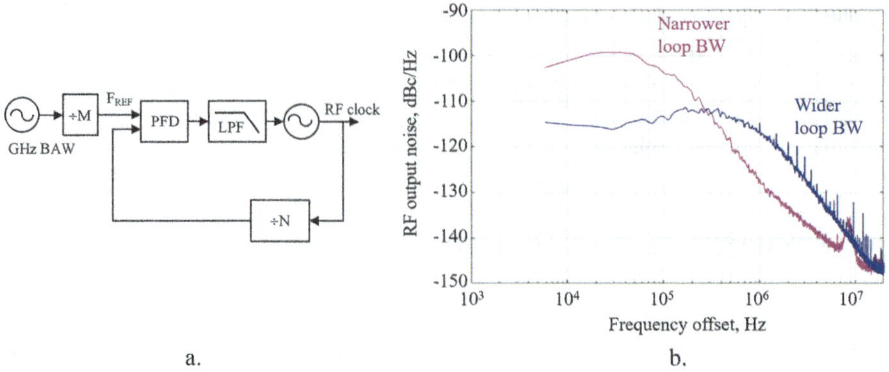

Fig. 8 (**a**) PLL block diagram. (**b**) Wider PLL bandwidth lowers the in-band noise

increasing the energy per bit needed. A BAW resonator has a Q that is roughly 10× lower than the crystal and operates at about 100× higher frequency. Since the startup time of an oscillator is proportional to the Q and inversely proportional to the frequency, a 1000× faster startup is expected. Therefore, replacing a crystal oscillator with a BAW oscillator enables a startup time under 1 μs, providing a net energy savings for short data packets in duty-cycled systems.

6.3 *Robustness*

A key consideration for a new type of frequency reference is how well it performs in harsh environments, where it will be exposed to shock, vibration, and stress. The BAW resonator was tested under mechanical shock conditions according to the military standard 833H, which involves a peak acceleration of 1500 g with a pulse duration of 500 μs on three different axes, with five repeated shocks. From the results in [15], the integrated BAW shows about a 4× better immunity to shock than the external crystal. This translates to about 1–2 ppm of frequency shift using a BAW versus 4–8 ppm for the crystal. The increased immunity of the BAW resonator is due to its much lower mass than a crystal. This can be a critical advantage for systems operating in harsh environments, for example, factory monitoring, industrial applications, power tools, and even smart golf balls. Any stress sensitivity needs to be added into the frequency error budget to make sure that the total is within specification.

6.4 *Security*

A final advantage of using a BAW oscillator as a clock source is that it improves the security of the system. SoCs can be compromised by glitching the clock [16]. For example, the intended operation of a CPU is to have an instruction loaded in at the first clock cycle, and then on the second clock cycle, another instruction is loaded and the first one is executed, and so on. However, if a crystal oscillator is used, a pin on the device is accessible to the outside world with the clock source present on it. If a glitch is injected on the crystal pin before the next rising edge of the clock, an instruction that has already been loaded may not finish executing and would be skipped. If that instruction was a password check or similarly critical, the security of the system would be compromised. This is a simplified example of a key point: that all of the encryption built into a design ultimately is no stronger than the security of the clock system. BAW oscillators can improve the security of an SoC by having the clock source completely internal with no exposed pins.

7 Other Uses for BAW Oscillators

So far, this chapter has described advantages and considerations to use BAW oscillators for reference clocks. Before concluding, a brief overview of three additional BAW oscillator use cases will be presented.

7.1 BAW to Replace RF Synthesizer

With careful frequency planning, BAW oscillators can entirely replace RF frequency synthesizers. For example, [17] from 2017 shows a multi-channel ISM band transmitter using a BAW oscillator. The PLL has been removed and with it any performance limitations due to crystal startup time or PLL loop bandwidth. A frequency plan has been developed that covers the full 2.4 GHz ISM band of 80 MHz with no PLL or fractional dividers. Instead, the BAW oscillator, with a tuning range of under 1%, is used in combination with a programmable integer divider, IQ generation circuitry, and polyphase filters to cover the full band, which is a range of 3.3%. The advantage over the traditional approach is that the TX can start transmitting within 2.3 μs after being enabled, without any time spent waiting on a crystal oscillator to start or a PLL to lock. This is significantly more power efficient for duty-cycled systems. Similar approaches are shown in [18, 19].

7.2 BAW as a Jitter Cleaner

BAW oscillators are also useful for applications outside generating reference clocks. They can also provide an advantage in clock jitter cleaning applications that consist of two series-connected PLLs, as shown in Fig. 9, used to generate a high-frequency low phase noise output clock. The first PLL takes in a noisy reference clock and

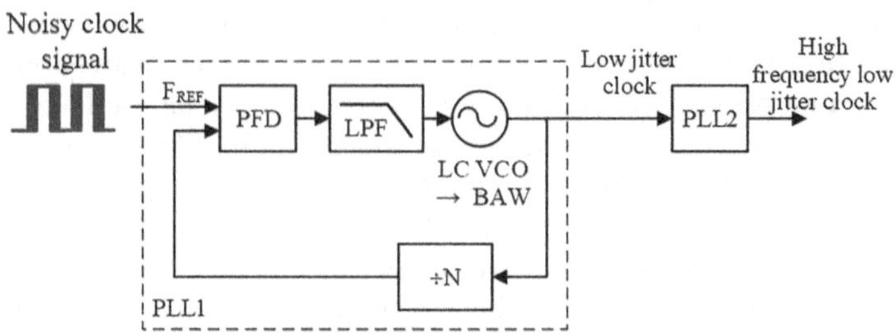

Fig. 9 BAW oscillator used to improve the performance of a jitter cleaner

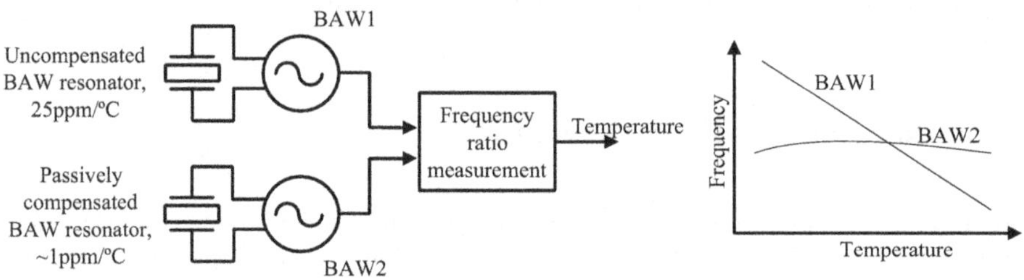

Fig. 10 Dual BAW structure used to implement a temperature sensor

outputs a cleaner, lower jitter, version of it. The second PLL then generates the high-frequency low jitter output. The noise at the output of the first PLL is dominated by the VCO. Typically, the VCO would be implemented with an on-chip LC-VCO which can have a tank $Q \sim 15$. If the VCO is replaced with a BAW oscillator, the tank Q becomes closer to 1000, and the VCO phase noise drops drastically. Therefore, the output clock from the first PLL becomes much lower noise, and the jitter cleaner is far more effective.

7.3 BAW as a Temperature Sensor

High-precision temperature sensors can be fabricated using a modification of the BAW technology [20]. As shown in Fig. 10a, the frequencies of two BAW-based oscillators are compared. One resonator, labeled BAW1, has the passive temperature compensation layers removed, while the second resonator, labeled BAW2, has the normal passive temperature compensation layers. Their frequencies as a function of temperature are shown in Fig. 10b. The temperature is proportional to the difference between the two frequencies. Higher accuracy can be achieved by measuring the frequencies over a longer time. Note this approach requires calibration in production.

8 Conclusions

Bulk acoustic wave resonators are now in volume production as frequency references. Billions of BAW resonators have been sold over the years as filters and duplexers, but only recently have they started to be used in clocking circuits. Although BAW resonators require both passive and active temperature compensation to reach frequency stability acceptable for communication protocols, they are an advancement over 100-year-old crystal technologies in many ways. They enable reduced size and cost, improved frequency stability compared to crystals at the temperature extremes, and are more robust to shock and vibration due to their lower mass, making BAW oscillators particularly interesting for harsh industrial environments. They provide advantages for duty-cycled systems by allowing lower average power consumption and give designers flexibility in choosing a PLL reference clock frequency to get the best performance for RF synthesizers. They also provide lower jitter from clock jitter cleaners. Finally, BAW technology provides increased security against clock glitch attacks, a feature which is becoming evermore important as more and more devices are being connected.

References

1. W.G. Cady, The piezo-electric resonator. IEEE Proc. Inst. Radio Eng. **10**(2), 83–114 (1922)
2. J. Jung et al., A 52MHz −158.2dBc/Hz PN @ 100kHz digitally controlled crystal oscillator utilizing a capacitive-load-dependent dynamic feedback resistor in 28nm CMOS, in *2022 IEEE International Solid-State Circuits Conference (ISSCC)*, vol. 65, (2022)
3. L. Xu et al., A 0.51nW 32kHz crystal oscillator achieving 2ppb Allan deviation floor using high-energy-to-noise-ratio pulse injection, in *2020 IEEE International Solid- State Circuits Conference (ISSCC)*, (2020)
4. A. Schaer, Behavior and current performance of SAW and BAW oscillators, in *IEEE European Microwave Conference*, (1976), pp. 249–253
5. https://www.ti.com/lit/wp/snoaa34/snoaa34.pdf. Accessed 4 Sept 2022
6. S. Salgar et al., Modeling and simulation of the thin film bulk acoustic resonator, in *Proceedings of the 2002 IEEE International Frequency Control Symposium and PDA Exhibition*, (2002)
7. E. Yen et al., Integrated high-frequency reference clock systems utilizing mirror-encapsulated BAW resonators, in *2019 IEEE International Ultrasonics Symposium*, (2019)
8. F. Vanhelmont et al., A 2 GHz reference oscillator incorporating a temperature compensated BAW resonator, in *2006 IEEE Ultrasonics Symposium*, (2006)
9. S. Razafimandimby et al., A 2GHz 0.25μm SiGe BiCMOS oscillator with flip-chip mounted BAW resonator, in *2007 IEEE International Solid-State Circuits Conference (ISSCC)*, (2007)
10. S. Rai et al., A 1V 600μW 2.1GHz quadrature VCO using BAW resonators, in *2007 IEEE International Solid-State Circuits Conference (ISSCC)*, (2007)
11. R. Thirunarayanan et al., Complementary BAW oscillator for ultra-low power consumption and low phase noise, in *2011 IEEE 9th International New Circuits and Systems Conference*, (2011), pp. 97–100
12. B. Bahr et al., Class-C BAW oscillator achieving a close-in FOM of 206.5dB at 1kHz with optimal tuning for narrowband wireless systems, in *2022 IEEE Radio Frequency Integrated Circuits Symposium (RFIC)*, (2022)

13. B. Bahr et al., Near-field-coupled bondless BAW oscillators in WCSP package with 46fs jitter, in *2021 IEEE Radio Frequency Integrated Circuits Symposium (RFIC)*, (2021)
14. D. Griffith et al., A +/−10ppm −40 to 125C BAW-based frequency reference system for crystal-less wireless sensor nodes, in *IEEE Symposium on Circuits and Systems*, (2017)
15. D. Griffith et al., An integrated BAW oscillator with <±30ppm frequency stability over temperature, package stress, and aging suitable for high-volume production, in *IEEE International Solid State Circuits Conference*, (2020)
16. https://wiki.newae.com/Tutorial_CW305-3_Clock_Glitching. Accessed 4 Sept 2022
17. P. Nadeau et al., Single-BAW multi-channel transmitter with low power and fast start-up time, in *ESSCIRC*, (2017)
18. K. Wang et al., A 1.8mW PLL-free channelized 2.4GHz ZigBee receiver utilizing fixed-LO temperature-compensated FBAR resonator, in *IEEE International Solid-State Circuits Conference Digest of Technical Papers*, (2014), pp. 372–373
19. R. Thirunarayanan et al., Reducing energy dissipation in ULP systems: PLL-free FBAR-based fast startup transmitters. IEEE Trans. Microw. Theory Tech. **63**, 1100–1117 (2015)
20. M. Heidarpour et al., A MEMS-assisted temperature sensor with 20-μK resolution, conversion rate of 200 S/s, and FOM of 0.04 pJK2. IEEE J. Solid State Circuits **52**(1), 185–197 (2017)

MEMS Oscillators Revolutionizing the Precision Timing Market

Kamran Souri and Sassan Tabatabaei

Abstract Frequency references are the heartbeat of most electronic systems. Over the last decade, MEMS-based frequency references have revolutionized the timing market by outperforming quartz crystal oscillators in many applications, where they can offer lower power, smaller size, better stability, programmability, and higher reliability. In this chapter, a MEMS-based temperature-compensated oscillator (TCXO) is presented which achieves a frequency stability of <±100 parts per billion (ppb) in the temperature range from −45 to 105 °C and an Allan deviation (ADEV) of $<2e^{-11}$ over 1 s averaging time. The result of an improved version of this TCXO is also presented where the frequency stability over temperature is improved by over tenfold, to <±10 ppb, which matches the performance of entry-level quartz-based OCXOs. Such oscillators are key building blocks in telecom, networking, and precision timekeeping systems. To enable such performance, SiTime's DualMEMS™ technology is used for temperature-to-digital conversion (TDC). This technology enables the design of high-precision temperature-compensated MEMS-based oscillators with superior performance for applications in breezy conditions and harsh environments. The TDC circuit is realized in a 0.18 μm CMOS process and achieves a resolution of 20 μK over a bandwidth of 100 Hz while consuming 19 mW of power, leading to a resolution FOM of 0.04 pJ°C2. Measurement results supporting the applications in harsh environments are presented, while a comparison with the high-performance quartz crystal TCXO solutions will be also provided.

K. Souri (✉)
SiTime, Rijswijk, The Netherlands
e-mail: ksouri@sitime.com

S. Tabatabaei
SiTime, Santa Clara, CA, USA

P. Harpe et al. (eds.), *Biomedical Electronics, Noise Shaping ADCs, and Frequency References*, https://doi.org/10.1007/978-3-031-28912-5_17

1 Introduction

Precision timing has become an integral part of developing emerging technologies in various application domains such as communication and enterprise, mobile and IoT, and automotive, industrial, and aerospace. The primary specifications include, but are not limited to, phase noise (PN), jitter, Allan deviation (ADEV), frequency stability over temperature, power consumption, and reliability. Moreover, the requirement for operation under environmental stressors such as airflow, rapid temperature changes, shock and vibration, poor power supply, and electromagnetic interference (EMI) imposes significant challenges for the timing devices which often lead to expensive solutions at the system level. Some examples of application requirements for frequency stability over temperature are as follows: consumer electronics such as wireless USBs needs $<\pm 30$ ppm frequency stability [1], while $<\pm 2.5$ ppm for commercial GPS [2] and $<\pm 0.1$ ppm for telecom application [3, 4] are needed. The stability requirement for applications, such as 4G/5G radio, base stations, small cell platforms, synchronous Ethernet (SyncE), and precision GNSS, is even more stringent, e.g., $<\pm 0.01$ ppm, which has historically been addressed by ovenized quartz crystal oscillators (OCXOs). The price to pay is the significantly higher cost, power consumption, larger package size, and reliability concerns. Recent advances in MEMS technology have established MEMS oscillators as an alternative to the legacy quartz crystal solution that had been dominant for several decades [5–14] in various applications. In particular, they have been rapidly gaining traction in the market due to the benefits they offer over quartz oscillators such as significantly faster time to market, leveraging semiconductor processes, and packaging, resulting in a smaller footprint [15, 16]. They also offer a wide range of features and programmability as well as a more attractive power consumption versus phase noise trade-off. Another advantage includes the superior performance in thermal transient tracking, which refers to maintaining frequency stability during fast thermal transient conditions. This requirement is gaining significance given the increase in power dissipation density and thus thermal dynamics in the electronic systems. The MEMS oscillators also exhibit higher immunity to shock and vibration, which makes them a suitable choice for applications with high stability requirements in harsh operating conditions such as mobile, industrial, automotive, and military applications, where the oscillator may be subject to substantial external vibration [17, 18].

2 MEMS OSC Architecture

As shown in Fig. 1, an example MEMS-based programmable oscillator package contains three elements: a MEMS die, a CMOS die, and a lead frame. As illustrated in Fig. 2, the MEMS die only carries resonators, while the CMOS die includes the electronics, such as the oscillator sustaining circuit, frequency synthesizer,

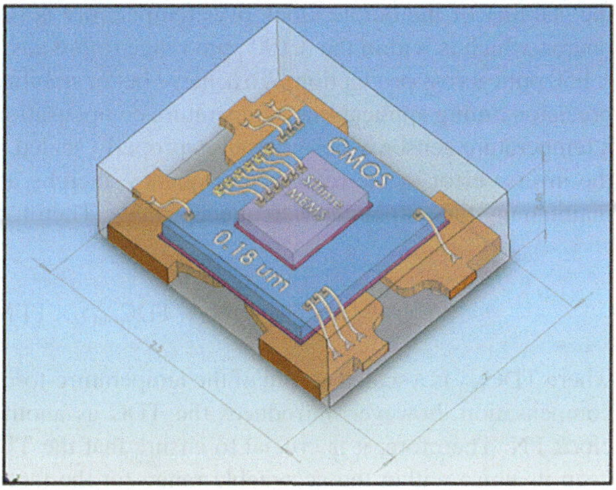

Fig. 1 MEMS-based programmable oscillator package

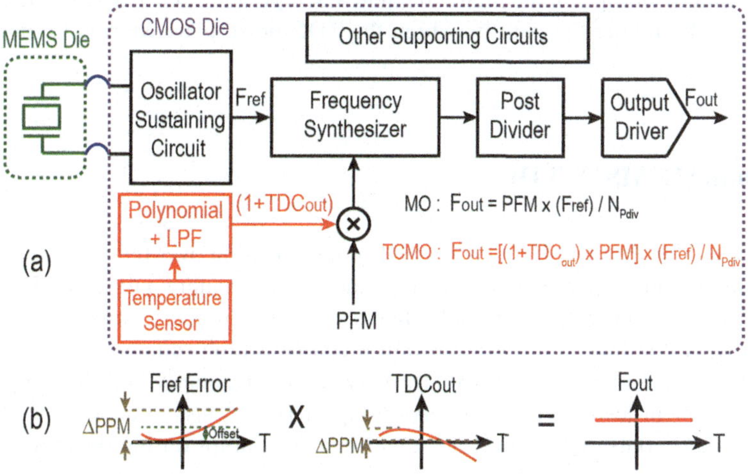

Fig. 2 Top (**a**): MEMS-based programmable oscillator in both XO and TCXO modes. Bottom (**b**): Reference frequency, TDC output, and output frequency variation over the temperature in TCXO mode

temperature sensor, post-divider, etc., that are necessary to sustain the oscillation of the MEMS resonator and to output a clock at the desired frequency. The output frequency (F_{out}) without temperature compensation can be expressed as:

$$F_{\text{out}} = F_{\text{ref}} \times \text{PFM}/N_{\text{Pdiv}} \tag{1}$$

where F_{ref} is the reference frequency, PFM is the programmable frequency multiplier, and N_{Pdiv} is the post-divider value. In this temperature-uncompensated mode, the stability of the output clock over temperature is set by that of the MEMS resonator, which is within the ± 100 ppm range (more advanced developments achieve $<\pm 50$ ppm across production). To achieve better stability, e.g., ± 0.1 ppm required in precision timing applications, temperature compensation is essential which employs a temperature sensor, whose output is properly scaled by a polynomial to generate the inverse error of the reference frequency, thereby adjusting the PFM value and maintaining the output clock frequency stable. Therefore,

$$F_{\text{out}} = F_{\text{ref}} \times (1 + \text{TDC}_{\text{out}}) \times \text{PFM}/N_{\text{Pdiv}} \tag{2}$$

where TDC_{out} is a scaled output of the temperature-to-digital converter (TDC). This compensation, however, introduces the TDC as another contributor to the output clock PN. Therefore, it is crucial to ensure that the TDC has enough resolution to keep its noise within the acceptable range for the targeted clock PN and jitter. As can be seen from Eq. (2), the higher the native temperature sensitivity of the MEMS resonator, the higher the gain by which the TDC noise is transferred to the output PN.

3 DualMEMS™ TDC

Apart from the stringent frequency stability specification over temperature, another key requirement for precision timing applications is to maintain the low ADEV, e.g., $<1\text{e}^{-10}$ at 1 s averaging time, under harsh environmental stressors. This imposes a significant challenge on the temperature compensation scheme. In quartz-based TCXOs, the existing physical distance between the resonator and the temperature sensing element leads to thermal lags during thermal transient conditions. This, in turn, results in outside-the-specifications frequency jumps during the rapid thermal transients. Besides, depending on the TDC noise and temperature sensitivity of the resonator, temperature compensation could impact the close-in PN (and thus ADEV) of the compensated output, due to the inherent noise of the TDC. To tackle this, the quartz crystal industry has been using OCXOs to ensure the ambient thermal conditions do not impact the oscillator circuit's operating temperature. Besides, by tuning the turnover point of the resonator, at which the temperature sensitivity is minimized, to be at the oven temperature, the requirement on the TDC noise is dramatically relaxed. This approach, although delivering superior performance, is expensive in terms of high power consumption which is needed to ovenize the package, cost, package size, and reliability since the device should operate at the oven temperature (which is set to higher than the upper range of the ambient temperature supported) during its entire lifetime.

To tackle the abovementioned TCXO constraints, a TDC architecture based on a DualMEMS™ resonator scheme was proposed by SiTime in 2016 [14]. As introduced in SiTime's Elite platform and shown in Fig. 3, two different resonators,

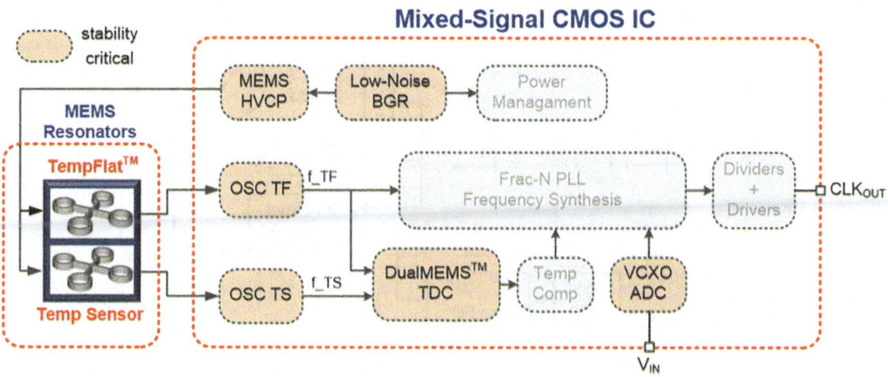

Fig. 3 SiTime's Elite platform MEMS-based programmable oscillator using DualMEMS™ TDC architecture. Two resonators, TempFlat and TempSense, are fabricated on the same MEMS die

namely, TempFlat and TempSense, are fabricated on the same die. As the name suggests, the TempFlat MEMS resonator has relatively flat temperature characteristics, with a maximum temperature coefficient $<\pm 1$ ppm/°C, while the TempSense resonator is intentionally designed to have strong temperature sensitivity, with a temperature coefficient of -7 ppm/°C. The two resonators are then connected to the CMOS die where two oscillator circuits OSC-TF and OSC-TS are used to extract the signals f_TF and f_TS, respectively, and sustain the oscillations. The signal f_TF is then fed to a fractional-N PLL as a reference, while the DualMEMS™ TDC circuit digitizes the MEMS resonator's die temperature by computing the ratio of the two frequencies, which is a strong function of temperature. The key observation is that in this scheme, there is almost no thermal lag between the reference resonator and the temperature sensing element, as they are both fabricated on the same die. This is a key advantage that enables the operation of this device in harsh environments, without the need for ovenization.

A second key advantage of the proposed TDC scheme is its natural fit to the temperature compensation of the MEMS oscillator reference, which is sometimes referred to as a noiseless compensation. Recall from the introduction, a significant challenge in designing TCXOs is to minimize the PN hit by the inherent noise of the temperature compensation path. Ideally, the PN of the TCXO should match that of the uncompensated oscillator if the noise of the temperature compensation path is negligible, in the absence of temperature fluctuations. Since the temperature coefficient of the TempSense resonator is significantly higher than the reference TempFlat resonator, for an optimal design as the reference clock PN decreases, the TDC PN contributions decrease by almost the same factor, ensuring close-in PN always is dominated by the reference clock. In other words, the noise from the temperature compensation path is negligible compared to that of the reference oscillator, hence the name noiseless compensation. Note this is a natural advantage of this scheme due to the higher sensitivity of the sensing element to temperature, whereas in other compensation architectures, e.g., using BJTs or thermistors, the

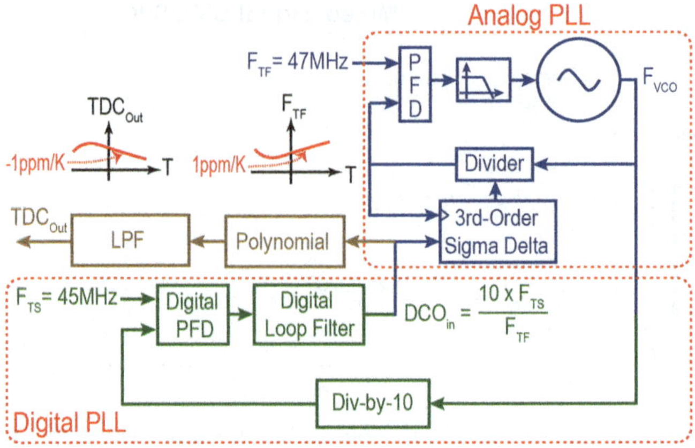

Fig. 4 Block-level diagram of the frequency ratio engine

PN impact due to the sensing element's noise PSD should be designed to be significantly (e.g., <12 dBc) below that of the reference, which is a non-trivial challenge, particularly for precision oscillators with low close-in PN and ADEV targets.

Figure 4 shows the block-level architecture of the proposed frequency ratio engine. It consists of an analog fractional-N PLL referenced to ClkTF and a third-order sigma-delta modulator, nested in a digital PLL. Assuming a fixed f_{TF}, the analog PLL behaves like a DCO since its output frequency is set by the sigma-delta input, which is a digital number. Hence,

$$f_{\mathrm{DCO}} = f_{\mathrm{TF}} \times \mathrm{DCO_{in}} \tag{3}$$

where $\mathrm{DCO_{in}}$ is the sigma-delta input value and f_{DCO} is the PLL output frequency. The digital PLL is referenced to ClkTS. In this configuration, running the VCO at higher frequencies is desired since it is used as part of the time-to-digital converter for the digital phase quantizer. A frequency divide value of 10 provides enough time resolution for the phase quantizer while keeping its power consumption in a reasonable range [14]. As a result, the DCO runs at:

$$f_{\mathrm{DCO}} = f_{\mathrm{TF}} \times 10. \tag{4}$$

Therefore, by combining (3) and (4), the sigma-delta input value is expressed as:

$$\mathrm{DCO_{in}} = \frac{10 \times f_{\mathrm{TS}}}{f_{\mathrm{TF}}} \tag{5}$$

Hence, the feedback loops force DCO_{in} to be always a scaled ratio of the input clock frequencies that vary over temperature. Thus, the temperature can be computed by post-processing the DCO_{in} value. Since this temperature sensor is aimed to compensate for the TempFlat MEMS frequency variation over temperature, as shown in Fig. 4, its output is properly scaled to exhibit the inverse characteristic of the f_{TF} error across temperature. To do so, DCO_{in} is applied to a TDC data path composed of a digital seventh-order polynomial correction, followed by a low-pass filter. The order of the polynomial is chosen based on the level of the nonlinearities of the two resonators and the target output frequency stability. Therefore, the polynomial coefficients are individually set for each device after characterizing the temperature dependence of the clocks over temperature. The low-pass filter cuts the noise of the TDC above the desired bandwidth.

4 Stability Signal Chain

Apart from the MEMS resonator's aging, various circuit-level mechanisms could also contribute to the frequency drift at the output. In order to achieve the frequency stability specification of $<\pm 100$ ppb, the so-called stability signal chain, shown in Fig. 3, was carefully designed to ensure negligible drift over time. This involves the two oscillators and the DualMEMS™ TDC, and the bias path used for MEMS biasing, i.e., the low-noise bandgap reference, and the high-voltage charge pump (HVCP) circuits. Chopping, common-centroid layout, and careful floor planning are extensively used along the stability signal chain to minimize the impact of various offset and drift sources. Another key reason for using chopping along the stability signal chain is to remove $1/f$ noise, which would otherwise impact the close-in and ADEV of the two oscillators.

Figure 5 shows the circuit diagram of the low-noise bandgap reference, which is an extension of the architecture proposed by Brokaw [19]. The two NPN transistors, Q_1 and Q_2, are biased at a 1:8 current density ratio, while the currents in the two branches are set equally. Using NPNs is preferred to avoid using an opamp to form the loop to set the current $I = \Delta V_{BE}/R_1$, thus no impact from the amplifier's noise. The native NMOS N_1 is used to provide the base currents and to ensure the collector current ratio between Q_1 and Q_2 is accurately defined. The target noise performance of the reference is a noise floor of <-140 dBV/Hz with a flicker corner of <10 Hz. The flicker corner is dominated by the PMOS current sources and the resistors since the flicker noise contribution of the NPNs is negligible. Also, proper sizing of the resistors was implemented to ensure negligible noise impact at 1 Hz. An advantage of the chosen architecture is the ratio R_2/R_1 is halved since the current of the two branches is added up and passed through R_2.

The flicker noise of the PMOS current sources, however, can't be mitigated just by brute force sizing, since the area cost will be prohibitively high. Instead, chopping was employed which effectively removes the flicker noise to negligible levels. However, this introduces another challenge: the chopping-induced ripples of

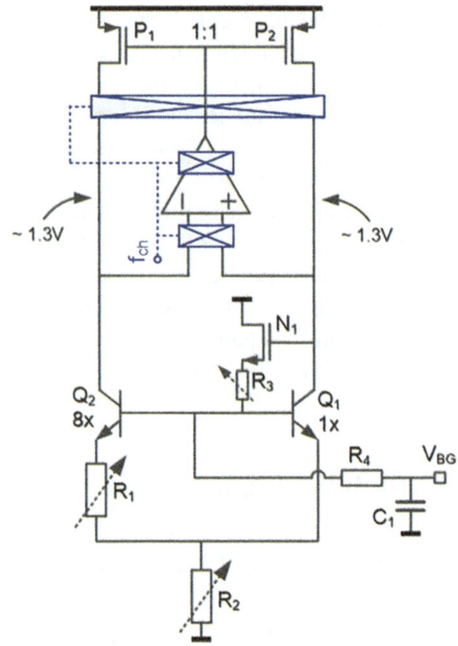

Fig. 5 Circuit-level diagram of the chopped bandgap reference

up to ±2 mVpp (at $f_{\text{chop}} = 1$ MHz) at the bandgap output, which propagates to the output by introducing unacceptably high spurs at the chopping frequency and its harmonics. To meet the stringent spur specification of <−95 dBc, the ripple at the bandgap output should be reduced to <1 uV, which is not possible by only increasing the size of the output filter. This was addressed by using an amplifier to equalize the voltage of the two collectors, thus significantly reducing the amplitude of the ripples to <±100 uV, before filtering.

Another key building block of the stability signal chain is the high-voltage charge pump (HVCP) used to generate a precision high voltage (10–12 V) for biasing the MEMS resonator. Since the resonator's frequency is modulated through its bias node, it's critical to ensure the noise and the aging and drift of the bias voltage are well defined and meet the specifications. This was addressed by designing a closed-loop HVCP, locked to the precision bandgap voltage, as shown in Fig. 6. An error amplifier in the feedback loop is used to feed the chain of ten charge pump stages. The output voltage is set by the feedback resistors: $V_{\text{bias\_MEMS}} = V_{\text{BG}} \times (1 + R_1/R_2)$. Again, chopping is used to mitigate the flicker noise and the offset drift of the error amplifier. The resistors R_1 and R_2 are carefully sized to ensure the flicker noise at the output is within the specification. To reduce the ripple amplitude at the bias node to <100 nV specification, a two-stage RC low-pass filter was used. A multi-phase clocking was used for the charge pump stages, reducing the ripple amplitude by spreading the switching energy over time, thus relaxing the filter requirements.

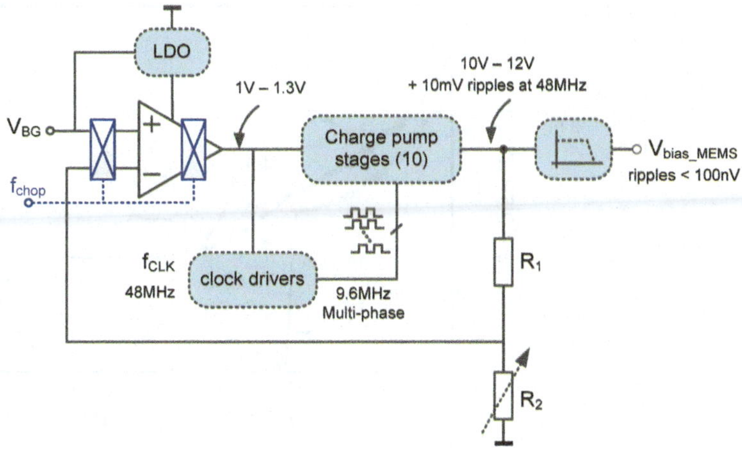

Fig. 6 Circuit-level diagram of the chopped closed-loop high-voltage charge pump (HVCP)

5 Measurement Results

There is no straightforward way to directly measure the noise of the TDC, given its very low noise floor. Instead, this was indirectly measured through the output PN []. The TDC data path was initially bypassed to make the TDC noise contribution observable in the frequency range from 200 Hz to 2 kHz. As shown in Fig. 7, its contribution was then amplified by 24 dB through the digital gain stage in the TDC data path. The indirectly measured TDC resolution for the DualMEMS™ TDC proves to be 20 μK over a bandwidth of 100 Hz, resulting in a resolution FOM of 0.04 pJK^2 [20].

The parts are packaged in ceramic and measured over the temperature range −55 to 105 °C. Figure 8 depicts the measured frequency stability of 108 parts from one lot. As shown, the stability after trimming is well within ±50 ppb. Another key specification for the high-precision timing market is the frequency versus temperature slope (*dF*/*dT*) which is shown to be ~1.2 ppb/°C for the same parts. To evaluate the operation under the airflow conditions, ADEV was measured in a temperature chamber at 25 °C with and without the fan being on. As shown in Fig. 9, ADEV for the MEMS TCXO is almost intact under the airflow condition and up to 100 s averaging time. However, for the quartz-based best-in-class TCXO under test, ADEV drastically degrades, up to 38× at 2 s averaging time, due to its suboptimal temperature compensation scheme.

Another test was done to evaluate the part's performance under fast thermal transients and to compare it against the quartz solution. As shown in Fig. 10, DUTs were first exposed to a running fan, placed at a short distance from the parts. The frequency error of the MEMS TCXO was measured to be <0.5 ppb, while that of the quartz-based TCXO jumps by >20 ppb. In the next step, the two parts were exposed to a heat air gun running at 50 °C. As shown, the MEMS TCXO degrades by only

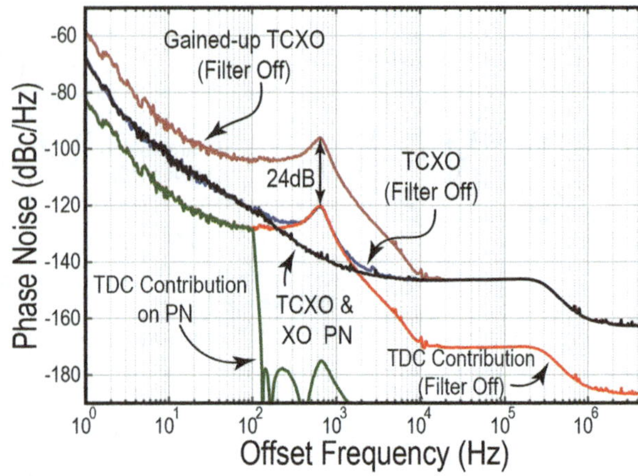

Fig. 7 Output clock phase noise in different modes of operations. The TDC noise was indirectly measured through PN and by gaining up its contribution in the compensation data path

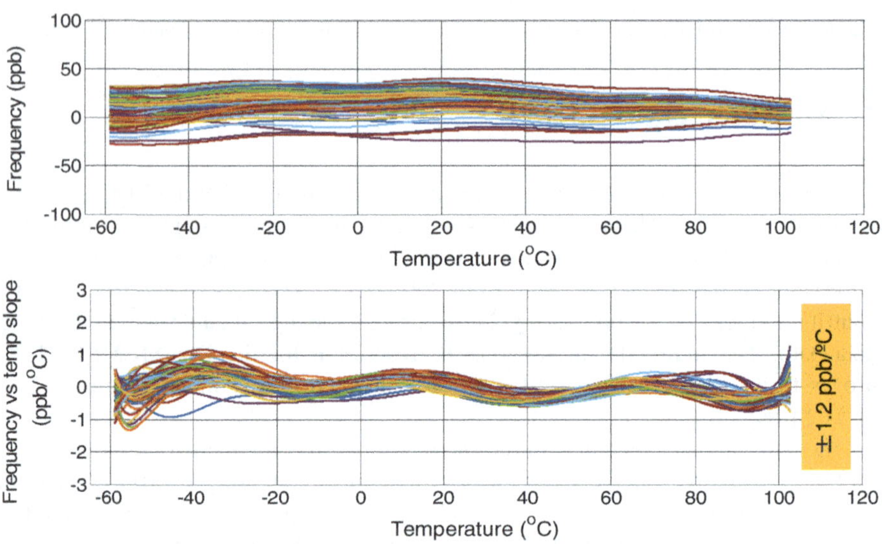

Fig. 8 Top: Measured frequency stability over temperature. Bottom: Measured frequency versus temperature slope (dF/dT). A number of 108 parts were measured from the same lot

<3 ppb, while the quartz-based TCXO jumps by >200 pb. The MEMS TCXO's superior performance during thermal transients is due to the high speed, high-resolution temperature compensation, as well as extremely tight thermal coupling between the reference and the temperature sensing resonators.

Another key advantage of MEMS resonators, compared to their quartz-based counterparts, is their robustness to shock and vibration. This is an inherent advantage due to their significantly smaller mass. Operation under shock and vibration in quartz resonators could result in higher phase noise, jitter, and possible malfunction of the device. To evaluate this, we compared the PN of the MEMS oscillator with

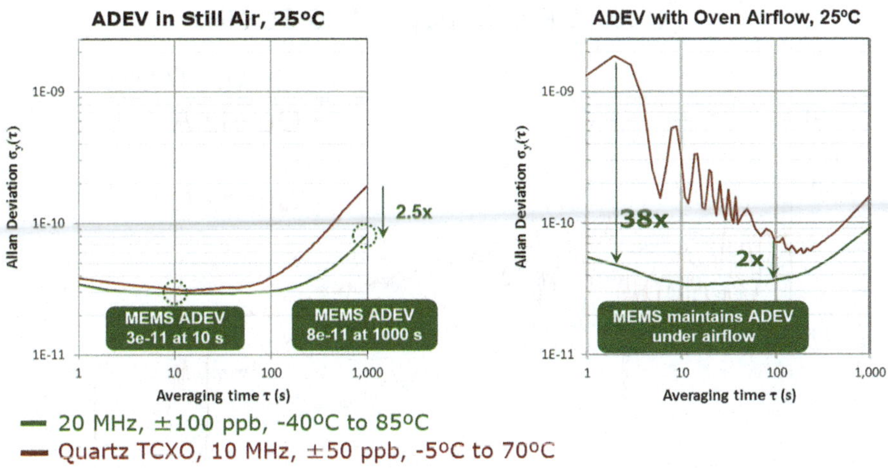

Fig. 9 ADEV measurement for MEMS (green) and quartz (red) TCXOs in still air and under airflow condition

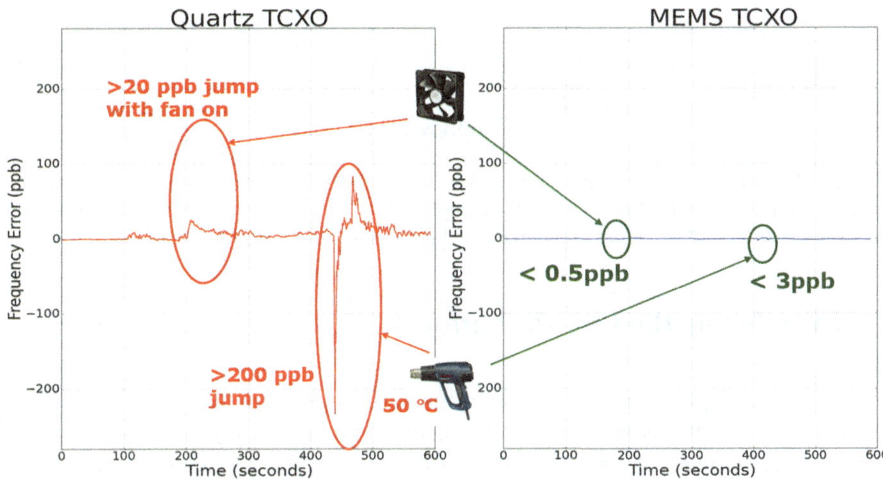

Fig. 10 Frequency error measurement for MEMS (right) and quartz (left) TCXOs under fast thermal transients. The MEMS TCXO performs better by almost two orders of magnitude

three different quartz-based parts while being exposed to random vibration within the frequency range of 15 Hz–2 kHz and with an amplitude of 7.5 g (rms). The result is shown in Fig. 11. The PN of the MEMS TCXO has hardly been impacted, while a clear jump of >20 dBc/Hz within the frequency range of the imposed random vibration is measured for all three quartz-based designs. This is a significant advantage for operation under harsh environments, which is becoming more critical in applications such as automotive, aerospace, and military.

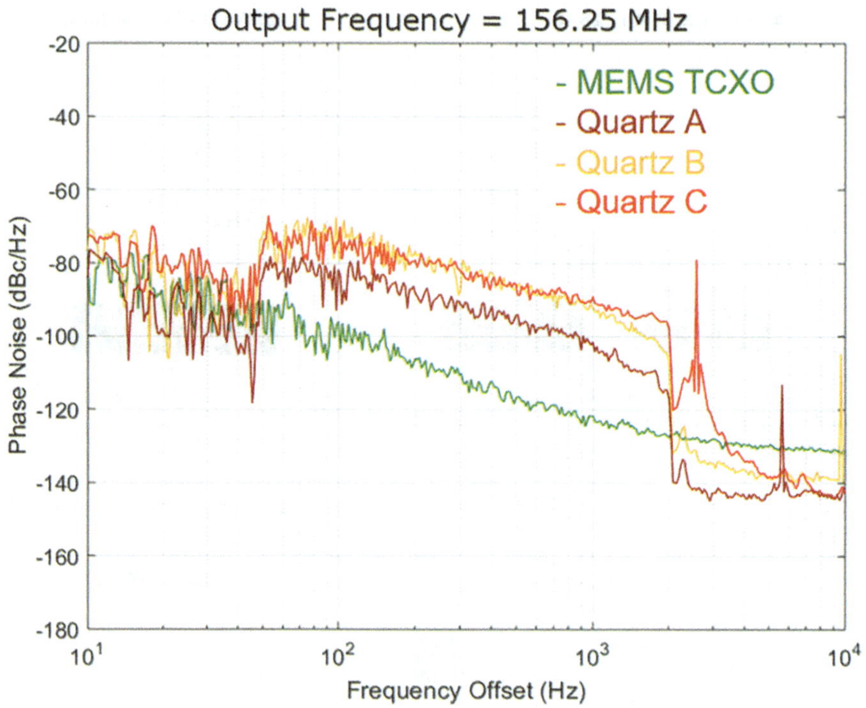

Fig. 11 Phase noise degradation under random vibration. The injected vibration noise in the frequency range of 15 Hz–2 kHz and with an amplitude of 7.5 g (rms). The PN for MEMS TCXO under test is almost intact, while the quartz-based devices see an impact of >20 dBc/Hz

6 Introducing the Elite-X Platform

While the Elite platform delivers superior performance for the precision timing market, the recent market traction for even more tight frequency stability has significantly grown. In an attempt to address such market demand, the Elite-X TCXO platform was developed, targeting a significant 10× improvement in frequency stability over temperature. This was achieved by reducing the stress sensitivity of the MEMS resonator, as well as a more tailored ceramic package design. As shown in Fig. 12 (left), 20 devices from 1 lot were measured in the socket and delivered frequency stability of ±3 ppb over the temperature range from −45 to 105 °C. More importantly, the same parts were soldered down and remeasured over the same condition. As shown in Fig. 12 (right), the result only slightly degraded while still delivering a superior frequency stability error of <±5 ppb. Measurements for the other three lots proved the consistency of the results. Such a result is unprecedented for TCXOs and matches the entry-level quartz-based OCXO performance. Clearly, the advantage of using the MEMS-based solution will be lower power consumption, smaller footprint, lower cost, and higher reliability.

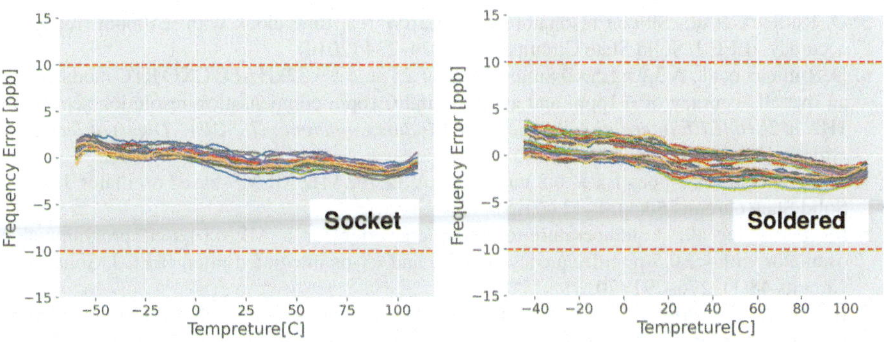

Fig. 12 Frequency stability over temperature for 20 devices from 1 lot. Measurements were initially done in a socket (left) and repeated after solder-down (right)

7 Conclusions

Recent developments in MEMS-based timing devices have proved them as the technology of choice for precision timing when compared to their legacy quartz-based counterparts. They leverage the existing mature, fast-growing semiconductor infrastructure to deliver the most optimum design for the resonator as well as the required supporting circuits. For a temperature-compensated design, the SiTime's Elite platform delivers frequency stability well below ±100 ppb, while it was shown the optimized design by the Elite-X family can improve this to ±10 ppb, an unprecedented performance for TCXOs. Using DualMEMS™ technology is a key enabler to deliver outstanding performance in the presence of environmental temperature stressors. It was shown that the precision timing performance of the device remains nearly intact under fast thermal transients, delivering entry-level OCXO performance at much lower power consumption, smaller package footprint, and guaranteed reliability. Last but not least, the inherent resilience to shock and vibration makes MEMS oscillators the natural choice for operation under harsh environmental conditions. Such performance promises a consistent MEMS domination in the category of precision timing in the future.

References

1. WirelessUSB™ crystal guidelines, AN19219 [Online]. Available http://www.cypress.com/file/134326/download
2. Datasheet, ASTXR-12 [Online]. Available http://www.abracon.com/Oscillators/ASTXR-12.pdf
3. Marc Wiess, Telecom Requirements for Time and Frequency Synchronization [Online]. Available http://www.gps.gov/cgsic/meetings/2012/weiss1.pdf
4. IEEE 1588™, Telecommunications Applications [Online]. Available http://www.nist.gov/el/isd/ieee/upload/tutorial-telcom.pdf

5. D. Ruffieux et al., Silicon resonator based 3.2μW real time clock with ±10ppm frequency accuracy. IEEE J. Solid State Circuits **45**(1), 224–234 (2010)
6. D. Ruffieux et al., A 3.2×1.5×0.8mm^3 240nA 1.25-to-5.5V 32kHz-DTCXO RTC module with an overall accuracy of ±1ppm and an all-digital 0.1ppm compensation-resolution scheme at 1Hz, in *2016 IEEE International Solid-State Circuits Conference (ISSCC) – Digest of Technical Papers*, (2016), pp. 208–209
7. S.Z. Asl et al., A 3 ppm 1.5 × 0.8 mm2 1.0 μA 32.768 kHz MEMS-based oscillator. IEEE J. Solid State Circuits **50**(1), 1–12 (2015)
8. M. Perrott et al., A temperature-to-digital converter for a MEMS-based programmable oscillator with <±0.5-ppm frequency stability and <1-ps integrated jitter. IEEE J. Solid State Circuits **48**(1), 276–291 (2013)
9. H.C. Lee, A. Partridge, F. Assaderaghi, Low jitter and temperature stable MEMS oscillators, in *2012 IEEE International Frequency Control Symposium Proceedings*, (2012), pp. 1–5
10. A. Kourani, E. Hegazi, Y. Ismail, RF MEMS reference oscillator platform with ±0.5ppm frequency stability for wireless handsets, in *International Symposium on Signals, Circuits and Systems (ISSCS)*, (2015), pp. 1–4
11. A. Partridge, H.C. Lee, We know that MEMS is replacing quartz. But why? And why now? in *2013 Joint European Frequency and Time Forum & International Frequency Control Symposium (EFTF/IFC)*, (2013), pp. 411–416
12. Electronic Design, MEMS Oscillators Replace Quartz Crystal Oscillators [Online]. Available http://electronicdesign.com/communications/mems-oscillators-replace-quartz-crystal-oscillators
13. SiTime's MEMS First Process, SiTime Corp [Online]. Available http://www.sitime.com/support2/documents/AN20001_MEMS_First_Process.pdf
14. M. Roshan et al., A MEMS-assisted temperature sensor with 20uK resolution, conversion rate of 200S/s and FOM of 0.04 pJK^2. IEEE J. Solid State Circuits **52**(1), 185–197 (2017)
15. SolidState Technology, MEMS resonators vs. crystal oscillators for IC timing circuits [Online]. Available http://electroiq.com/blog/2009/01/mems-resonators-vs-crystal-oscillators-for-ic-timing-circuits/
16. Digi-Key, MEMS Oscillators Challenge Quartz Crystals in RF Applications [Online]. Available http://www.digikey.com/en/articles/techzone/2014/oct/mems-oscillators-challenge-quartz-crystals-in-rf-applications
17. SiTime, Shock and Vibration Performance Comparison of MEMS and Quartz-based Oscillators [Online]. Available http://www.sitime.com/support2/documents/AN10032-Shock-Vibration-Comparison-MEMS-and-Quartz-Oscillators.pdf
18. SiTime, Resilience and Reliability of Silicon MEMS Oscillators [Online]. Available http://www.sitime.com/support2/documents/AN10045-SiTime-Resilience-Reliability-MEMS-Oscillators.pdf
19. P. Brokaw, A simple three-terminal IC bandgap reference. IEEE J. Solid State Circuits **sc-9**(6), 388–393 (1974)
20. K.A.A. Makinwa, Temperature Sensor Performance Survey [Online]. Available http://ei.ewi.tudelft.nl/docs/TSensor_survey.xls

Fast Startup and Fully Differential Crystal Oscillator

Wim Kruiskamp

Abstract Crystal oscillators are essential circuits in low-power wireless systems. The de facto standard circuit for a crystal oscillator is the three-point Pierce oscillator. Despite the simplicity and good performance of this circuit, it has some drawbacks: large required capacitance, slow and inefficient startup, and single-ended operation, hence poor common-mode rejection. This chapter discusses these drawbacks and presents an alternative circuit to address these drawbacks. The proposed switched capacitor crystal oscillator requires almost four times less capacitance, includes fast and efficient startup, and is fully symmetrical and fully differential.

1 Introduction

Low-power wireless systems typically use a crystal oscillator to generate an accurate low-noise reference clock for the radio. The crystal oscillator typically spends most of its time in a low-power sleep state and only briefly wakes up during the receiving and transmitting of data by the transceiver. Fast startup and low startup energy are therefore important aspects for such an oscillator.

Traditionally, the crystal oscillator is implemented as a three-point Pierce oscillator, described by [1] and depicted in Fig. 1. In this oscillator, the crystal is connected between the input and output of a self-biased inverter, with load capacitors from either crystal terminal to ground. The simplicity and good performance of this circuit have made it the de facto standard in many RF systems for decades. This circuit however suffers from a few important drawbacks.

Firstly, the total capacitance required is four times higher than for a topology where the load capacitor is directly across the crystal. These capacitors usually dominate the total area, adding cost to the circuit.

W. Kruiskamp (✉)
Renesas Electronics Corporation, 's-Hertogenbosch, The Netherlands
e-mail: wim.kruiskamp.uj@renesas.com

P. Harpe et al. (eds.), *Biomedical Electronics, Noise Shaping ADCs, and Frequency References*, https://doi.org/10.1007/978-3-031-28912-5_18

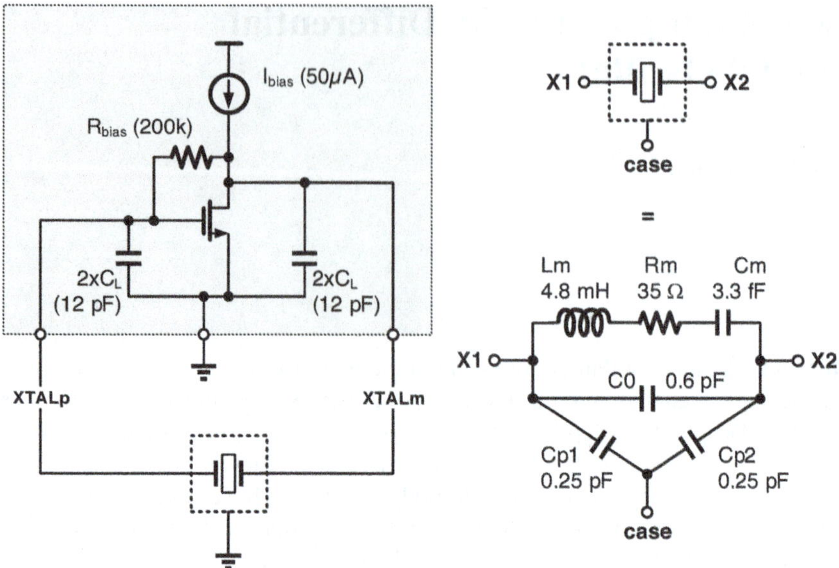

Fig. 1 Typical three-point Pierce oscillator with an equivalent circuit of a typical 40 MHz crystal

Fig. 2 Measured waveform of a typical 32 MHz Pierce oscillator during startup

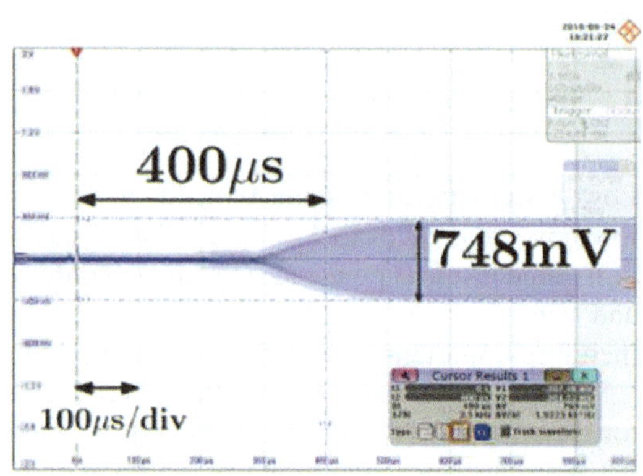

Secondly, the circuit is a linear circuit starting from noise. The high-quality factor of the crystal makes that it takes many periods, typically in the order of ten thousand, before the oscillating amplitude is at its desired level (see Fig. 2). During steady-state operation, when the oscillation amplitude is large, the inverter operates in class-C mode and is very efficient. During startup however, when the oscillation amplitude is still small, the inverter operates in class-A mode and is extremely inefficient. This makes that the startup of a Pierce crystal oscillator not only takes a lot of time but also consumes a lot of energy. In low-power wireless systems, like Bluetooth Low Energy, the startup of the crystal oscillator can be a significant part of the total energy consumption, especially if the transmit and receive periods are short.

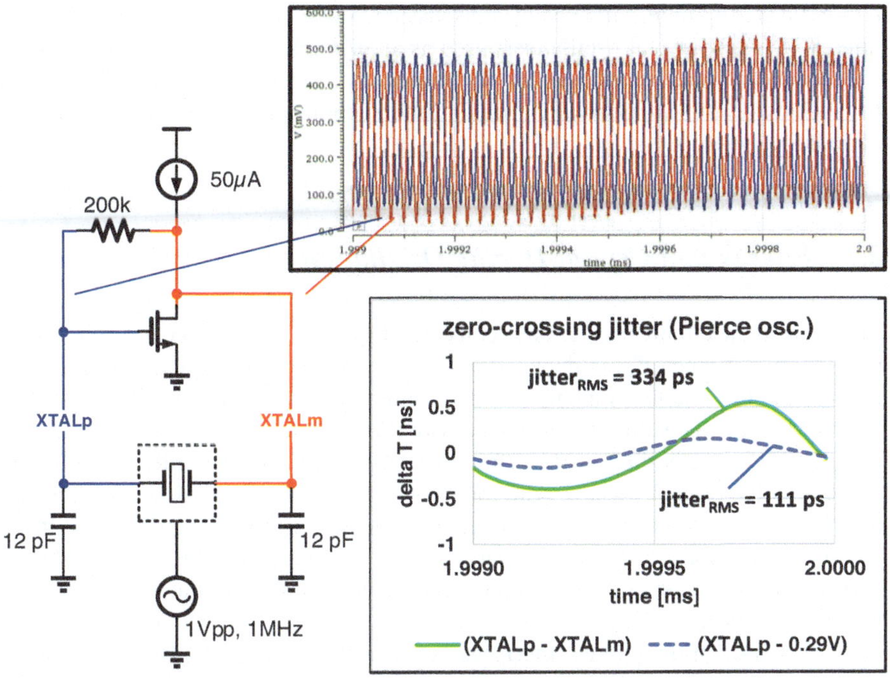

Fig. 3 Simulated jitter of the typical 40 MHz single-ended Pierce oscillator of Fig. 1, with a 1 Vpp, 1 MHz signal applied on the crystal case

Thirdly, the three-point Pierce oscillator is a single-ended circuit. Any common-mode noise is amplified by the inverter and converted to differential noise. This makes the three-point Pierce oscillator very sensitive to common-mode crosstalk, for instance, coupled in via ESD diodes. This is quantified by the simulation of Fig. 3. A 1 MHz signal with an amplitude of 1 Vpp is applied to the crystal case. This signal is coupled to the crystal pins by the parasitic case capacitance Cp1 and Cp2. With single-ended probing on XTALp, the resulting jitter is equal to 111 ps. With differential probing across the crystal, the resulting jitter is three times as high and is equal to 334 ps. Since the typical jitter performance of a well-designed crystal oscillator is in the sub-ps range, common-mode distortions of a millivolt will already have a noticeable effect. A fully symmetrical and fully differential crystal oscillator is therefore desired, especially when used in a noisy environment.

In this chapter, a fully differential crystal oscillator is proposed, which overcomes all three listed drawbacks.

2 Fast Startup

While the high-quality factor of a crystal resonator allows excellent phase noise, it also implies a long startup time. In this section, we will first look at the theoretical minimum startup time, followed by a discussion how to reach this minimum startup time. A 40 MHz crystal oscillator is used as an example. The theory is general for all frequencies. It also applies to low-frequency crystal oscillators, e.g., 32,768 kHz, but such oscillators are generally used for time-keeping, and therefore startup is less of an issue.

2.1 Energy in a Crystal During Oscillation

The equivalent circuit of a crystal consists of an LRC branch in parallel with a parasitic capacitance C_0 as depicted in Fig. 4. The capacitance C_m is only a few femtofarad for a stable crystal up to tens of femtofarad for a less stable crystal and is more or less independent of the frequency of the crystal. C_0 represents the parasitic capacitance between the crystal terminals and is usually in the order of a pF. The inductor value typically scales with the frequency to the power minus two, and the resistor typically scales with the frequency to the power minus one. In an oscillator circuit, a load capacitor C_L is connected parallel to the crystal to tune the frequency to the exact frequency and at the same time to increase the oscillation energy for a given amplitude at the crystal pins.

During oscillation, energy alternates between the inductor L_m on the one hand and the capacitors C_m, C_0, and C_L on the other hand. The resistor R_m models the losses in the crystal and is at the same time the only noise source within the crystal. An active circuit, not shown in Fig. 4, must restore the losses to get a stable steady-state oscillation.

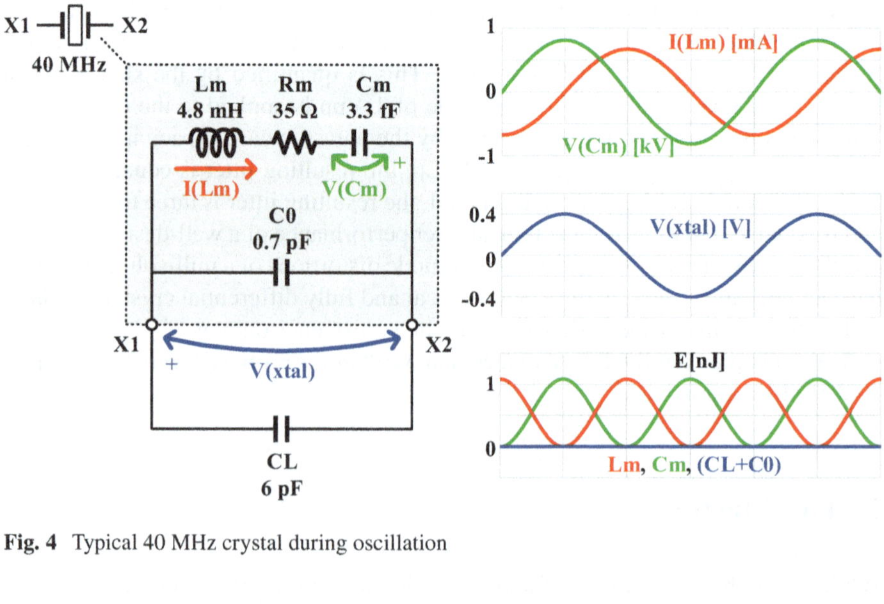

Fig. 4 Typical 40 MHz crystal during oscillation

The harmonic current $I(L_m)$ generates a voltage $V(\text{xtal})$, exactly 90 degrees out of phase with itself, by flowing through C_0 parallel to C_L. The same current will also flow through C_m, generating a $(C_0 + C_L)/C_m$ larger voltage across C_m. A typical oscillation amplitude at the crystal pins of 400 mV therefore corresponds to an 800 V amplitude across C_m. Since the energy in a capacitor is equal to $½CV^2$, the energy in C_m is $(C_0 + C_L)/C_m \sim 2000$ times more than the combined energy in C_0 and C_L. 99.95% of the energy therefore stays within the LRC branch of the crystal. The large harmonic energy and high internal amplitude make it possible to have a stable and low-noise signal. The downside however is that this energy must be added to the crystal during startup, which takes time and costs a lot of extra energy when not done in an efficient way.

The energy stored in the crystal, as a function of amplitude at the crystal pins, is equal to:

$$E_{\text{xtal}} = \frac{(C_L + C_0)^2}{2\,C_m} \cdot (V_{\text{xtal}})^2 \tag{1}$$

The power dissipation during steady state is due to the equivalent resistor R_m

$$I_{\text{xtal}} = V_{\text{xtal}} \cdot 2\pi \cdot (C_L + C_0) \cdot f_{\text{xtal}} \tag{2}$$

$$P_{\text{xtal}} = \frac{R_m}{2} \cdot (I_{\text{xtal}})^2 = \frac{R_m}{2} \cdot (V_{\text{xtal}} \cdot 2\pi \cdot (C_L + C_0) \cdot f_{\text{xtal}})^2 \tag{3}$$

For our typical 40 MHz crystal, this equals to $E_{\text{xtal}} = 1.09$ nJ, $I_{\text{xtal}} = 0.67$ mA, and $P_{\text{xtal}} = 7.9$ µW.

2.2 *Theoretical Minimum Startup Time*

The fastest way to start a crystal oscillation, for a given maximum voltage that can be applied to the crystal pins, is by applying a block wave to the crystal with a maximum amplitude and a polarity synchronized with the direction of the current through the equivalent inductor in the crystal. This optimum startup is depicted in Fig. 5.

The initial state is minus the supply voltage across the crystal. No current is flowing, and the voltage across C_m is plus V_{dd}. The first step is to change polarity: a step from minus V_{dd} to plus V_{dd} across the crystal pins. That step of two times V_{dd} will appear across the inductor, and the inductor current will start to rise. This current will charge C_m, which in its turn will reduce the voltage across the inductor. The inductor voltage will become negative, reducing its current, and eventually the current becomes zero. At that moment, the polarity is changed again: a step of

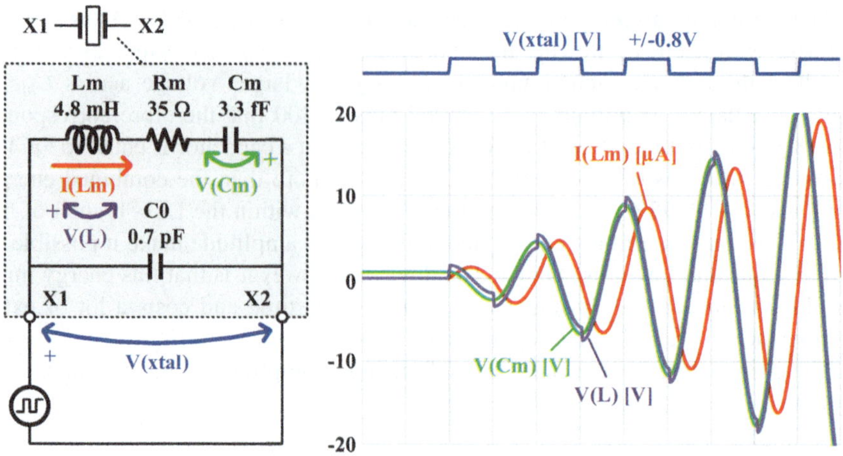

Fig. 5 Startup by synchronized block wave

minus two times V_{dd}. This step again appears across the inductor, which voltage will change from minus two times V_{dd} to minus four times V_{dd}. This increased voltage will make the current change faster, and the current will swing low and eventually become zero again. By keep toggling the crystal voltage any time the inductor current is zero, the internal voltages will grow linearly with two times V_{dd} per toggle, resulting in a growth of four times V_{dd} per period.

The optimal startup time, ignoring the effect of R_m, can be expressed by the wanted amplitude V_{xtal} at the crystal pins and the maximum voltage of the block wave V_{dd}:

$$T_{\text{startup}} = \frac{1}{f_{\text{xtal}}} \cdot \frac{V_{\text{xtal}}}{4\ V_{\text{dd}}} \cdot \frac{C_L + C_0}{C_m} \tag{4}$$

For our 40 MHz example, this equals to only 6.3 μs to reach a 0.4 V amplitude. The energy consumption during this startup time consists of the energy which is stored in the crystal (1) and the energy dissipated in C_0 (5) and R_m (6) during startup. Note that no dissipation in C_L is considered during startup, since C_L can be disconnected during startup and connected during steady state only.

$$E_{C0} = T_{\text{startup}}\ 2\ f_{\text{xtal}}\ 2\ C_0\ {V_{\text{dd}}}^2 = V_{\text{xtal}}\ V_{\text{dd}}\ C_0\ \frac{C_L + C_0}{C_m} \tag{5}$$

$$E_{Rm} = \frac{R_m}{6}\ {I_{\text{xtal}}}^2\ T_{\text{startup}} = R_m\ \frac{\pi^2}{6}\ \frac{{V_{\text{xtal}}}^3}{V_{\text{dd}}}\ \frac{(C_L + C_0)^3}{C_m} f_{\text{xtal}} \tag{6}$$

For our 40 MHz example, this equals to an energy loss equal to $E_{C0} = 0.45$ nJ plus $E_{Rm} = 0.02$ nJ while storing $E_{\text{xtal}} = 1.08$ nJ in the crystal: a theoretical efficiency of 70%. The energy associated with the (dis)charging of C_0 can be reduced by stepwise charging techniques as presented by Lechevallier in [2]. The simplest example of this is to first short the crystal terminals before applying the new voltage level across the crystal, reducing the energy dissipation E_{C0} by 50%.

2.3 Synchronizing the Block Wave to the Crystal Current

In order to add energy to the crystal, the phase of the block wave needs to be (significantly) within 90 degrees of the phase of the current through the inductor of the crystal. To start a crystal oscillator with an additional free-running oscillator therefore requires that oscillator to match the crystal frequency within at most 0.1% but preferably better than 100 ppm. This is a very challenging or maybe even unrealistic accuracy for an on-chip oscillator, especially including temperature variations and aging.

A more practical approach [3–5] is to use a free-running oscillator only for the first tens of periods. After this initial startup, the free-running oscillator is either periodically synchronized with the crystal oscillation [3] or replaced by a fast-locking PLL [4, 5]. This approach relaxes the accuracy requirement of the free-running oscillator by an order of magnitude. This is still challenging, but not impossible as demonstrated by Verhoef [3], Megawer [4], and Jung [5]. A downside of this approach is that it assumes that the crystal is not oscillating at the time the startup begins and the phase of the crystal oscillation is defined by the first toggle of the block wave. However, in case there is still energy left in the crystal at the beginning of startup, the phase of the block wave and the harmonic oscillation can be anything, and there is no guarantee that the block wave is enforcing the harmonic oscillation. Worst-case it can even damp the harmonic oscillation, making it impossible to lock to it after the first tens of periods. For a robust solution, it is therefore essential that the block wave is synchronized to the harmonic oscillation from the beginning. The approach of Lechevallier [6] realizes this by monitoring the voltage across the switches driving the crystal during startup as depicted in Fig. 6.

As mentioned earlier, the ideal switching moment is when the motional current through the equivalent inductor is zero. The zero-crossing can be detected by monitoring the voltage drop over the on-resistance of the switches in the H-bridge, assuming that the phase of the current through the switch I_{sw} is identical to the phase of the current I_L. This however brings contradicting requirements for the on-resistance of the switches: It has to be low-impedance in order to avoid phase shift in combination with the parasitic capacitance of the crystal, yet it has to be high-impedance to get a decent amplitude across the switch out of the, initially small, harmonic current.

To avoid such a compromise, the driving phase can be separated from the detection phase as depicted in Fig. 7 [7, 8]. The high-side switches S1 and S3

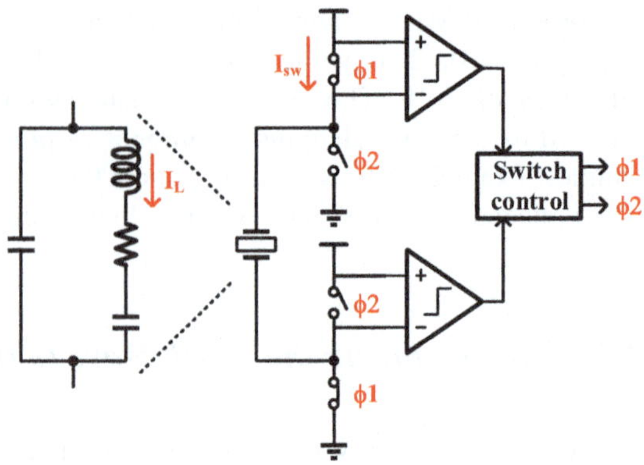

Fig. 6 Self-timed energy injection by monitoring the H-bridge voltage

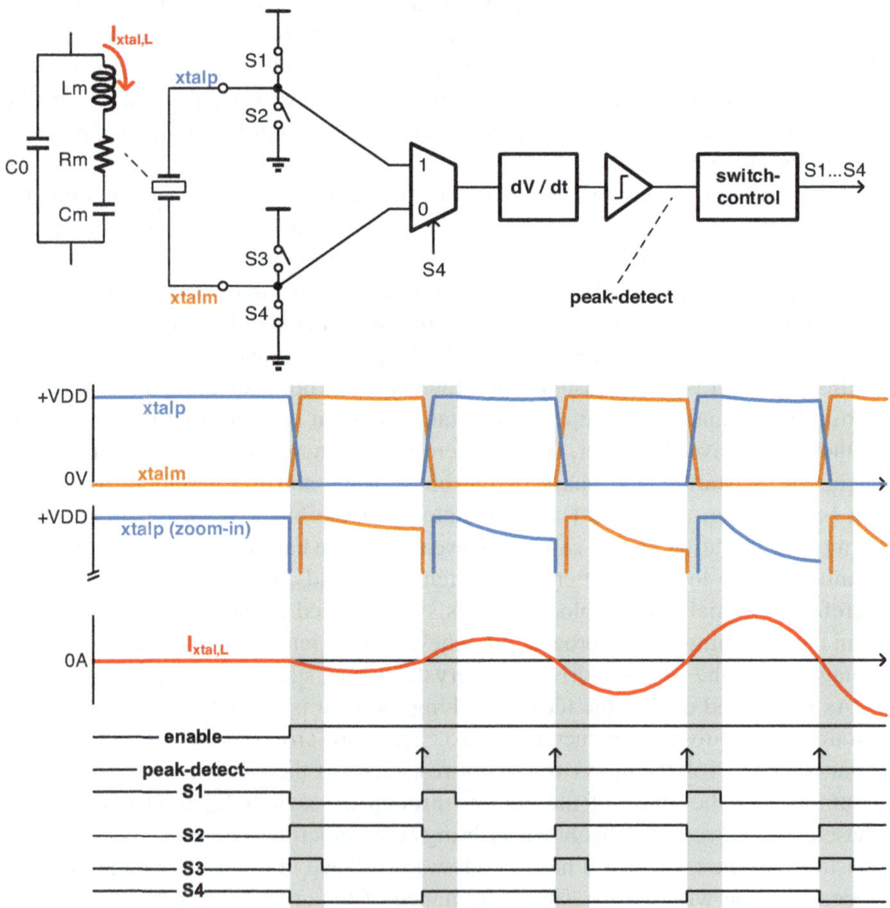

Fig. 7 Time-multiplexing the drive phase and the detection phase

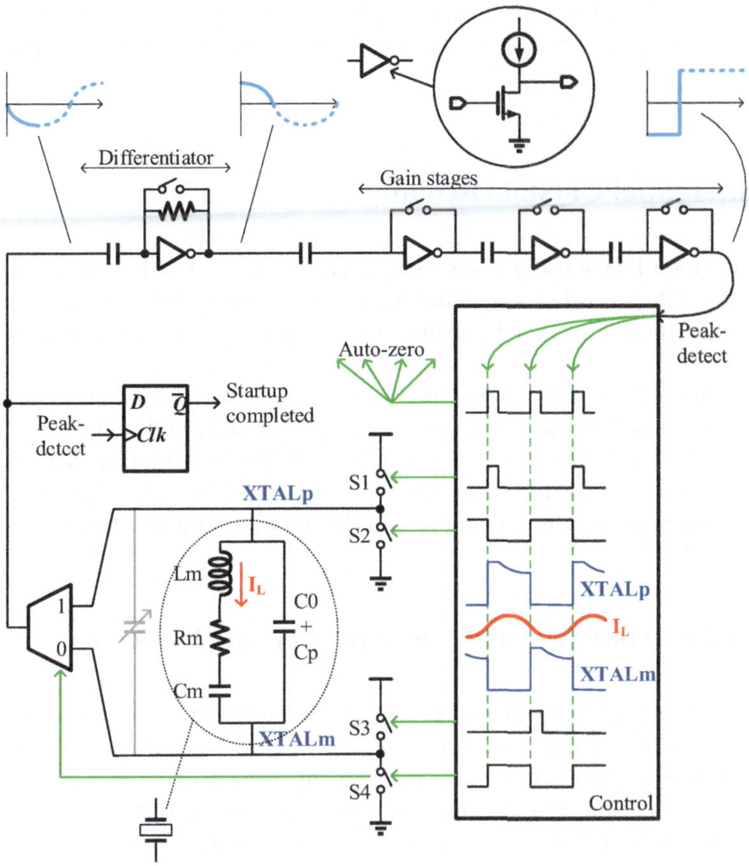

Fig. 8 Fast startup implementation

are only closed for the first part of each half-period (the drive phase). The voltage will remain on the parasitic capacitance C_0 for the remainder of the half-period (the detection phase), and the now opened high-side switch will not reduce nor influence the oscillating voltage across the crystal anymore. The inductor current will create a voltage, exactly 90 degrees out-off phase with itself, across the (parasitic) capacitance between the crystal pins. By differentiating and amplifying that voltage, the ideal moment to change the polarity of the H-bridge can be determined.

The signal during startup is a single-ended signal since one side of the crystal is connected to ground. The peak detection is implemented with the circuit depicted in Fig. 8. The differentiator and the gain stages are implemented with common-source amplifiers. The current sources provide isolation to noise on the supply. The differentiator and each gain stage are auto-zeroed during each drive pulse as the sine wave during the first cycles is very small. Startup is completed when the output of the multiplexer is below half the supply voltage at the rising edge of the peak detect signal, indicating that the differential peak-to-peak amplitude is equal to the supply voltage.

3 Differential Crystal Oscillator

The three-point Pierce oscillator is a single-ended circuit. Unlike with the differential LC oscillators as often seen in RF circuits, it is not possible to just duplicate and cross-couple the single-ended circuit to make a differential version as depicted in Fig. 9.

Unlike the LC oscillator, there is no DC path through a crystal. In fact, this circuit will act as a latch, with one of the outputs equal to ground and the other output equal to the supply. The circuit in Fig. 9 can only work if a DC path is added parallel to the LRC branch to make the oscillator DC-stable. There are only a few examples of such a differential crystal oscillator [9–11], but it never became a popular circuit.

3.1 Differential Switched Capacitor Crystal Oscillator

In this section, an alternative approach for a differential crystal oscillator is presented. During startup as described in Sect. 2.3, the voltage across the crystal is the summation of a large block wave and a growing sine wave, depicted in Fig. 10. If we would continue this process, the sine wave would eventually become almost rail-to-rail. For power-saving reasons, among others, it is usually preferred to operate the crystal oscillator in steady-state operation with a smaller than rail-to-rail amplitude. This can be achieved by adding a small block wave to the sine wave, just enough to keep the amplitude of the harmonic oscillation constant.

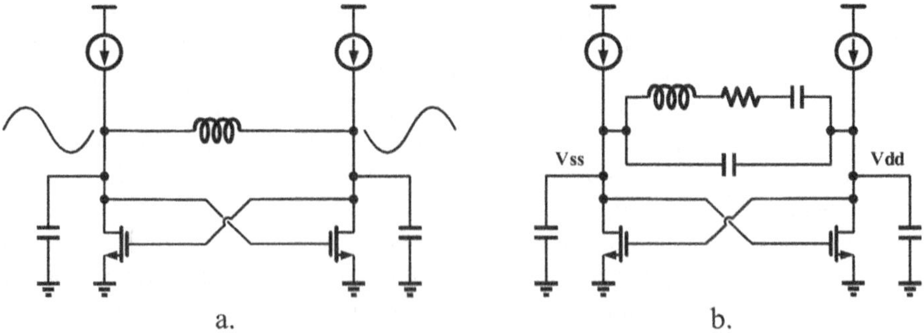

Fig. 9 (**a**) LC oscillator. (**b**) Non-working differential crystal "oscillator"

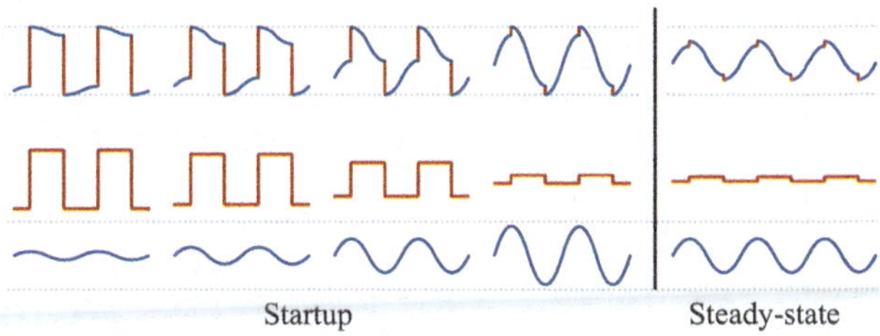

Fig. 10 Crystal waveforms: block wave plus sine

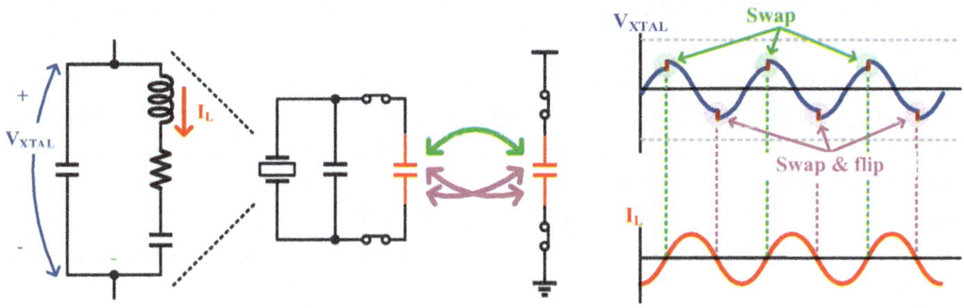

Fig. 11 Switched capacitor concept

The desired steady-state waveform, depicted on the top right of Fig. 10, can be achieved by the switched capacitor circuit of Fig. 11 [8, 12]. The load capacitor C_L is split into a fixed capacitor and a smaller capacitor, connected via switches. A similar capacitor at the side is charged to the supply voltage. At each peak of the crystal voltage, when the current through the inductor is zero and all energy is stored in capacitors, the currently connected switched capacitor is replaced by the fully charged similar capacitor. This will instantly increase the voltage across the crystal, adding energy to the system. At each valley of the crystal voltage, the currently connected switched capacitor is again replaced by a similar capacitor, which this time is charged negatively to the supply voltage. This will again increase the voltage across the crystal, adding energy to the system. The steady-state amplitude settles to a value where the added power exactly compensates the power dissipation in the crystal.

Note that this circuit is fully symmetrical, and the load capacitor is connected directly across the crystal, reducing the total required capacitance to almost a factor 4 compared to the Pierce oscillator.

3.2 PSRR Improvement by Correlated Sampling

A risk of the presented switched capacitor circuit from the previous section is supply noise, sampled by the switched capacitor and added to the crystal. This noise results in an offset of the crystal voltage $V\text{xtal}$, causing jitter when converted to a digital

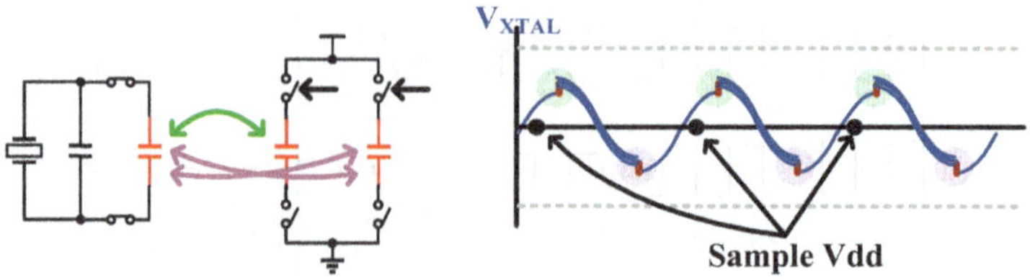

Fig. 12 Correlated sampling

clock by a comparator. This jitter can be reduced by using correlated sampling as illustrated in Fig. 12.

The two capacitors at the side are sampled simultaneously to the supply voltage: one is used at the next peak, and the other is used at the next valley. Supply noise added at the peak is in this way largely cancelled by subtracting a similar error at the valley. This reduces the jitter at the rising edge, but not at the falling edge, of the sine wave. Fortunately, many PLLs in radio systems only use the rising edge of the reference clock. The switching scheme of Fig. 12 can be implemented with the topology depicted in Fig. 13.

4 Measurement Results

The switched capacitor crystal oscillator with fast startup is implemented in a 22 nm process. As depicted in Fig. 14, the 8 pF varicap and the switched capacitors dominate the total area, although the required capacitance is almost four times smaller as would be required for a Pierce oscillator.

The oscillator was evaluated with a 40 MHz crystal (Murata, RCGB40M000F1S2FR0). No calibration was required for the circuit other than trimming the varicap to a frequency of exactly 40 MHz.

4.1 Fast Startup

The measured voltage across the crystal during startup is depicted in Fig. 15. The top curve nicely shows the linear growth in amplitude of the harmonic oscillation on top of the block wave. Zoomed in at the beginning of the startup, the sine wave is not yet visible, but nevertheless the block wave is equal to 40 MHz, indicating that it is synchronized with the harmonic oscillation. Further down in the startup procedure, the sine wave becomes clearly visible, with the edges of the block wave nicely aligned with the peaks of the sine. The startup ends when the crystal voltage drops below half the supply voltage, indicating that the peak-to-peak amplitude is equal to the supply voltage.

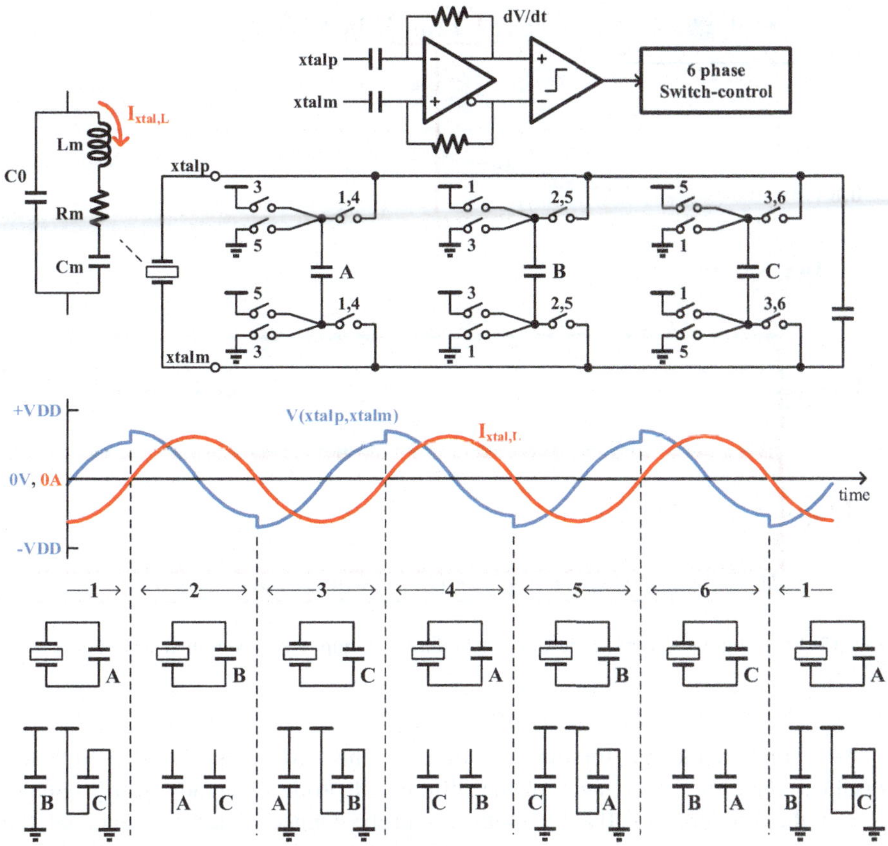

Fig. 13 Switched capacitor crystal oscillator circuit

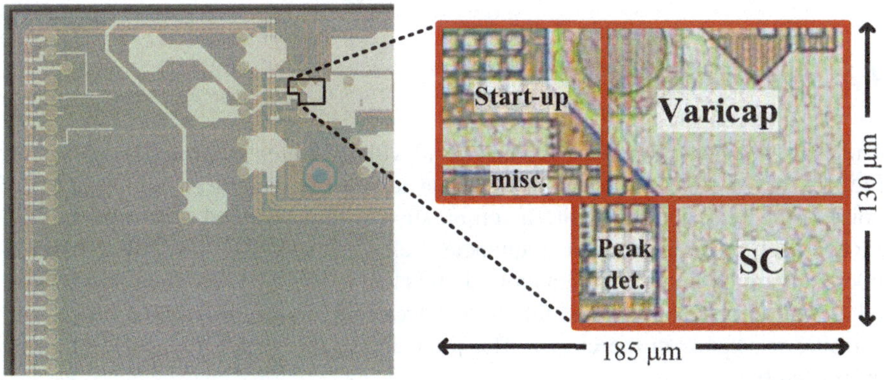

Fig. 14 Die microphotograph

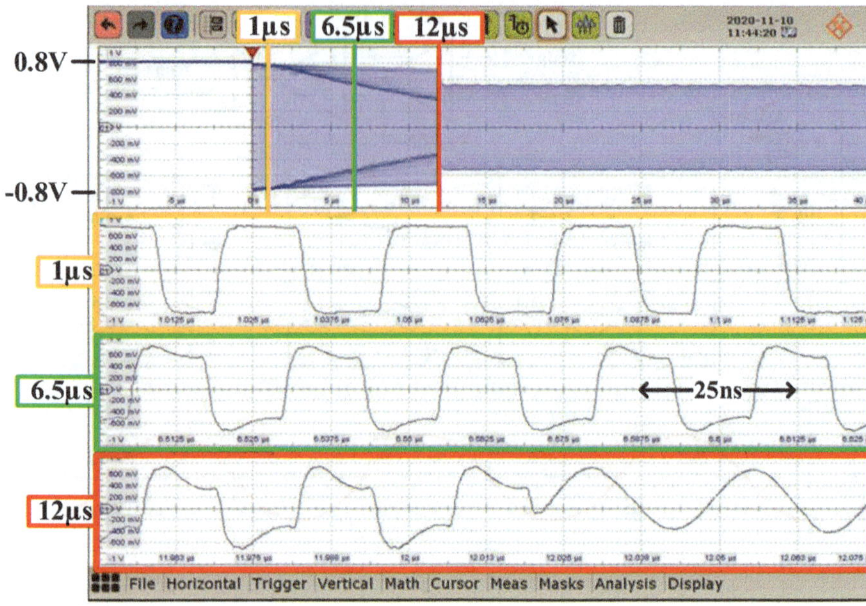

Fig. 15 Measured waveform across the crystal during startup, with zoom in at 1 μs, 6.5 μs, and 12 μs

Note that in the measurement of Fig. 15, the varicap is almost completely disabled since the capacitance of the differential probe is already nearly enough to trim the frequency to 40 MHz. Without a probe connected across the crystal, the varicap will be connected in steps during the startup procedure, to reduce the losses associated with the (dis)charging of the varicap by the block wave signal.

4.2 *Phase Noise*

The phase noise, measured across the crystal with a differential probe, is depicted in Fig. 16. At low offset frequency from the carrier, the phase noise is dominated by $1/f$ noise of the peak detect circuit, in combination with the second-order effect of the LRC branch in the crystal for frequencies below 10 kHz. The −153dBc/Hz noise floor is assumed to be a combination of switched capacitor kT/C noise and various other white noise sources like the peak detect comparator, differential probe, and parasitic resistance in the circuit. The jitter due to the switched capacitor can be calculated by:

$$\text{jitter}_{\text{rms}} = \frac{\sqrt{\overline{v_n^2}}}{dV_{\text{xtal}}/dt} = \frac{\sqrt{\frac{kT}{(C_L+C_0)}}}{2\pi\ f_{\text{xtal}}\ V_{\text{xtal}}} \tag{7}$$

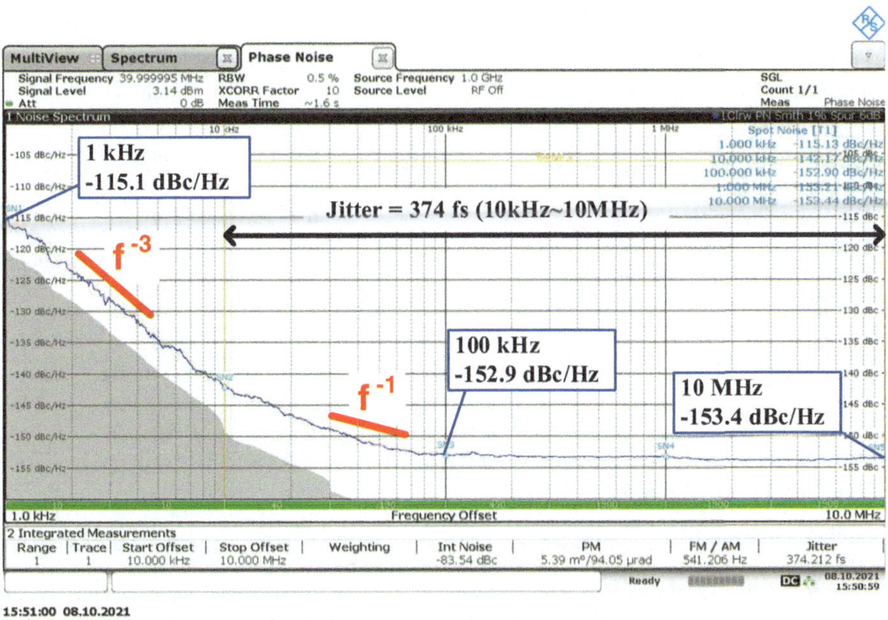

Fig. 16 Measured phase noise

For our 40 MHz switched capacitor crystal oscillator, this equals to 250 fs, which is below the measured jitter, integrated from 10 kHz to 10 MHz, of 374 fs.

4.3 Performance Summary (Table 1)

5 Conclusions

A fast startup technique is presented which doesn't require an accurate startup oscillator and which overcomes the difficulty of measuring a small current through a low-Ohmic switch by separating the drive phase and the measurement phase in time. This startup technique will especially be useful in low-power radio systems in which the receive and transmit periods are short.

Also presented is a switched capacitor crystal oscillator as a fully differential alternative for the single-ended Pierce oscillator. The main benefit of this oscillator over the traditional Pierce oscillator is the significantly improved CMRR, especially important for circuits that operate in a noisy environment. Furthermore, the fact that the load capacitor is connected directly across the crystal reduces the required on-chip capacitance by almost a factor 4.

Table 1 Main performance of the circuit

Parameter		Value	Unit
Technology		22	nm
Supply voltage		0.8	V
Core area		0.021	mm^2
Frequency		40	MHz
Load capacitance C_L		6	pF
Differential amplitude		0.8	Vpp
Steady-state power consumption		55	μW
Startup time		12	μs
Startup energy[a]		8	nJ
Phase noise	@1 kHz	−115.1	dBc/Hz
	@10 kHz	−142.2	dBc/Hz
	@100 kHz	−152.9	dBc/Hz
	@1 MHz	−153.2	dBc/Hz
RMS jitter [10 kHz–10 MHz]		374	fs
CMRR[b]		−67.4	dBc/1Vpp

[a]The startup energy is derived by a combination of measurements and simulations
[b]CMRR is defined as the level of spurious tone with a 1 Vpp 1 MHz signal applied on the crystal case. As a reference, the simulation result of the typical Pierce oscillator (Fig. 2) shows a spurious tone of −34 dBc with single-ended probing on XTALp, mainly defined by the ratio Cp1 divided by 2 C_L, and an even higher spurious tone of −25 dBc with differential probing across the crystal pins, due to the gain of the inverter. The measured spurious tone of −67.4 dBc is therefore a 33–42 dB improvement in CMRR over a typical Pierce oscillator

References

1. E.A. Vittoz, M.G.R. Degrauwe, S. Bitz, High-performance crystal oscillator circuits: Theory and application. JSSC **23**(3), 774–783 (1988)
2. J.B. Lechevallier, H.S. Bindra, R.A.R. van der Zee, B. Nauta, Energy efficient Startup of crystal oscillators using stepwise charging. IEEE J. Solid State Circuits **56**(8), 2427–2437 (2021)
3. B. Verhoef, J. Prummel, W. Kruiskamp, R. Post, A 32MHz crystal oscillator with fast start-up using synchronized signal injection, in *ISSCC*, (2019), pp. 304–305
4. K.M. Megawer et al., A 54MHz crystal oscillator with 30× start-up time reduction using 2-step injection in 65nm CMOS, in *ISSCC*, (2019), pp. 302–303
5. J. Jung et al., A single-crystal-oscillator-based clock-management IC with 18× start-up time reduction and 0.68ppm/°C duty-cycled machine-learning-based RCO calibration, in *ISSCC*, (2022), pp. 58–60
6. J.B. Lechevallier, R.A.R. Van Der Zee, B. Nauta, Fast & energy efficient startup of crystal oscillators by self-timed energy injection. IEEE JSSC **54**(11), 3107–3117 (2019)
7. M.W. Kruiskamp, B. Verhoef, Crystal oscillator with fast start-up. U.S. Patent 10 951 166 (2021, March 16)
8. W. Kruiskamp, A fully differential 40 MHz switched-capacitor crystal oscillator with fast start-up, in *Proceedings of the 48th European Solid-State Circuits Conference (ESSCIRC)*, (2022), pp. 397–400
9. D. Ruffieux, A high-stability, ultra-low-power quartz differential oscillator circuit for demanding radio applications, in *Proceedings of the 28th European Solid-State Circuits Conference (ESSCIRC)*, (2002), pp. 85–88

10. Y. Chang et al., A differential digitally controlled crystal oscillator with a 14-bit tuning resolution and sine wave outputs for cellular applications. IEEE JSSC **47**(2), 421–434 (2012)
11. Y. Rajavi, M.M. Ghahramani, A. Khalili, A. Kavousian, B. Kim, M.P. Flynn, A 48-MHz differential crystal oscillator with 168-fs jitter in 28-nm CMOS. IEEE JSSC **52**(10), 2735–2745 (2017)
12. M.W. Kruiskamp, Switched capacitor crystal oscillator. U.S. Patent 10 965 250 (2021, March 30)

Index

P. Harpe et al. (eds.), *Biomedical Electronics, Noise Shaping ADCs, and Frequency References*, https://doi.org/10.1007/978-3-031-28912-5

M

N

O

P

The manufacturer's authorised representative in the EU is Springer Nature Customer Service Centre GmbH, Europaplatz 3, 69115 Heidelberg, Germany. If you have any concerns regarding our products, please contact ProductSafety@springernature.com

Printed and bound by CPI Group (UK) Ltd, Croydon, CR0 4YY
28/09/2026
02224075-0016